FATIGUE
Environment and Temperature Effects

SAGAMORE ARMY MATERIALS
RESEARCH CONFERENCE PROCEEDINGS

Recent volumes in the series:

20th: Characterization of Materials in Research: Ceramics and Polymers
Edited by John J. Burke and Volker Weiss

21st: Advances in Deformation Processing
Edited by John J. Burke and Volker Weiss

22nd: Application of Fracture Mechanics to Design
Edited by John J. Burke and Volker Weiss

23rd: Nondestructive Evaluation of Materials
Edited by John J. Burke and Volker Weiss

24th: Risk and Failure Analysis for Improved Performance and Reliability
Edited by John J. Burke and Volker Weiss

25th: Advances in Metal Processing
Edited by John J. Burke, Robert Mehrabian, and Volker Weiss

26th: Surface Treatments for Improved Performance and Properties
Edited by John J. Burke and Volker Weiss

27th: Fatigue: Environment and Temperature Effects
Edited by John J. Burke and Volker Weiss

28th: Residual Stress and Stress Relaxation
Edited by Eric Kula and Volker Weiss

FATIGUE
Environment and Temperature Effects

Edited by

John J. Burke

Army Materials and Mechanics Research Center
Watertown, Massachusetts

and

Volker Weiss

Syracuse University
Syracuse, New York

PLENUM PRESS • NEW YORK AND LONDON

Library of Congress Cataloging in Publication Data

Sagamore Army Materials Research Conference (27th: 1980: Bolton Landing, N.Y.)
   Fatigue, environment and temperature effects.

   Includes bibliographical references and index.
   1. Materials—Fatigue—Congresses. I. Burke, John J. II. Weiss, Volker, 1930—   III.
Title.
TA418.38.S23 1980   620.1'126   82-12392
ISBN 0-306-41101-6

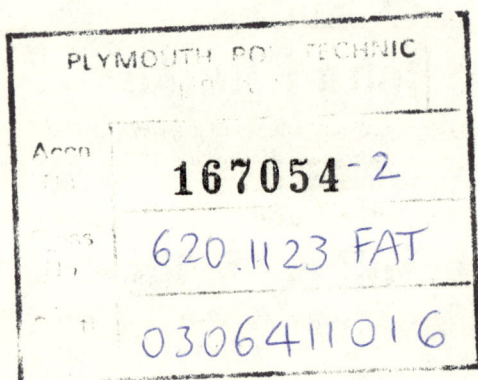

Proceedings of the 27th Sagamore Army Materials Research
Conference, held July 14–18, 1980, at the Sagamore Hotel,
Bolton Landing, Lake George, New York

© 1983 Plenum Press, New York
A Division of Plenum Publishing Corporation
233 Spring Street, New York, N.Y. 10013

Printed in the United States of America

PREFACE

The Army Materials and Mechanics Research Center in coop-
eration with the Materials Science Group of the Department of
Chemical Engineering and Materials Science of Syracuse University
has been conducting the Annual Sagamore Army Materials Research
Conference since 1954. The specific purpose of these conferences
has been to bring together scientists and engineers from academic
institutions, industry and government who are uniquely qualified
to explore in depth a subject of importance to the Department of
Defense, the Army and the scientific community.

These proceedings entitled, FATIGUE - ENVIRONMENT AND TEMPER-
ATURE EFFECT, address the overview of temperature and environmental
effects of fatigue, room temperature environmental effects, high
temperature and environmental effect - mechanisms, high temperature and environmental effect - mechanisms, materials and
design-engineering applications.

We wish to acknowledge the assistance of Messrs. Joseph
Bernier and Dan McNaught of the Army Materials and Mechanics
Research Center and Helen Brown DeMascio of Syracuse University
throughout the stages of the conference planning and finally the
publication of the book.

The continued active interest and support of these conferences
by Dr. E. Wright, Director of the Army Materials and Mechanics
Research Center, is appreciated.

The Editors

Syracuse University
Syracuse, New York

CONTENTS

SESSION I

OVERVIEW OF TEMPERATURE AND ENVIRONMENTAL EFFECTS OF FATIGUE

Overview of Temperature and Environmental Effects of
    Fatigue of Structural Metals 1
  L. F. Coffin

SESSION II

ROOM TEMPERATURE ENVIRONMENTAL EFFECTS

Corrosion Fatigue Crack Propagation 41
  F. P. Ford

Surface Reactions and Fatigue Crack Growth 59
  R. P. Wei and G. W. Simmons

Determination of Prefracture Damage and Failure Prediction
    in Corrosion-Fatigues Al-2024-T4 by X-Ray Diffraction
    Methods 71
  T. Takemoto, S. Weissman, and I. R. Kramer

SESSION III

ROOM TEMPERATUE ENVIRONMENTAL EFFECTS

ΔK Thresholds in Titanium Alloys – the Role of Microstructure,
    Temperature and Environment 83
  C. J. Beevers and C. M. Ward-Close

Dislocation Distribution in Plastically Deformed Metals . . . 103
  I. R. Kramer, R, Pangborn, and S. Weissman

SESSION IV

HIGH TEMPERATURE AND ENVIRONMENTAL EFFECT – MECHANISMS

The Effect of Microstructure on the Fatigue Behavior of Ni
    Base Superalloys 119
    S. D. Antolovich and N. Jayaraman

Creep Crack Growth 145
    S. Floreen

Temperature Dependent Deformation Mechanisms of Alloy 718
    in Low Cycle Fatigue 163
    T. H. Sanders, Jr., R. E. Frishmuth, and G. T. Embley

Deformation Induced Microstructural Changes in Austenitic
    Stainless Steels 183
    J. Moteff

Fatigue and Fracture Resistance of Stainless Steel Weld
    Deposits After Elevated Temperature Irradiation 195
    J. R. Hawthorne

SESSION V

HIGH TEMPERATURE AND ENVIRONMENTAL EFFECT – MECHANISMS

High-Temperature Static Fatigue in Ceramics. 221
    R. N. Katz, G. D. Quinn, and E. M. Lenoe

Environment, Frequency and Temperature Effects on Fatigue
    in Engineering Plastics. 231
    R. W. Hertzberg and J. A. Manson

SESSION VI

MATERIALS

Creep-Fatigue Effects in Structural Materials Used in
    Advanced Nuclear Power Generating Systems 241
    C. R. Brinkman

The Effect of Environment and Temperature on the Fatigue
    Behavior of Titanium Alloys 265
    J. A. Ruppen, C. L. Hoffmann, V. M. Radhakrishnan,
    and A. J. McEvily

Creep-Fatigue-Effects in Composites 301
    N. S. Stoloff

                           SESSION VII

              DESIGN - ENGINEERING APPLICATIONS

Thermal Fatigue Analysis. 329
    D. F. Mowbray and G. G. Trantina

Life Prediction for Turbine Engine Components 353
    T. Nicholas and J. M. Larsen

A Kinetic Model of High Temperature Fatigue Crack Growth . . 377
    J. J. McGowan and H. W. Liu

Design-, Operation-, and Inspection-Relevant Factors of
      Fatigue in Crack Growth Rates for Pressure Vessel
      and Piping Steels 391
    W. H. Cullen and F. J. Loss

INDEX 411

OVERVIEW OF TEMPERATURE AND ENVIRONMENTAL EFFECTS ON FATIGUE OF

STRUCTURAL METALS

L. F. Coffin

General Electric Company
Corporate Research and Development
Schenectady, N. Y.   12301

INTRODUCTION

This introductory lecture of the 27th Sagamore Army Materials
Research Conference is intended to provide a general overview of
the effects of temperature and environments on the fatigue behavior
of structural alloys. A broad view, it is hoped, will compliment
the more specialized subjects to be treated in the next three days.
Included there will be such topics as the role of environment on the
fatigue process at room and elevated temperatures, the fatigue performance of specific materials, largely metals but also ceramics and
engineering plastics and engineering applications, all to be dealt
with in some detail.

To begin this overview, since the concern is with fatigue, that
is, with the phenomenon of crack initiation and growth under cyclic
loading, it is useful to consider this process of crack initiation
and growth in more detail. With reference to Figure 1 we show three
aspects of the process, over which most of the current discussions
of fatigue take place. Since fatigue is well recognized to be
associated with cyclic plastic strain, the cracking process will
initiate at that point in a body subjected to cyclic loading where
the cyclic plastic strain is most severe. This may be a notch, a
surface irregularity or some other stress concentration, or, if the
body is of uniform geometry, where some discontinuity exists in the
material such as an inclusion, void, or some other abnormality in
the microstructure. Under the action of the cyclic loading, a
plastic zone is created as shown in Figure 1(a) from which cracks
will initiate and grow. The growing crack propagates first through
the plastic zone created by the notch and thus grows in it a largely
plastic environment. It is identified as a short crack, defined

1

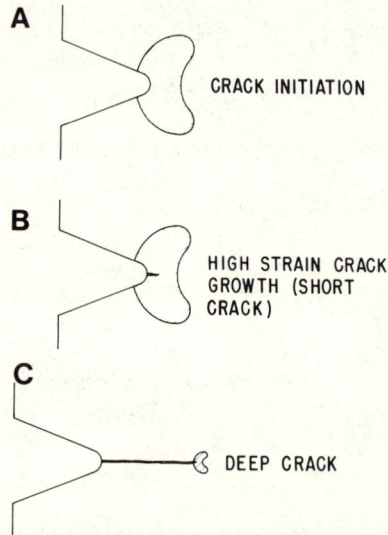

Fig. 1 Processes and Plastic Zones for Crack Initiation and Growth.

as that crack length which is small relative to the plastic zone.
The conditions here are special, as we shall see later, and characterization and quantification of this form of crack growth is not
fully resolved from current research. Eventually the crack passes
beyond the plastic zone of the notch and grows on its own. Here the
plastic zone is determined by the crack geometry and loading rather
than by that of the notch.

One distinguishing feature of Figure 1(c) over Figure 1(b) is
that the plastic zone of the notch may accommodate a greater degree
of plastic deformation than that produced by the long crack since
the notch is capable of plastic deformation in compression due to
the upsetting (strain reversal) when the loading on the body is
removed, while the plastic zone of the crack can result from tensile
loading only, since upsetting is largely prevented by crack closure.
A second feature of the high strain crack growth regime, Figure
1(b) is the mode of crack growth. Due to the reversible nature of
the plastic strain, a preferred form of crack growth is by shear,
i.e., by Mode II or Mode III (using fracture mechanics terminology),
rather than by normal or planar growth, Mode I. These matters need

greater discussion and resolution.

To assist in the determination of material property information for the regimes shown in Figure 1, we refer to Table I. It is common to investigate the fatigue crack initiation process in notches using a smooth uniaxially loaded specimen and to relate the information obtained from crack initiation in strain controlled smooth specimen tests to that of the notch.[1,2] An alternative approach is to obtain fatigue crack initiation information from notched geometries such as uncracked compact tension tests and to relate this information to smooth specimen performance.[3-5] This later approach is particularly useful in cases where environments preclude the use of strain-control techniques. For the short crack or the elastic-plastic problem, a variety of techniques can be used, including the defected notch,[6-7] the compact tension configuration[8] or smooth specimens where surface growth of microcracks are monitored.[9] When the crack is deep and can be well characterized by linear elastic fracture mechanics methodology, a variety of standardized geometries can be employed as indicated in Table I.

Crack initiation and crack growth measurement represent important parts of the material property studies and some of the more common techniques for observing and measuring crack growth are summarized in Table II. Surface replica measurement of micro-cracks can be very useful in determining microstructural damage.[9] Crack initiation defined in terms of microcracks, that is, in terms of an engineering size crack, can be observed in the course of conducting standard smooth specimen low cycle fatigue tests by load changes or changes in hysteresis loop shape.[10] For notch geometries, AC or DC electrical potential methods are found to be useful,[5,11] while compliance techniques for determining crack growth in standardized test specimen geometries are well developed.[12] When cracks are short, visual techniques are commonly used,[7] but electrical potential methods have also been developed for such cases.[6,13] Observation of crack growth in deep cracks has been fairly well standardized through the use of visual, compliance and electrical potential methods.[11]

With the availability of the appropriate laboratory specimen and crack monitoring technique, the next question to consider is the controlling parameter of interest in the particular regime of cracking. Here the loading is broadened to include both monotonic and cyclic loading as well as a concern for rate effects. With reference to Table III it will be noted that a wide variety of parameters such as stress σ, strain ε, strain $\dot{\varepsilon}$, frequency ν, stress intensity factor K and path independent line integral J are useful in representing the static and cyclic crack initiation and growth processes for fracture.

Finally, with reference to Table IV are given some of the commonly accepted means for representing the crack initiation and

Table I

SPECIMEN GEOMETRIES

Smooth Surface Notch	Short Crack	Deep Crack
SMOOTH SPECIMEN UNIFORM OR HOUR GLASS	DEFECTED NOTCH	CT (WOL)
	SHARPLY NOTCHED	EDGE OR CENTER CRACK
BLUNT NOTCH CT	SMOOTH SPECIMEN	DEFECTED SMOOTH SPECIMEN

Table II

TECHNIQUES FOR OBSERVING AND MEASURING CRACKS

Smooth Surface Notch	Short Crack	Deep Crack
REPLICAS	VISUAL	VISUAL
LOAD CHANGES	ELECTRICAL POTENTIAL	COMPLIANCE
HYSTERESIS LOOP		ELECTRICAL POTENTIAL
COMPLIANCE		
ELECTRICAL POTENTIAL		

Table III

CONTROLLING PARAMETER

	Smooth Surface Notch	Short Crack	Deep Crack
Cyclic	$\Delta\varepsilon_p$, $\Delta\varepsilon_e$	$\Delta\varepsilon_p$, J OR PSEUDO ΔK	ΔK OR K_{MAX}
Monotonic	ε_p, σ	σ_{NET}, K OR J	σ_{NET}, K
Rate Effects	ν OR $\dot\varepsilon$	ν, $\dot\varepsilon$, J	ν

growth phenomenology. For smooth specimens it is common to use
either the plastic strain range $\Delta\varepsilon_p$ or the total strain range $\Delta\varepsilon$ vs.
N_f the cycles for crack initiation. In the case where mean stresses
may exist, an alternate form proposed by Smith et. al.[14] is useful,
namely $\sigma_{max}\varepsilon_a E$ vs. N_f when ε_a is $\Delta\varepsilon/2$. For notches the Neuber notch
analysis procedures result in the parameter $\sigma_a\varepsilon_a E$ vs. N_f.[1,2] For
monotonic loading at high temperature the expression $\varepsilon = F(\sigma,\dot{\varepsilon})$ is
useful. No means for representing the high temperature time dependent
fatigue life prediction phenomenology is given here and the reader
is referred to other sources[15,16] for this extensive topic.

A variety of methods have been proposed for representing the
short crack phenomenology including that of El Haddad, et.al,[17-18]
where the crack length a used in stress intensity factor determinations is replaced by $a + a_o$ where a_o is introduced to provide a
finite stress intensity factor when the crack length becomes van-
ishingly small. Solomon[7] has shown that the form $d\ln a/dN = F(\Delta\varepsilon_p,\nu)$
is useful in describing the growth of elastic-plastic crack growth,
while Dowling and Begley[8] show that elastic-plastic cyclic crack
growth can be represented by using J or $\frac{\Delta K^2}{E}(1-\nu^2)$.

For deep cracks the functional relation between ΔK and da/dN is
commonly used, while K_{max} is useful when considering corrosion fatigue. Creep crack growth or stress corrosion crack growth can also
be represented by K, J or $\dot{J}$.

With these generalities for characterizing and evaluating both
crack initiation and crack growth behavior behind us, let us move
into consideration of real material behavior at elevated temperature.

Damage Processes In Time Dependent Fatigue

The application of cyclic plastic strain to structural components at elevated temperature can produce a number of damaging
processes which affect the microstructure and the resulting fatigue
resistance. These damage processes lead to premature failure by
fatigue when compared to fatigue failure under conditions of time
independency, that is, where frequency, strain rate and environmental
effects are absent. The topic of elevated temperature fatigue damage
was reviewed in more detail in various papers[19,20] where the important damage categories were discussed including substructure, cyclic
strain aging, grain boundaries, environment, wave shape, and plastic
instability. Of primary interest in the present paper are the damage
processes arising from grain boundaries, from the environment and
from wave shape effects.

Environmental damage is viewed as a synergistic phenomenon with
cyclic plastic deformation and occurs both at a free surface and an
exposed crack tip. Here brittle protective oxide films are ruptured,

Table IV

MEANS OF REPRESENTING RESULTS

Smooth Surface Notch	Short Crack	Deep Crack
$\Delta\epsilon$ OR $\Delta\epsilon_P$ VS. N_F	EL HADDAD ET AL	$DA/DN = F(\Delta K)$ OR $F(K_{MAX})$
NEUBER: $\sigma_A \epsilon_A E$ VS. N_F	D LN $A/DN = F(\Delta\epsilon_P, \nu)$	
SMITH ET AL: $\sigma_{MAX} \epsilon_A E$ VS. N_F	$DA/DN = B \, \Delta \, J^N$	
$\epsilon = F(\sigma, \dot{\epsilon})$	$DA/DT = F(\dot{J})$	$DA/DT = F(K)$

exposing fresh, nascent metal, and chemical attack proceeds until protective films are reformed. With fatigue, this process of straining, film rupture and attack occurs repeatedly, leading to localization of the cyclic strain and an abundance of the reaction products in the strained region.[21] Grain boundaries are selectively attacked because of chemical segregation existing at the boundaries and the ready oxidation of these grain boundary species. At high frequencies time is insufficient for local chemical reaction and cracking reverts to its normal transgranular mode.[22] At low frequencies or long hold times in tension, chemical processes lead to acceleration of micro- and macrocrack growth rates. It is important to note that, unlike grain boundary cavity formation which is an interior damage process at high temperature, environmental damage occurs mostly at the surface where the environment is present. Some environmental effects can also be expected when grain boundary diffusion by the environmental species can occur.[21]

The wave shape can also produce grain boundary damage.[20] This form of damage is commonly called creep-fatigue. Here damage develops by process of cavity nucleation and growth and of triple point cracking and intergranular cracking results. These processes are similar to those more generally studied in creep under static loading. Veevers and Snowden[23] have reviewed the status of this topic in fatigue. The grain boundary damage processes that are of importance here are those occurring on boundaries normal to the direction of applied load where cavities are found to form or where triple point cracks are observed.

Effect Of Air Environment On Fatigue Crack Initiation At High Temperature

It is well known that most fatigue cracks examined in low-cycle fatigue at high temperature are filled with oxides.[21,24] We have found that the oxidation processes are highly localized at fatigue nucleation sites and along the fracture path. Figure 2 shows the enhancement of the local oxidation at slip-band nucleation sites with decreasing frequency for A286, an iron-base superalloy tested at 593^0C[25]. Further studies revealed that crack propagation from these initiation sites was largely intergranular, particularly at low frequencies, and that the cracks were filled with oxides.[25]

There are various nucleation mechanisms arising from the interaction between the environment and cyclic strain. The most likely cause for this interaction and the degrading effect it has on fatigue resistance is repeated film rupture due to plastic strain. Since fatigue nucleation is commonly a result of highly localized slip, protective films are continually disrupted with repeated cyclic strain. The enhanced locally oxidized region then acts as a notch to further concentrate the strain, and the development of a fatigue crack is ensured. From this viewpoint, it is clear that the specific

Fig. 2   Crack Initiation at Notch Root Surface Along Inclined
         Polishing Marks and Accompanying Localized Oxidation-Direc-
         tionally Solidified A286 Tested in Air at 593°C and 0.1 cpm.
         Heat Treatment 3.[22]

features of the deformation, such as coarse slip (which occurs during
strain aging) or grain boundary sliding (with occurs at low strain
rates and high temperatures), can be expected to amplify the film
rupture and local oxidation process.

The importance of environment in influencing high-temperature
low-cycle fatigue behavior can best be determined by testing a material in both air and high vacuum. Conventional balanced low-cycle
fatigue tests were performed in vacuums of 10^{-8} torr with striking
results.[25,26] Figure 3, a typical example, shows the comparative
behavior in air and vacuum of A286 at 593°C, together with room-
temperature vacuum test results. Note that the significant effects
of frequency of cycling when testing in air are not found in vacuum;
also note the large difference in lifetimes between the highest
frequency air test and the vacuum test. It is also interesting to
observe that the effect of the test temperature is minimal in a
vacuum environment. Additionally, it was found that crack propagation
was transgranular. Tests of this type have been performed on several
materials at temperatures where significant degradation in fatigue
resistance in air occurs, with results similar to those obtained
for A286.

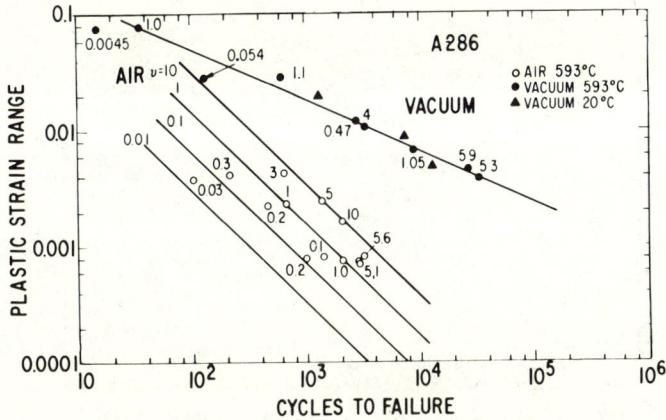

Fig. 3 Plastic Strain Range vs. Fatigue Life for A286 in Air and Vacuum at 593°C. Numbers Adjacent to Test Points Indicate Frequency (cpm); Solid Lines are Regression Analysis.[25]

Plastic strain range-life data for several materials were obtained and are plotted in Figure 4. Included are low-cycle fatigue test results reported many years ago for room-temperature air[27] (open points), together with a number of high-temperature tests run either in high vacuum or in a highly purified argon atmosphere (solid points). The test results in Figure 4 are normalized to the plastic strain range for failure at one cycle in order to establish a common base for comparison of the slope by eliminating ductility differences. Lines corresponding to various exponents β (the Coffin-Manson exponent) are also drawn, showing that the test results fit within a scatter band $0.45 < \beta < 0.60$, and the bulk of the data fits between $0.50 < \beta < 0.55$. Note that temperatures as high as 1150°C are employed in some of the experiments. Cycling frequency for all these data is 0.2 cpm or greater. These findings provide further endorsement for the Coffin-Manson relationship, but, more importantly, they show the significance of the environment in elevated-temperature low-cycle fatigue. An additional feature of these tests is that all fractures are transgranular.

Effect Of Frequency And Structure

To investigate the effect of cyclic frequency, fully reversed loading, notched bar fatigue tests of A286 were conducted with three specific aging heat treatments designed to alter the γ' size and homogeneity of deformation. These heat treatments are identified elsewhere.[22]

The results of these notched bar tests at 593°C on specimens with $K_t = 3$ testing at $\Delta\sigma/2 = 414$ MPa^2 (60,000 psi) at three fre-

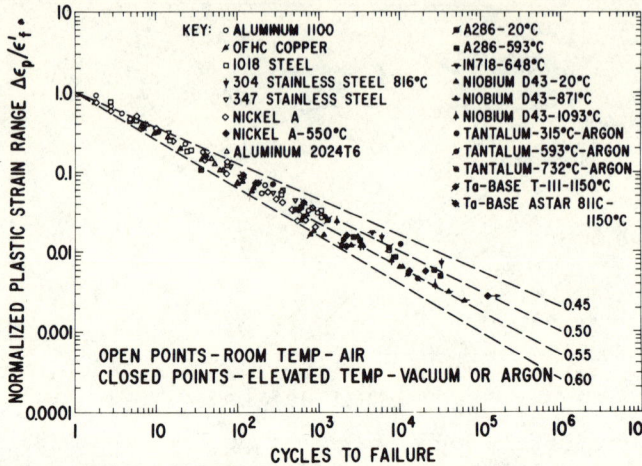

KEY: o ALUMINUM 1100   ⚹ A286-20°C
     ⚹ OFHC COPPER   ▲ A286-593°C
     ▢ 1018 STEEL   ↛ IN718-648°C
     ▼ 304 STAINLESS STEEL 816°C   ▲ NIOBIUM D43-20°C
     ▽ 347 STAINLESS STEEL ▲ NIOBIUM D43-871°C
     ◇ NICKEL A   ↟ NIOBIUM D43-1093°C
     ◆ NICKEL A-550°C   ▲ TANTALUM-315°C-ARGON
     △ ALUMINUM 2024T6   ⚹ TANTALUM-593°C-ARGON
                            ▪ TANTALUM-732°C-ARGON
                            ⚹ Ta-BASE T-III-1150°C
                            ⚹ Ta-BASE ASTAR 811C-
                                       1150°C

OPEN POINTS - ROOM TEMP - AIR
CLOSED POINTS - ELEVATED TEMP - VACUUM OR ARGON

Fig. 4   Summary Plot of Plastic Strain Range vs. Cycles to Failure
         for Severals Metals in Room-Temperature Air and High-Temper-
         ature Vacuum or Argon. Plastic Strain Range Normalized to
         Fatigue Ductility.[26]

quencies are shown in Figure 5. The dashed line for the standard
treatment represents a regression analysis of results of tests under
a variety of notches and stress levels. A strong frequency dependence
of life is apparent below about 5 cpm for all heat treatments. However, the lives are substantially different at a given frequency
for the different heat treatments. A double aging treatment (#3) in
particular shows much less frequency sensitivity than the single
lower temperature treatments.

     Since for A286 failure at low frequencies was intergranular,
elimination of transverse grain boundaries might be expected to increase the life in the low-cycle regime. Accordingly, two notched
specimens with K_t = 3.0 were prepared from remelted stock which had
been directionally solidified. Clearly a substantial improvement
was obtained as seen in Figure 5.

     If the transverse grain boundaries are retained but the envi-
ronment is removed for A286, the high vacuum test point on Figure 5
at ν = 1.6 x 10^{-3} Hz indicates a further improvement. This confirms
that most of the frequency dependence observed in air tests is related
to the influence of oxidation processes at the crack tip. The failure
of the directionally solidified specimens tested in air to last as
long as the vacuum test may be attributed in part to the cast structure in which transverse dendrite boundaries become preferred crack
propagation sites.

     Tests were also performed at higher frequencies (up to 30 cpm).
From the view that, at high frequencies, there would be little time
for environmental attack, curves were constructed to indicate a

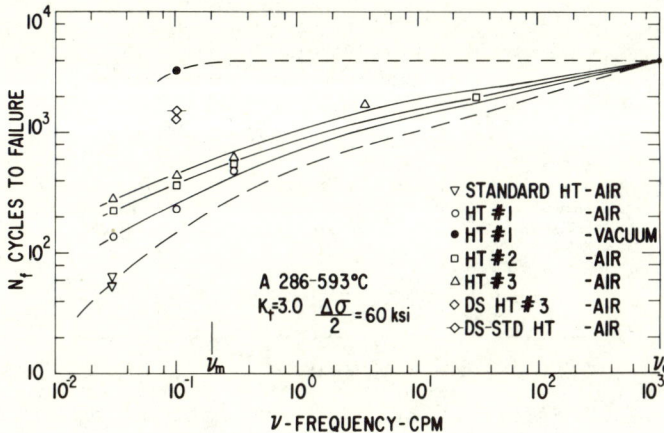

Fig. 5   Effect of Frequency on Life of Notched Fatigue Bars of
A286 at 593°C in Air and Vacuum.[22]

common convergence point at 1000 cpm. Above this frequency the material can be expected to be environmentally insensitive and independent of frequency. For frequencies approaching this convergence
point the fracture surfaces would be expected to be largely transgranular at this stress level. Fractographs of the 30 cpm test
supported this view.

A Damage Model For Frequency Effect

      The results presented above for balanced loop tests can be used
in support of a damage model proposed by Solomon and Coffin[28] and
extended by Woodford and Coffin[22] to account for the role of frequency
and environment in high temperature fatigue (Figure 6). The model
assumes equal and linear tensile and compressive going ramp rate
cycling so that frequency and strain rate are directly related. It
also assumes that any creep damage produced at low strain rates in
tension is balanced by equal recovery of damage in compression. Hence
the damage processes which occur are a result of environmental
effects. Woodford and Coffin[22] found a change in fracture mode from
transgranular to intergranular in air at 304 stainless steel as the
frequency of testing was decreased. However, in vacuum, failure was
always transgranular even at a frequency of 1.6×10^{-4} Hz. Similar
results were found for A286.[28] In the model it is suggested that
three frequency regimes exist in air which can be identified with
different damage processes. At high frequencies ($\nu > \nu_e$) the damage
process is independent of frequency and failure is by the normal
transgranular fatigue mechanism. In the intermediate frequency range
($\nu < \nu < \nu_e$) the air environment is capable of interacting with the
cracks and there is a change of fracture mode from transgranular to
intergranular with decreasing frequency. At lower frequencies
($\nu < \nu_m$) microstructural instabilities may occur and intergranular

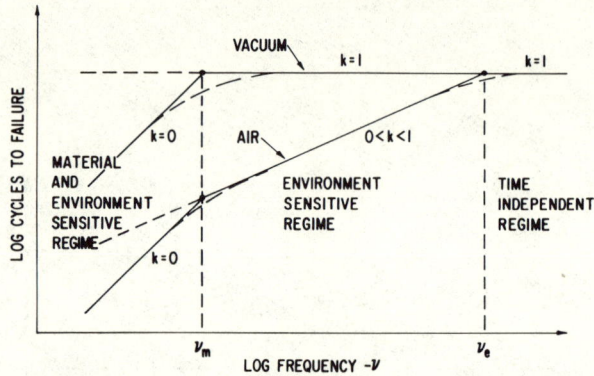

Fig. 6 Model of Effect of Frequency on Fatigue Life at Constant
Plastic Strain Range at Elevated Temperature.[22]

fracture may result. Figure 6 reflects this behavior.

Effect Of Wave Shape On Fatigue Behavior In Air And Vacuum

Over the years, a large amount of testing experience has shown
that waveshape had an important influence on life. Strain-hold time
studies on austenitic stainless steels[29] and materials of similar
strength and ductility reveal that tensile strain holds are the most
damaging mode for equivalent periods (Figure 7). On the other hand,
for the cast nickel-based superalloys, compressive strain-hold tests
are most damaging.[30,31] Some tests on carbon and low-alloy steels
show that frequency, but not waveshape, influences the life.[32] A
discussion of these waveshape effects is given elsewhere.[15,19]

In tensile strain-hold time tests on AISI 304 stainless steel,
[29,33] significant difference in fracture morphology and in life are
found when compared to combined tensile and compressive holds.
Fractures for the former were largely intergranular while additions
of compression hold times of shorter duration cause largely transgranular fracture and increased life. Similar findings have been
reported for AISI 304 stainless steel.[34] Internal cavities have
also been noted in 20/25/Nb/Cr stainless steel at 750^0C[35] and on a
1 CrMoV steel at 565^0C.[36]

Other types of unbalanced loop tests reveal similar degradation
of life and changes in fracture morphology. These include fully
reversed creep at constant stress[37,38] of AISI 316 stainless steel
at 704^0C, strain range partitioning waveshapes such as CP loops,[15]
thermal mechanical tests on A286[39] at 595^0C. The common feature of
all of the above tests was the significant decrease in life and the
appearance of intergranular cracking and interior cracking and
cavity formation when the time for while tensile stresses were applied
exceeded the time spent under compressive stress in a given cycle.

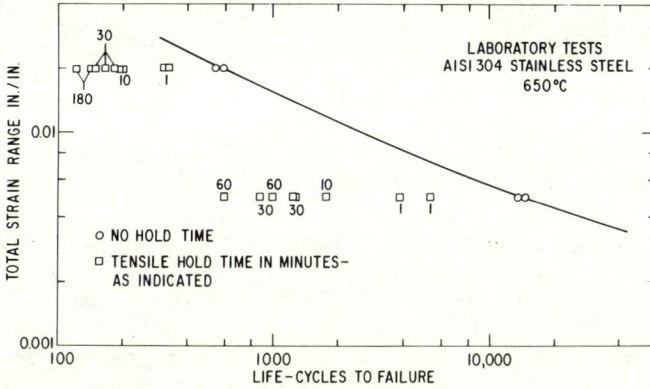

Fig. 7 Effect of Hold Time on Life of 304 Stainless Steel at 650°C. After Berling and Conway.[29]

In order to separate wave shape effects from environmental damage, a series of unequal ramp rate tests was undertaken.[15,40] Tests were performed both in air and high vacuum on AISI 304 stainless steel at 650 and 810°C with a plastic strain range of 0.02 and on A286 at 593°C with plastic strain ranges of 0.01 and 0.005. Tests were run at the same overall frequency, fast-slow, equal, and slow-fast. Results are summarized in Figure 8 for A286. Of special interest is the fact that slow-fast cycles are more damaging and fast-slow cycles less damaging than equal tension- and compression-going frequencies for both strain ranges. Also noted but not tabulated was the loop mean stress, which was negative for the slow-fast cycle and positive for the fast-slow cycle. Of particular interest were the high vacuum results, which indicated a relatively small increase in life for the slow-fast tests and greater increases

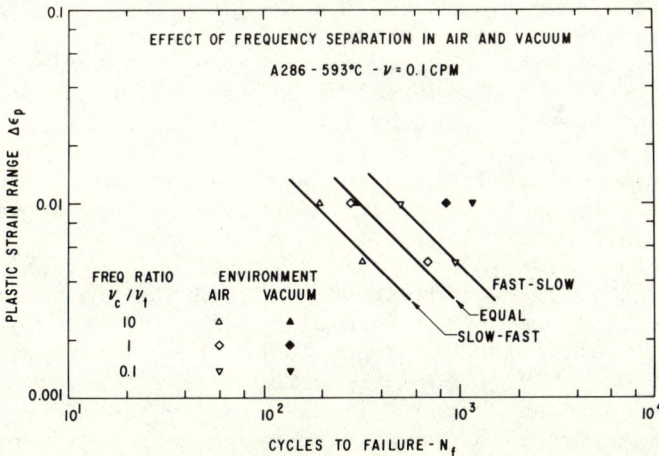

Fig. 8 Effect of Loop Imbalance on Fatigue Life of A286 at 593°C in Air and Vacuum; ν = 0.1 cpm.[15,40]

for the equal and fast-slow tests, all of the same period. Although
further confirmation is required, these preliminary results suggest
that wave shape effects can be more or less damaging independent of
the environment.

Similar results are found for the 304 stainless steel.[20] At
650 and 810°C in air, the slow-fast sequence is substantially more
damaging than the fast-slow. Also of interest are the high vacuum
results, where it is found that the slow-fast test in vacuum is
nearly as damaging as the corresponding air test, while for balanced
loops (equal times for tension and compression going) the vacuum
causes a nearly seven-fold increase in life over air. Furthermore,
the fracture surface for the vacuum slow-fast test is intergranular,
as in the air test, while with equal rates in vacuum, the cracking
is transgranular. Figure 9 and 10 show the microstructure of failed
slow-fast tests at 650°C in air and in vacuum with a plastic strain
range of 0.02. Note, as indicated, the intergranular nature of the
cracking in both environments.

Damage Model For Unbalanced Loop Testing

A further paper[20] modified the model described earlier to account
for unbalanced hysteresis loops and concomitant creep damage. Emphasis was given to the unequal strain rate results under the assumption that this form of wave shape is a rational standard for
damage evaluation. A similar picture can be presented for strain
hold time test results. A qualitative damage picture was proposed,
based on evidence obtained from equal and unequal ramp rate tests on
OFHC copper at 400°C and on 304 stainless steel at 650°C.[20] The slow-
fast type of cycle introduces creep damage in addition to environmental damage. There is general agreement that intergranular cavitation and triple-point cracks in monotonic creep at high temperatures are produced by tensile and not compressive axial stresses[41]
and that they are also enhanced by lowering the strain rate.[42] Other
evidence[43] suggests the reversal of stress in fatigue would tend to
heal any creep damage caused by the tensile stress.

Thus in slow-fast testing surface and interior grain cavity
growth can occur, the growth rate depending on the temperature,
strain range, the overall frequency and the relative forward and
reverse strain rates. A schematic view of the progressive grain
boundary cavity growth is described in Figure 11 for each phase of
the slow-fast loading. This form of damage can assist in the crack
initiation and early growth processes by causing intergranular
cracking but probably is more important in the interior as Tomkins
and Wareing[35] have suggested. It also should be more or less independent of external environment, except as diffusion of environmental species into cavities might impede collapse. Thus, in slow-
fast crack initiation in air three contributing factors are operative: creep cavity damage, environment, and cyclic strain damage.

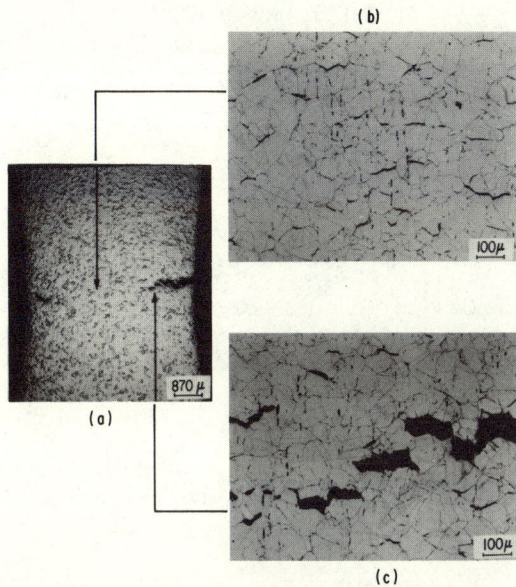

Fig. 9 Microstructure of Failed Slow-Fast Strain-Rate Test on AISI 304 Stainless Steel at 650°C in Air.[20]

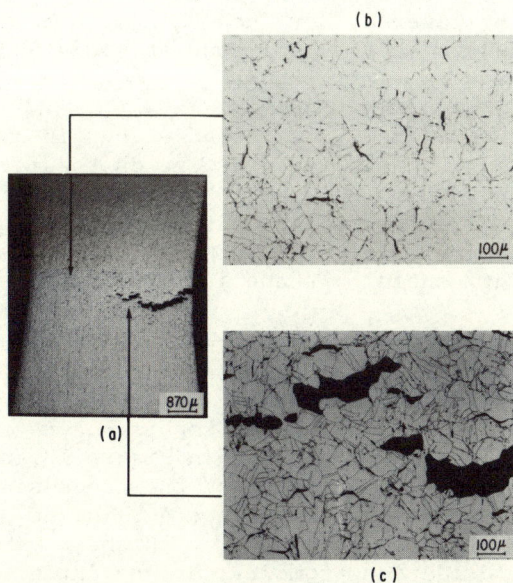

Fig. 10 Microstructure of Failed Slow-Fast Strain-Rate Test on AISI 304 Stainless Steel at 650°C in Vacuum.[20]

SLOW-FAST UNBALANCED LOOP

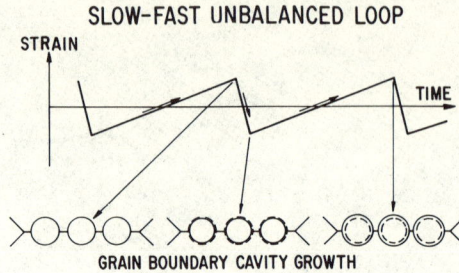

Fig. 11 Model of Cavity Growth for Slow-Fast Wave Shape Showing
Progressive Growth Ratchetting for Each Phase of the
Cycle.

When slow-fast tests are conducted in vacuum, the environmental
damage contribution is removed but intergranular cracking and grain
boundary cavity damage persists.

Considering the fast-slow type of hysteresis loop, failure has
been observed to occur by the propagation of a single transgranular
crack. Often the tensile-going strain rate is high so that the crack
is only exposed to the environment for a short period of time and
this is likely to lead to a normal transgranular fatigue crack. As
far as creep processes are concerned, the compressive strain rate is
low so that if the tensile ramp introduces intergranular damage there
is time during the reversal for such damage to be healed. A schematic
view of the process of cavity collapse at grain boundaries is shown
in Figure 12.

Since no creep damage is expected either in air or vacuum,
environmental damage in air would depend on the forward strain rate
only and lives in vacuum would be longer than in air tests, other conditions being equal. Initiation and early growth processes are expected to be transgranular. Unfortunately, the fast-slow tests results are also influenced by shape instability effects.[15,44]

Equal tensile and compressive ramp cycles produce a transgranular
or intergranular main crack depending on the frequency. Intergranular
surface cracks and secondary intergranular interior cracks near the
fracture have been observed in OFHC copper but not bulk damage. According to the schematic view shown in Figure 13, no cavity growth is
anticipated on grain boundaries since the processes of growth and collapse during each phase of the cycle are balanced and no net changes
are anticipated. This is consistent with the environment controlled
fatigue regime. Oxidation promotes surface intergranular initiation
and further growth takes place along the grain boundaries which are
the easiest diffusion paths for oxygen. The damage processes in
equal ramp rate regimes are identified as environmental and time-

FAST-SLOW UNBALANCED LOOP

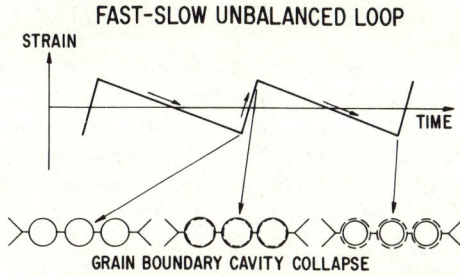

Fig. 12 Model of Cavity Collapse for Fast-Slow Wave Shape Showing
        Progressive Cavity Collapse Ratchetting for Each Phase
        of the Cycle.

independent fatigue in air, while in vacuum damage is due only to
time-independent fatigue. Fatigue cracks would be transgranular in
vacuum at all frequencies, unless, time-dependent metallurgical
effects influence the behavior. In air, the degree to which the
fracture is intergranular depends on the frequency or strain rate,
(assuming a given material and temperature).

Observations on a High Strength High Temperature Alloy

     Recent experience on wave shape and environmental studies on
a high strength, high temperature alloy requires some modification
of the above model when the alloy and grain size composition is such
that high strength is achieved at high temperature. Studies[45] were
performed on a high thermal conductivity alloy, MZC copper (nominal
composition 0.06% magnesium, 0.15% zirconium, 0.40% chromium, balance
copper) solution treated for optimum strength at temperature. This

EQUAL RAMP RATES

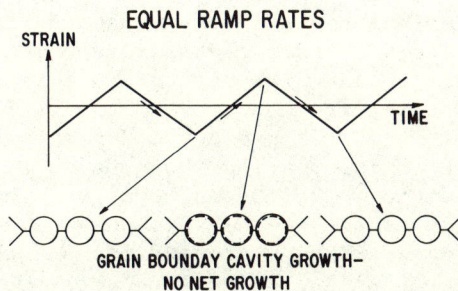

Fig. 13 Model of Cavity Behavior for Equal Wave Shapes Showing No
        Progressive Growth or Collapse During Each Phase of the
        Cycle.

alloy is considered as the subskin for a composite water-cooled turbine bucket or nozzle. Testing was conducted in a specially constructed argon gas chamber to provide an appropriate inert environment. Two specific effects on fatigue behavior were evaluated: one
considering the role of wave shape in which unequal forward and reverse ramp rates were imposed, and the other treating the question
of the effect of gaseous environment on the material's fatigue resistance. In the case of the latter, it was found that testing in
high vacuum gave an improved low cycle fatigue resistance over a
purified argon atmosphere, while fatigue results in air were almost
equivalent to those in argon (Table V). This surprising result was
interpreted to be associated with a lowering of grain boundary surface energy (all fracturing was intergranular) and a loss of protective oxide films in atmospheres with trace quantities of oxygen.
In the case of air, severe oxidation led to crack blunting which
served as a deterrent to crack growth.

Little difference was found between fatigue tests of equivalent periods when tests with slow forward (tension going) ramp rates
and fast reverse (compression going) rates (S-F) were compared with
tests of equal ramp rates (E-E), while large extensions in life resulted from fast forward and slow reverse (F-S) cycles (Table VI).
This is in contrast with the OFHC copper results where severe losses
in fatigue resistance develop in S-F cycling. Metallographic studies
showed no bulk grain boundary damage for MZC copper in contrast to
severe bulk cavity damage at grain boundaries for OFHC copper. The
wave shape damage process for MZC copper was interpreted as an environmental one, the degradation depending on the tension going
time.

The coarse grain size of this material, seen most clearly in
Figures 14 and 15, in conjunction with the alloy's strength at
temperature, are probably responsible for the intergranular fractures found in all tested specimens. As indicated earlier, it is
unusual to encounter intergranular cracking in high-vacuum balanced
loop (E-E) testing or in F-S tests. The intergranular nature of the
fracture process must be assumed to be inherent for MZC copper with
this processing history because of the vacuum results. It may also
be generic to coarse grain, high strength alloys at high temperature. It provides an environmental sensitivity not found when
transgranular crack growth is present. Since crack tip growth is
related to the work of deformation whose major components are surface energy and plastic work, it can be argued that, when grain
boundary cracking is found, the plastic work component is small
relative to transgranular crack propagation and the grain boundary
surface energy is controlling. Thus the presence of absorbed
films on the fracture surface at the crack tip and the concomitant
surface energy change has a high leverage on crack growth.

TABLE V

MZC Copper Test Results

370°C 0.2 CPM

Effect of Environment

Spec.	Atmosphere	Wave Shape	N_f Cycles
4L	Vacuum	E-E	> 1670
11L	Argon	E-E	200
12L	Air	E-E	560
10L	Argon	F-S	> 910
16T	Air	F-S	1000
Cu 5	Argon	S-F	300
4T	Air	S-F	230

TABLE VI

MZC Copper Test Results
370°C 0.2 CPM
Effect of Wave Shape

Spec.	Wave Shape	Atmosphere	N_f cycles	$\Delta\sigma$ @ $N_f/2$ ksi	$+\sigma$ peak @ $N_f/2$ ksi
Cu 5	S-F	Argon	300	51.8	26.0
11L	E-E	Argon	200	60.8	28.3
9L	E-E	Argon	270	56.9	32.3
10L	F-S	Argon	910	53.4	25.8
4T	S-F	Air	230	54.9	28.0
12L	E-E	Air	560	52.3	28.5
12T	E-E	Air	300	54.0	30.1
16T	F-S	Air	1000	54.5	29.8
4L	E-E	Vacuum	1670	53.1	26.7

Fig. 14 Longitudinal section of MZX copper specimen 12L (E-E tested in air).

Fig. 15 Higher magnification of specimen 12L showing fatigue crack.

Effect Of Hold Times In Ni-Base Superalloys At Elevated Temperatures –
Crack Initiation

    Some years ago, it was observed that strain controlled tests
on cast R'80 involving strain-hold periods and $A_\epsilon = \infty$ gave longer
lives with tensile hold times as compared to compressive hold times
or equal tensile and compressive hold times.[30] These results are
shown in Figure 16. Concurrently it was found that continuously-
recorded hysteresis loops showed a progressive mean stress shift
until an equilibrium loop was established, the direction of the shift
depending on which hold period was employed. Thus a compressive
mean stress was found for tensile strain hold periods, and a tensile
mean stress resulted from compression strain holds.

    Similar results are found in the case of René 95, a high strength
disk alloy. Tests were conducted for the Air Force Materials Laboratory (AFML) using strain-controlled loading and hourglass specimens
of cast and wrought René 95 at 1200^0F with one and ten minute hold
periods in tension and compression.[46] The A-ratio here was also ∞.
Figure 17 is constructed from these data and shows the strain-life
results for the several tests performed. Included with each test
point in parenthesis is the tensile/compressive peak stresses in ksi
units. Although the differences in life are as described above for
one minute hold periods, this life difference is seen to be much
greater for ten minute hold periods. Similarly, the mean stress
shift is observed to be much greater for the longer hold period.
From these results, tensile strain hold time periods cannot be considered as particularly damaging to cast and wrought René 95, other
than for effects that could be construed to be the normal degradation
due to decreasing cyclic frequency.[47] In fact, Figure 17 shows the
tensile hold fatigue lines to out last those having compression strain
holds. In this connection it is interesting to note in Figure 16
that tensile hold time tests result in longer lives for cast René 80
than continuous cycling tests of the same frequency.[30] In the AFML
work,[46] one minute and 10 minute hold time tests produced lifetimes
about equal to 20 CPM tests.

    This hold time effect can be explained if one introduces a
simplified view as shown in Figure 18. Here a stress-strain loading
with an A_ϵ ratio of ∞ is assumed. Considering the initial loading
to be largely elastic, upon reaching the strain hold point 1, inelastic stress relaxation occurs, such that the unloading results in
a compressive stress of 0. Again, assuming nearly elastic loading,
further stress relaxation at 1 causes further compression at 0.
Continued cycling causes further lowering of the stress-strain loop.
Concurrently, inelastic plastic strain begins to develop as the
unloading path approaches 0. The magnitude of this plastic strain
increases with increasing loop shift.

The loop stabilizes when $(\varepsilon^r_{in})_1 = (\varepsilon^p_{in})_o$ and a compressive mean stress

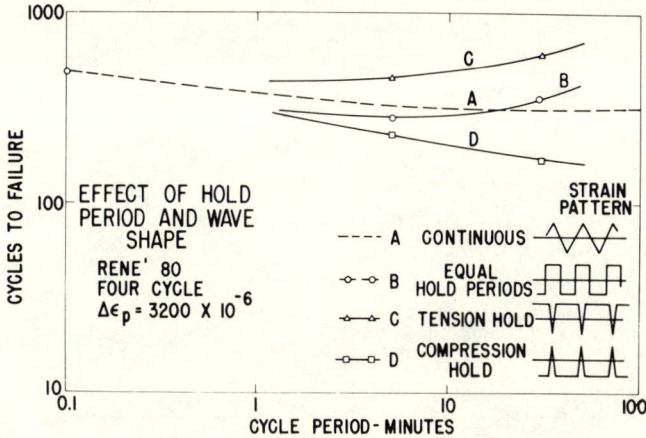

Fig. 16 Effect of Hold Period and Wave Shape on Fatigue Life
of Cast René 80 at 871°C, $\Delta\varepsilon p = 3200 \times 10^{-6}$.[30]

is so introduced. For compressive strain hold periods, the reverse
action develops, and a tensile mean stress results. It is argued[30]
that the compressive mean stress accompanying a tensile hold period
is beneficial to fatigue resistance, while compressive hold periods
and tensile mean stress acts to reduce the fatigue life.

Effect of Hold Times on Fatigue Crack Growth in Superalloys at
Elevated Temperatures

     Load controlled crack growth test results have been reported[48]
for several alloys (including Waspaloy, wrought Astroloy, HIP Astroloy, in 100 and NASA IIB-7). Figure 19 and 20 summarize their
findings at 1200°F for a frequency of 0.33 Hz (Figure 19) and for
a 15-minute hold period (Figure 20). Note the large increases in
cyclic crack growth for these alloys when a hold period is inserted.
It would appear that the large increase in cyclic crack growth with
tensile hold times is a phenomenon quite general to this class of
alloys.

     The acceleration in cyclic growth rates under load controlled
conditions in nickel base superalloys is a matter of some concern
when considering possible applications to disk alloys, where temperatures can approach 1200°F and where extended tensile dwell periods
are encountered. However, the earlier reported set of observations
should first be considered to provide further perspective on this
concern, namely, the results from tensile dwell periods from smooth
specimen strain controlled tests also controlled tests also conducted
at 1200°F. The wide difference in performance between smooth and
pre-defected specimens is striking and some consideration of the
reason for this large and important geometry effect on hold time
cyclic life is in order. The extension in life for the smooth

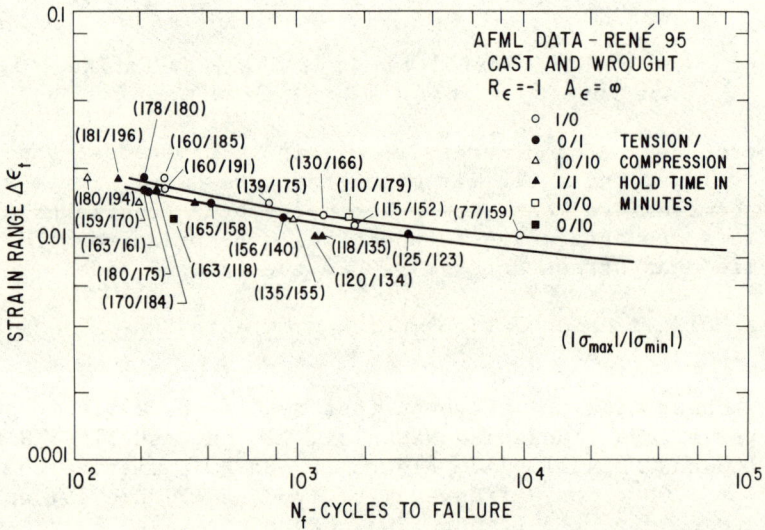

Fig. 17 Air Force Material Laboratory Hold Time Test Results on Cast and Wrought René 95, 1200°F, $A_\epsilon = \infty$.[46]

specimen is presumably related to the contribution of those cycles required to initiate a fatigue crack from some inherent defect size characteristic of the microstructure and to propagate that crack to the starting defect size used in the pre-defected tests. A direct calculation of this contribution is not possible at present since smooth specimen hold time data is currently available for $A_c = 1$ tests under strain control. However, an interesting deduction can be made if one assumes no difference in smooth specimen hold time tests for $A = \infty$ and $A = 1$ at 1200^0F. Since tensile strain dwell periods are not particularly damaging for smooth specimens at this temperature, the acceleration of crack growth with dwell time can only mean that the initiation life of the smooth specimen tests are delayed. Expressed analytically, we have,

$$N_i^c + N_p^c = N^c \tag{1}$$

and

$$N_i^h + N_p^h = N^h \tag{2}$$

Here the superscript c refers to continuous cycling, the superscript h to hold time, subscript i for the process of initiation and early growth of a crack from its initiating site, and p for the process of growing the crack from its early growth size to

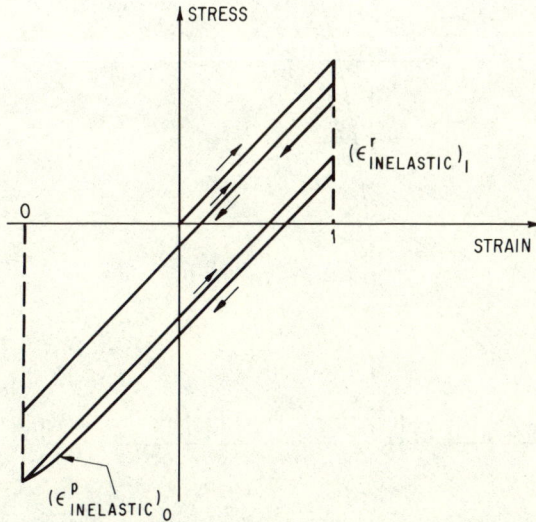

Fig. 18   Schematic of Hysteresis Loop Shift for High Strength Nickel-Base Alloys at Elevated Temperature with Tensile Strain Hold Times.

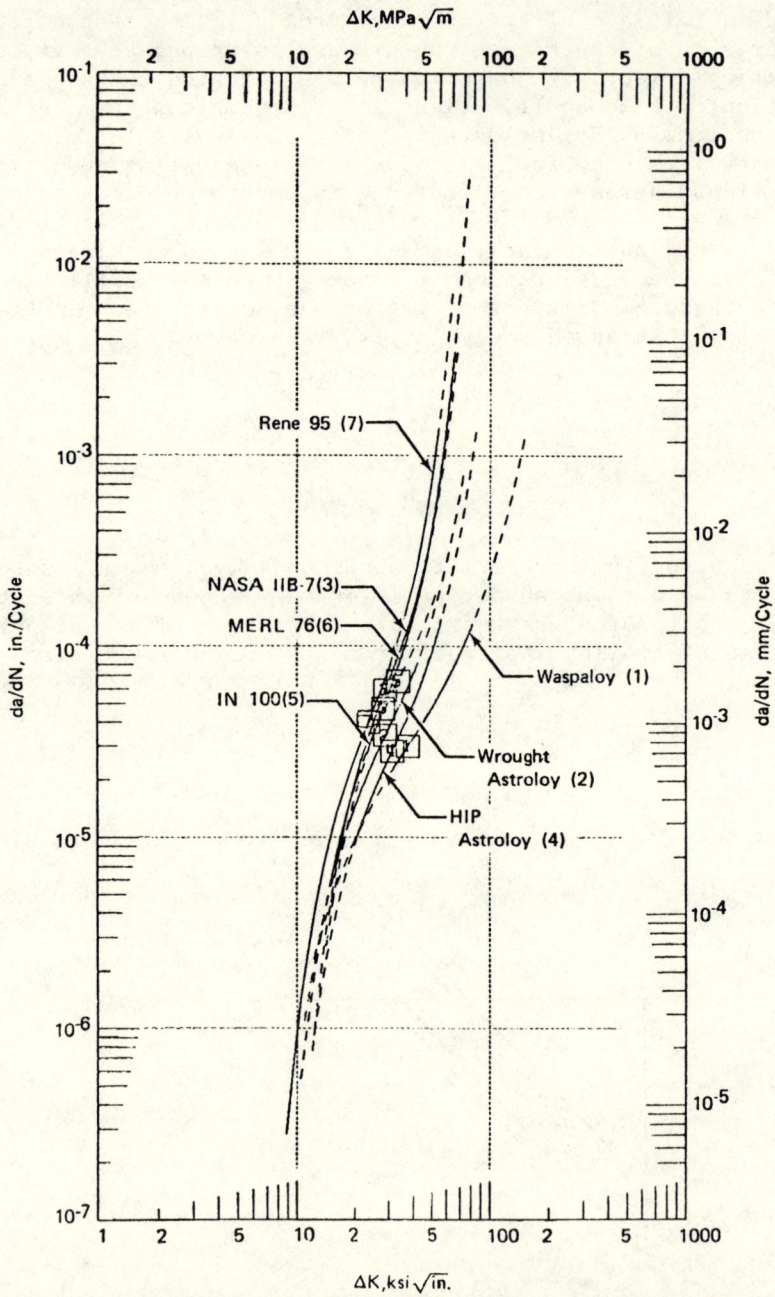

Fig. 19   Crack Growth Rate of Several Alloys at 0.33 H_z,   R = 0.05, 1200^0F.[48]

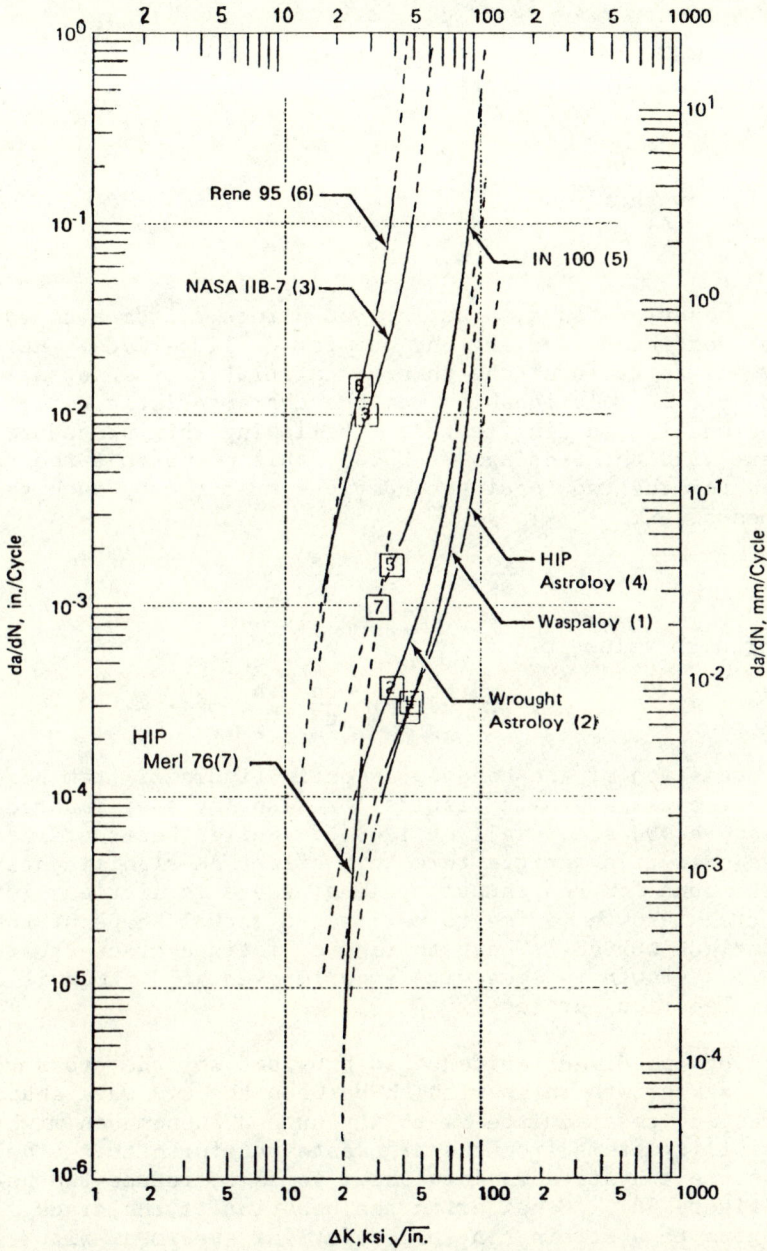

Fig. 20   Crack Growth Rates for Several Five Alloys with 900-sec
Dwell, R = 0.05, 1200°F.[48]

failure.   If, as assumed above, for smooth specimens

$$N^c = N^h \tag{3}$$

while from crack growth tests

$$N_p^c > N_p^h \tag{4}$$

it must follow that

$$N_i^h > N_i^c \tag{5}$$

    This behavior can also be derived directly from a da/dN vs. ΔK
curve for continuous cycling and one for dwell periods. Here the
life (cycles to failure) can then be calculated by integration of
the da/dN vs. ΔK relationship from the characteristic defect size
to that producing sudden fracture. Following this procedure and
consistent with the findings of Eq. 5, this means that the da/dN vs.
ΔK curves for the two loading conditions must cross, such that for
low values of ΔK,

$$(\frac{da}{dN})^c > (\frac{da}{dN})^h \tag{6}$$

while for high values of ΔK,

$$(\frac{da}{dN})^c < (\frac{da}{dN})^h. \tag{7}$$

    A suggestion of the behavior seen in Figure 21 from Reference
48, where the crack growth results for Waspaloy are compared for
0.33 Hz and a 900 sec. dwell at 1200^0F. While these data tend to
suggest a hold time fatigue threshold effect ΔK significantly greater
than that found for continuous cycling, there is little solid data
in this crack growth regime to verify the actual shape of the da/dN
vs. ΔK fatigue curves. The gathering of fatigue crack growth data
in these low growth regimes with superimposed hold times is an
extremely laborious process.

    Although no direct evidence is provided for the crossover in
fatigue crack growth rates with ΔK between the two wave shapes, it
is interesting to speculate as to why such a phenomenon might occur.
One possibility stems from the arguments put forth that a hold time
acts to cause a hysteresis loop shift so as to reduce the hold time
stress (Figure 18). Considering the behavior at the crack tip to
be analogous to a strain controlled fatigue specimen, with tensile
hold period the local tensile stress progressively relaxes, and the
compressive stress produced on unloading increases. Thus at low
values of ΔK, crack closure is enhanced and crack growth is retarded.
At higher stress intensities, this effect is overwhelmed by the
environmentally-enhanced static crack growth.

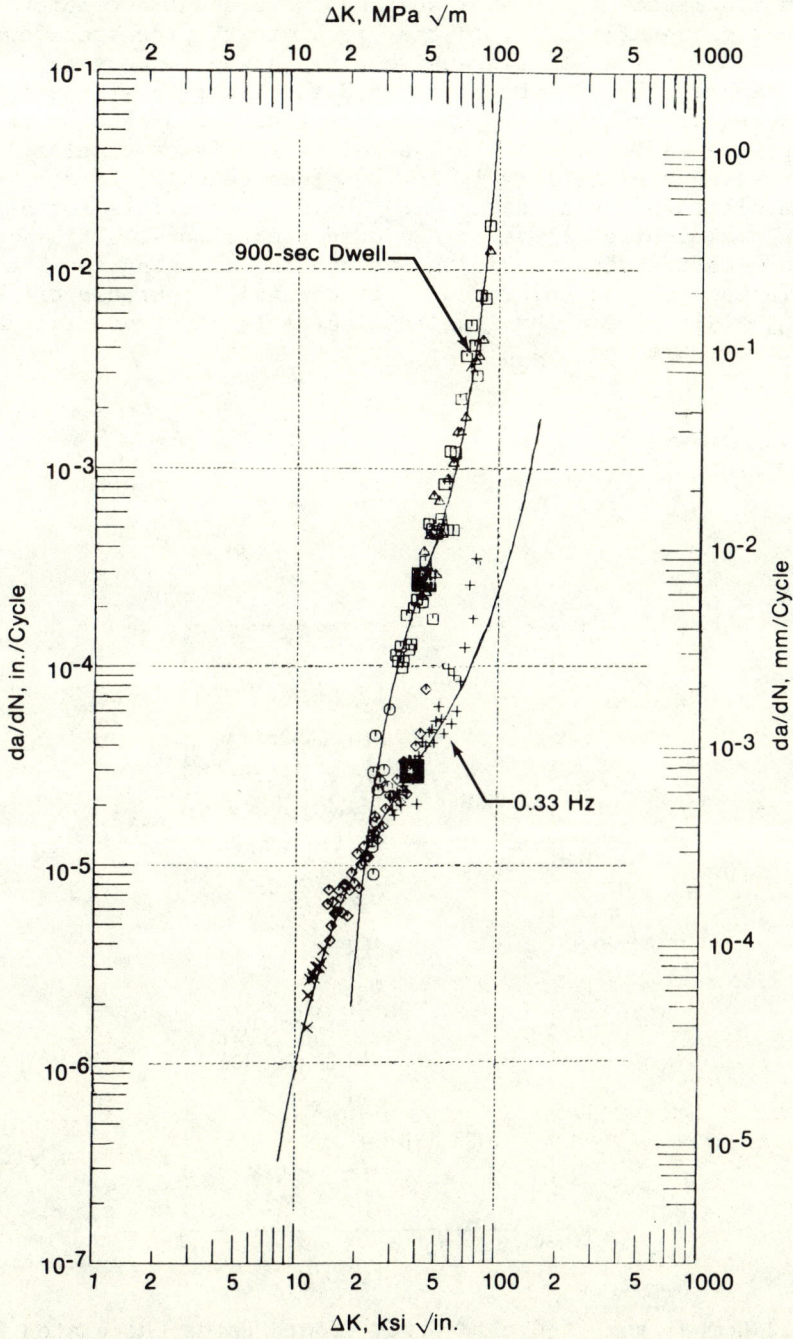

Fig. 21   Comparison of Waspaloy 900-sec Dwell and 0.33 Hz Crack
Growth Rates at R = 0.05, 1200°F. [48]

   The crack growth behavior seen by the present and referenced
data has a close similarity to the crack growth processes found for
ferrous alloys tested in aggressive aqueous environments. Here
static crack growth can occur provided the limiting stress intensity
factor K_{ISCC} is exceeded. Cyclic crack growth can also occur, accelerated by the environment. This acceleration is particularly marked
as the frequency of cycling is lowered (see below). It does not seem
unreasonable to propose that a K_{ISCC} value also exists for high
strength nickel-base alloys at elevated temperature. This K_{ISCC}
would then define the transition between the damaging tensile hold
time effects due to accelerated environmentally enhanced crack growth
when $K_{max} > K_{ISCC}$, and the beneficial effects of stress relaxation
and crack closure when $K_{max} < K_{ISCC}$.

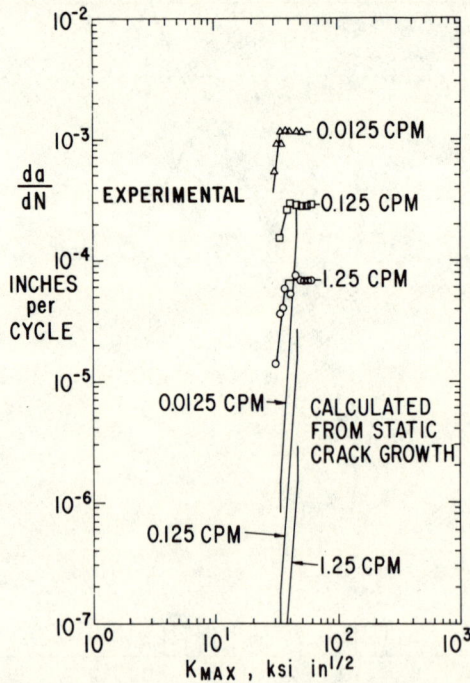

Fig. 22 Comparison of Cyclic Crack Growth Rates Calculated from
       Static Crack Growth Data with Experimentally Determined
       Cyclic Crack Growth Rates.[50]

Further Considerations on Fatigue Crack Growth At Elevated Temperatures

The previous discussion on the acceleration of fatigue crack
growth in deeply cracked nickel-base superalloys with dwell periods
under load at elevated temperature raises a number of questions with
respect to the nature of this acceleration. For example, should
this phenomenon be regarded as stress corrosion, where the growth
is the result of environmental damage at the crack tip, or as creep
crack growth where the crack acceleration is the result of time-
dependent damage arising from creep processes. In this section we
shall consider two widely different material-environment-temperature
systems in which the effect of cyclic frequency on crack growth has
been investigated using standardized crack growth testing procedures,
and where growth comparisons have been made on the basis of static
crack growth information. One of these systems involves the fatigue
crack growth of SA333 Grade 6 carbon steel in 8 ppm oxygen water at
1500 psi and 550°F, at three frequencies - .0208, .00208, and
.000208 Hz,[49,50] the other, the fatigue crack growth of conventionally
aged In 718 in air at 650°C at three frequencies 1, 0.1 and 0.01
Hz.[51]

Figure 22 shows the crack growth results for the carbon steel
in oxygenated water plotted vs. K_{max}. Elsewhere[50] it was shown
that K_{max} gave a more useful representation of the crack growth
results for this system than ΔK, particularly when results using
different R-values were included. Essentially it was found that
environmental crack growth acceleration took place above K_{ISCC} and
growth was influenced more by K_{max} than by ΔK. A characteristic
of the environmentally-accelerated crack growth behavior of carbon
steel in oxygenated high temperature water is a rapid acceleration
in growth rate da/dN above K_{ISCC} followed by a plateau regime, as
shown in Figure 22. Further, as the frequency of cycling is lowered
the da/dN level of the plateau is increased such that, for the
frequency range covered, a ten-fold decrease in frequency caused
nearly a four-fold increase in growth.

Also shown in Figure 22 are predictions of cyclic crack growth
at the three frequencies of interest as determined from the view
that the growth can be derived from static crack growth information.
Static crack growth testing had been performed on this steel in the
environment and was found to be represented by the relationship

$$\frac{da}{dt} = AK^n = 8.9 \times 10^{-14} K^{5.1} \text{ in/min} \qquad (8)$$

The Wei-Landes[52] linear superposition was applied to these data to
determine whether this approach was useful in explaining the plateau cyclic crack growth. Details of the actual analysis are

given elsewhere.[50] Figure 22 gives the results. Clearly the measured fatigue crack growth is underpredicted for all three frequencies. However, at the lowest frequency and the highest value of
K_{max}, the predictions and the experimental results converge. Presumably cyclic effects not accounted for by the above analysis are
acting to accelerate the growth rate over that calculated.

Now, considering the second system mentioned above, that of
conventionally aged In 718, crack growth tests results obtained are
shown in Figure 23. Here, in contrast to the representation of Figure
22, these results are given in terms of da/dT vs. K or ΔK. The figure
contains both static results (using K) and cyclic results (in terms
of ΔK). As indicated by the authors, the fatigue curves are separate
and distinct from the creep results, except at the lowest frequency
when the growth rates are comparable. Although the representation
is different for the two systems, in each case the da/dN growth rate
is underpredicted by the creep model. Interestingly, both systems
show about a three to four-fold increase in cyclic growth rate for
a ten-fold decrease in frequency. Clearly the behavior of each is
the same, namely that the fatigue and creep crack growth processes
have similar responses despite the fact that we are considering two
widely different high temperature crack growth situations.

Creep-Environment Crack Growth Interaction

What would be helpful is a more realistic model for describing
the actual time-dependent deformation and fracturing processes at
the crack tip. It would be particularly helpful if one could monitor

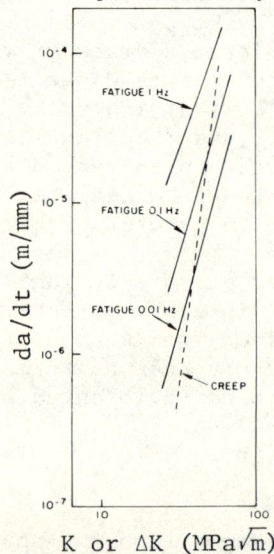

Fig. 23   Comparison of Fatigue and Creep Crack Growth Rates.
Conventionally Aged Material Tested in Air.[51]

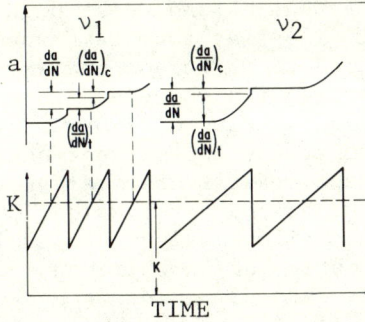

Fig. 24 Model Behavior for Crack Length vs. Time For Two Different Frequencies.

the growth on a nearly continuous basis to provide information as to what could be called the cyclic-dependent and the time-dependent growth during a loading cycle so that some of the complexities could be better sorted our. The possibilities are shown in Figure 24 wherein sawtooth wave forms of K vs. time are described for two different frequencies, ν_1 and ν_2, where $\nu_1 > \nu_2$. Also shown schematically in the figure is the crack length vs. time response for each frequency on a cycle-by-cycle basis. For each frequency the increment of growth da/dN is indicated, as is the cycle and time-dependent components $(da/dN)_c$ and $(da/dN)_t$, respectively. The separation of the growth per cycle into these two components as well as the specific shape of the $(da/dN)_t$ component as a function of time will provide a significant diagnostic technique for the determination of the crack growth processes. Further, it will be of great assistance in the modelling procedure. For example, the specific cycle-by-cycle growth as a function of K can be found and related to K_{ISCC} or some other K vs. da/dt static crack growth function. An experimental crack growth monitoring technique for accomplishing this is described in a recent report.[13]

It is worthwhile to consider the deformation that occurs in the plastic zone of a growing crack during a low frequency or dwell period test. It has already been shown that wave shapes which cause imbalance in the hysteresis loop can cause grain boundary cavity damage in smooth specimen tests, Figures 9 and 10. Certainly this effect can be expected to be operative in the crack tip plastic zone, accentuated by the high hydrostatic tension in this region. This form of damage can also be expected to depend on the number of loading cycles, since the earlier discussions described the damage as a ratchetting process, the grain boundary cavities growing progressively with each cycle.

Another aspect of the crack tip deformation process is the acceleration of creep that can occur with cycle plastic deformation. This was demonstrated dramatically by Feltner[53] for high purity aluminum at $78^\circ K$ under repeated loading at high stress (R=0) and by Benham[54] for structural alloys at room temperature. Thus cyclic creep can also be expected to influence the extent of the plastic zone and the strain distribution therein. This process is particularly significant when considering environmental damage. Vermilyea[55] has proposed a model for stress corrosion cracking in which cracking proceeds in specific steps. First chemical attack removes material at the crack tip until halted by passivation. Creep then takes place to rupture the passivated film causing more chemical attack and passivation, the process continuing in this manner. The mechanism is shown schematically in Figure 25. Since the cyclic creep process described above is not viewed as a thermally activated one, it is possible that this form of creep acceleration is operative at all temperatures and frequencies in the two systems under discussion (Figures 22 and 23).

A final point with respect to the creep-environment crack growth interaction problem relates to the role of gaseous atoms, notably oxygen diffusing into the crack tip region to cause intergranular cracking. Woodford,[56] Bricknell and Woodford[57] describe experiments on nickel base superalloys and on nickel following high temperature air exposure in which environmental damage has resulted. The extension of these views, whereby oxygen penetrates the grain boundary to finite depths, even at normal service temperatures, to embrittle the crack tip must be carefully considered.

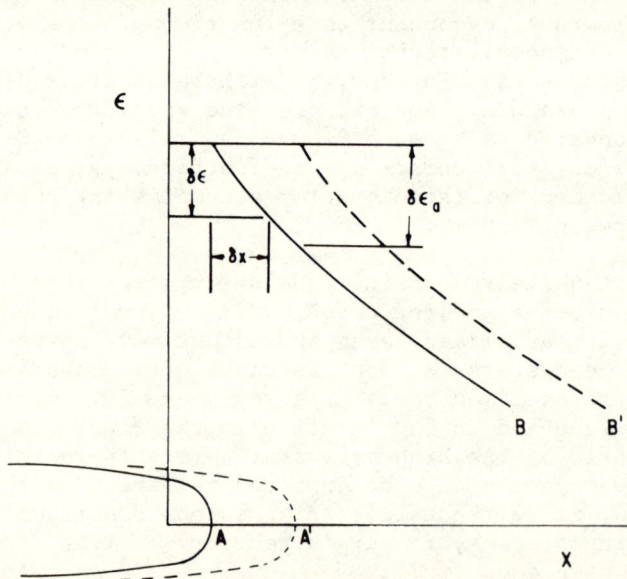

Fig. 25   Schematic Strain-Distance Relationships (Curves B, B')
          Ahead of a Stress Corrosion Crack Initially at Position
          A with a Second Position at A'.[55]

Table VII

SOME CHALLENGING QUESTIONS

Smooth Surface, Notch	Short Crack	Deep Crack
o Wave shape effects	o Method of representation	o Threshold behavior
o Long time extrapolation	o Modes of growth	K variations along front Closure Reaction products
o Environmental effects	o Closure	o Hold time and frequency effects
o Role of prior exposure	o Environment	o Time vs. cyclic dependent growth
o Role of microstructure	o Time vs. cyclic dependent growth	o Mechanisms of environmental damage
	o Treatment of defects	o Mechanisms of time dependent damage

SUMMARY

This complex subject can best be summarized by returning to our starting point with reference to Table VII, that is, by identifying the cracking problem at elevated temperature in terms of the three regimes: smooth surface or notch, short cracks and deep cracks. Under each category are listed many of the items covered in this paper, but some have not been covered. The items are self-explanatory. This list is a long one and many of the problem areas identified are complex and difficult. Some of these topics may be clarified by the papers you will hear during the next three days. However, for each topic removed from the list, a new one can be proposed to take its place. As we learn more, we ask more questions. For those of us active in the field the list is reassuring in that we will have plenty to do for some time to come. I am sure we will meet again on this subject.

REFERENCES

1. T. H. Topper, R. M. Wetzel and J. D. Morrow, "Neuber's Rule Applied to Fatigue of Notched Specimens", Journal of Materials, 1969, 4, 200-209.
2. J. D. Morrow, R. M. Wetzel and T. H. Topper, "Laboratory Simulation of Structural Fatigue Behavior", Symp. on Effects of Environment and Complex Load History on Fatigue Life, Amer. Soc. Test. and Mat., 1970, ASTM STP 462, 74-91.
3. J. M. Barsom and R. C. McNicol, "Effect of Stress Concentration on Fatigue Crack Initiation in HY-130 Steel", ASTM STP 559, Amer. Soc. Test. and Mat., 205-224.
4. W. G. Clark, Jr., "An Evaluation of the Fatigue Crack Initiation Properties of Type 403 Stainless Steel in Air and Steam Environments", ASTM STP 559, Amer. Soc. Test. and Mat., pp. 205-224.
5. T. A. Prater and L. F. Coffin, "The Use of Notched Compact Tension Tests for Crack Initiation Design Rules in High Temperature Water Environments", General Electric Report, 81CRD013, Feb. 1981.
6. R. P. Gangloff, "Electrical Potential Monitoring of Fatigue Crack Formation and Growth from Small Defects", ASTM Symp. on Fatigue Crack Growth Measurement and Data Analysis, Nov. 1979,
7. H. D. Solomon, "Low Cycle Fatigue Crack Propagation in 1018 Steel", Journal of Materials, Vol. 7, No. 3, 1972, pp. 299-306.
8. N. E. Dowling and J. A. Begley in Mechanics in Crack Growth, ASTM STP 590 (1976) p. 83.
9. M. F. Henry, "Crack Initiation and Early Growth in Low Cycle Fatigue - A Progress Report", General Electric Report, 72C100 (1972), Recommended Practice E606-80.

10. Annual Book of Standards Part 10, ASTM, Philadelphia, PA (1980).
11. C. J. Beevers, et al., The Measurement of Crack Length and Shape During Fracture and Fatigue, EMAS Ltd., Cradley Heath, Warley, UK, 1980.
12. Tentative Practice, E647-78T, Annual Book of Standards, Part 10, ASTM, Philadelphia, PA (1980).
13. T. A. Prater and L. F. Coffin, "Part Through and Compact Tension Corrosion Fatigue Crack Growth Behavior of Carbon Steel in High Temperature Water", General Electric Report, 81CRD159, August 1981.
14. K. N. Smith, P. Watson and T. H. Topper, "A Stress-Strain Function for the Fatigue of Metals", Journal of Metals, Vol. 5, No. 4, Dec. 1970, pp. 767-778.
15. L. F. Coffin, Jr., et al., Time-Dependent Fatigue of Structural Alloys, Oak Ridge National Laboratory, ORNL 5073 (June 1977).
16. L. F. Coffin in Methods of Predicting Material Life in Fatigue, 1979 Winter Annual Meeting, ASME, New York, 1979, pp. 1-24.
17. M. H. El Haddad, T. H. Topper and T. N. Topper, ibid. pp. 41-56.
18. M. H. El Haddad, K. N. Smith and T. H. Topper, Journal Eng. Materials and Technology, Vol. 101 (1979) pp. 42-46.
19. L. F. Coffin, Jr., Fracture 1977 I, Intern. Conf. Fracture 4, Waterloo, Ontario, June 1977, pp. 263-292.
20. D. Sidey and L. F. Coffin, Jr., Symposia on Fatigue Mechanisms, ASTM STP 675 (1979).
21. C. J. McMahon and L. F. Coffin, Jr., Met. Trans. 1, 3443 (1970).
22. D. A. Woodford and L. F. Coffin, Jr., 4th Bolton Landing Conf., Claitors Pub. Div., Baton Rouge, 421 (1974).
23. K. Veevers and K. V. Snowden, Journal Austral. Inst. Met. 20, 201 (1975).
24. G. J. Hill, Intern. Conf. on Thermal and High Strain Fatigue, Monograph and Report Series 32, pp. 312-27, The Metals and Metallurgy Trust, London (1967).
25. L. F. Coffin, Jr., Proceedings of International Conf. on Fatigue: Chemistry, Mechanics and Microstructure, NACE, pp. 590-600, 1972.
26. L. F. Coffin, Jr., Metall. Trans. 3, 1777-88 (1972).
27. L. F. Coffin, Jr., and J. F. Tavernelli, Trans. Metall. Soc., AIME 215, 749 (1959).
28. H. D. Solomon and L. F. Coffin, Jr., Fatigue at Elevated Temperatures, STP-520, pp. 112-22, ASTM, 1973.
29. J. T. Berling and J. B. Conway, Proc. 1st Int. Conf. on Pressure Vessel Technology, Delft, Holland, Part 2, pp. 1233-46 (September 29-October 2, 1969).
30. D. C. Lord and L. F. Coffin, Jr., Metall. Trans. 4, 1657 (1973).
31. W. Ostergren, J. Test. Eval. 4, 327-339 (1976).
32. M. M. Leven, Exp. Mech. 13, 353 (1973).
33. C. F. Cheng, et al., Fatigue at Elevated Temperatures, STP-520, pp. 355-364, ASTM (1973).

34. S. Majumdar and P. S. Maiya, ASME/CSME Pressure Vessel and
     Piping Conf., Montreal (June 1978).
35. B. Tomkins and J. Wareing, Met. Sci. J. 11, 414 (1977).
36. E. G. Ellison and AJF Paterson, Inst. Mech. Eng. 190, 333 (1976).
37. G. R. Halford, Cyclic Creep-Rupture Behavior of Three High Tem-
     perature Alloys, NASA-TN-D-6039 (May 1971).
38. S. S. Manson, Fatigue at Elevated Temperatures, STP-520, pp.
     744-82, ASTM (1973).
39. K. D. Sheffler, Vacuum Thermal-Mechanical Fatigue Testing Of
     Two Iron Base High Temperature Alloys, NASA-CR-134524 (1974).
40. L. F. Coffin, Jr., 1976 ASME-MPC Symp. on Creep-Fatigue Interac-
     tion, MPC-3, Amer. Soc. Mech. Eng., Dec. 1976.
41. R. Dutton, Mat. Res. and Standards, 9, No. 4, 11 (1969).
42. A. J. Perry, J. Materials Science, 9, 1016 (1974).
43. G. Wigmore and G. C. Smith, Metal Science J. 5, 58 (1971).
44. L. F. Coffin, Thermal Fatigue of Materials and Components, Amer.
     Soc. Test. and Mat., ASTM STP 612, 227 (1976).
45. L. F. Coffin, "The Effect of Environment and Wave Shape on the
     Low Cycle Fatigue Behavior of MZC Copper at $370^{0}C$", General
     Electric Report, 81CRD115, June 1981.
46. Life Prediction Techniques For Analyzing Creep-Fatigue Inter-
     action in Advanced Nickel-Base Superalloys, Wright State
     University, July 1976.
47. L. F. Coffin, "The Effects of Frequency on the Cyclic Strain and
     Fatigue Behavior of Cast René 80 at $1600^{0}F$", Met. Trans., I5,
     1053 (1974).
48. B. A. Cowles, D. L. Sims and J. R. Warren, "Evaluation of the
     Cyclic Behavior of Aircraft Turbine Disk Alloys", NASA Report
     CR159409, Oct. 1978.
49. T. A. Prater, F. P. Ford and L. F. Coffin, Metals Science, pp.
     424-432, Aug.-Sept. 1980.
50. T. A. Prater and L. F. Coffin, "Crack Growth Studies on a Carbon
     Steel in Oxygenated High-Pressure Water at Elevated Tempera-
     tures", General Electric Co. Report 81CRD067, April 1981.
51. S. Floreen and R. H. Kane, Fatigue of Engineering Materials and
     Structures, Vol. 2, No. 4, 1979, pp. 401-412.
52. R. P. Wei and J. D. Landes, Materials Research and Standards,
     Vol. 9, July 1969, p. 25.
53. C. E. Faltner, in Mechanisms of Fatigue in Crystalline Solids,
     Acta Metallurgica, Vol. 11, July 1963, pp. 817-828.
54. P. Benham, Met. Revs., Vol. 3, 1958, p. 11.
55. D. A. Vermilyea and R. B. Diegle, Corrosion, Vol. 32, 1976, pp.
     26-29.
56. D. A. Woodford, Metallurgical Transactions, Vol. 12A, pp. 299-
     308, 1981.
57. R. H. Bricknell and D. A. Woodford, Metallurgical Transactions,
     Vol. 12A, 1981, pp. 425-433.

CORROSION FATIGUE CRACK PROPAGATION

F.P. Ford

General Electric Research and Development Center
Schenectady, NY, U.S.A.   12301

There is a cogent argument for the proposition, illustrated in Figure 1, that there is a spectrum of rate-determining steps in environmentally-controlled subcritical crack propagation encompassed by the phenomena 'corrosion-fatigue' (C.F.), 'hydrogen-embrittlement' (HE) and 'stress-corrosion cracking' (SCC). The spectrum between the latter two phenomena was proposed by Parkins[1] and ranges from one limit where the cracking mechanism is dominated by strength and metallurgical parameters, epitomized by the hydrogen embrittlement of high strength steels in hydrogen gas, to the other limit where the cracking mechanism is dominated by the electrochemical processes in aqueous environments and is represented by systems such as mild-steel in nitrate or brass in ammoniacal solutions where failure is by active-path, slip-dissolution or (possibly) brittle film processes. Unfortunately many technologically important systems cannot be classified unambiguously at one of these limits, with the resulting academic debate as to the precise advancement mechanism or whether there is a combination of two distinct mechanisms.

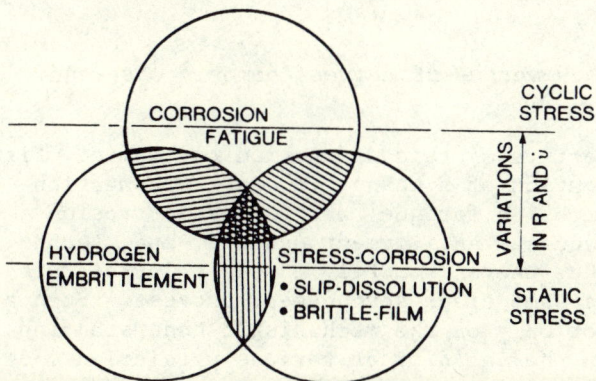

Figure 1. WEN diagram illustrating interaction of modes of subcritical crack growth under static load and under dynamic or cyclic load.

It can be argued that the mechanistic separation of corrosion-fatigue and the static modes of failure (SCC and HE) must be semantical since the operational R (= K_{min}/K_{max}) spectrum from -1 to 1 spans these two loading conditions. However, a distinct threshold condition has been observed in some alloy/environment systems above which environmental enhancement of fatigue crack growth da/dN is encountered, which corresponds to the condition where K_{max} during cyclic loading is equal to K_{Iscc} (static), the stress intensity value for the onset of subcritical crack growth under static loading conditions (Figure 2a). This observation has led to conclusions that corrosion-fatigue crack propagation rates can be predicted by summing the time-dependent crack growth under the appropriate static load in the environment, to that observed under fatigue loading in inert environments, i.e., the Wei-Landes superposition theory.[2] However, in other systems involving lower strength alloys in aqueous environments where slip-dissolution models are possible, it is apparent that environmental enhancement of the fatigue crack growth da/dN is observed at K_{max} values below K_{Iscc} (static) (Fig. 2b). These observations have led to categorizations of corrosion-fatigue [3] such as 'true-corrosion fatigue' (where $K_{max} \leq K_{Iscc}$ (static) and 'stress-corrosion fatigue' (when $K_{max} \geq K_{Iscc}$ (static)) with the implication that there are changes in the mechanism of corrosion fatigue crack propagation in these two K_{max} situations.

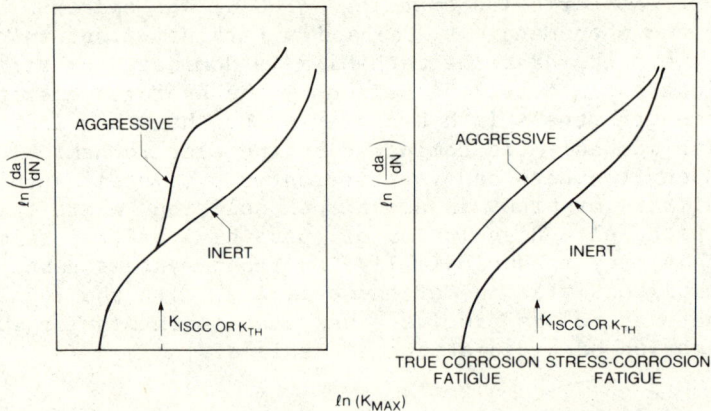

Figure 2.   Schematic da/dN versus ΔK curves commonly observed.

It is the general objective of this discussion to examine, first, the validity of this implication of a change in crack propagation mechanism between 'true-corrosion fatigue' and 'stress-corrosion fatigue' in ductile alloy/aqueous environment systems, and, second, the alternative proposal that there is merely a change in the rate-determining reaction of the same crack advancement process. Such a discussion has an impact not only on the mechanistic understanding of the process, but also on the validity of various prediction models

such as superposition(2) or process competition(4) and on design
criteria for components which (hopefully) are operating at low ΔK
and/or high R values.

Working Hypothesis

As a working hypothesis it is assumed that the time-dependent
crack advancement due to environmental effects alone is related to
the electrochemical reactions at the crack tip associated with either
dissolution (the slip-dissolution model), oxide-growth and -rupture
(the brittle-film model) or hydrogen-ion reduction (which would be
followed by hydrogen atom absorption and metal-metal atom bond rupture by some undefined hydrogen embrittlement mechanism). An examination of the factors involved[5] in both these oxidation (slip-
dissolution and brittle-film) and reduction (hydrogen embrittlement)
governed mechanisms indicates that there are common features (Fig. 3).
The crack growth step in both is electrochemically controlled; in
the oxidation theories the crack advance may be faradaically related
to the dissolution or oxide growth rates, whereas in the several
hydrogen embrittlement models the crack propagation rate is related
to the time required to achieve a critical hydrogen concentration
in front of the crack tip and this may be governed by the potential
dependent hydrogen adatom coverage on the crack-tip surface.

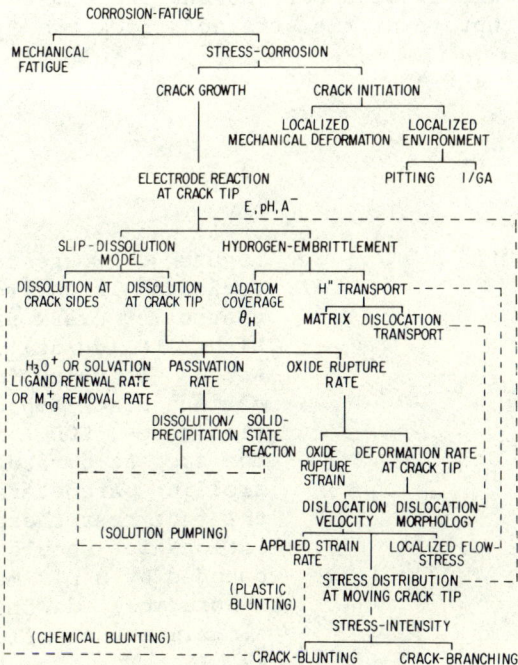

Figure 3. Interrelationship of parameters relevant to cracking in ductile alloy/aqueous environment systems.

However, there are further governing factors which are common to the crack advancement mechanisms. For instance, assuming that the strain at the crack-tip ruptures a thermodynamically stable and protective film (usually an oxide), both the mean dissolution rate and hydrogen adatom coverage at the crack tip will be dependent on the bare surface reaction rates, on the rate of passivation, and on the rate of oxide rupture by either creep under static loading conditions or on the applied strain rate under, for instance, corrosion-fatigue conditions. The similarity in the possible rate-determining steps between the two environmentally controlled advancement mechanisms and the myriad of interconnecting relationships between these possible rate-determining steps (associated, for example with solution pumping, chemical blunting, hydrogen transport by dislocations, etc.) make is difficult to distinguish unambiguously between the oxidation and reduction related mechanisms,[6] but a quantitative definition of these __common__ rate determining steps, rather than the advancement reaction itself, does make an analysis of the theoretical limits in crack-propagation possible for both static and dynamic straining conditions.

It is the specific objective of this discussion to examine the capability of the slip-dissolution model to derive quantitatively the crack velocities and morphologies in the observed spectrum of behaviour between 'stress-corrosion' and 'corrosion-fatigue' (Fig. 4), and to determine whether the K_{Iscc} value can be depressed through dynamic straining to the low values ($\sim \cdot 1$ $Ksi^{-3/2}$) that would be predicted[7] thermodynamically if a bare surface were maintained through continuous oxide rupture at the strained crack tip.

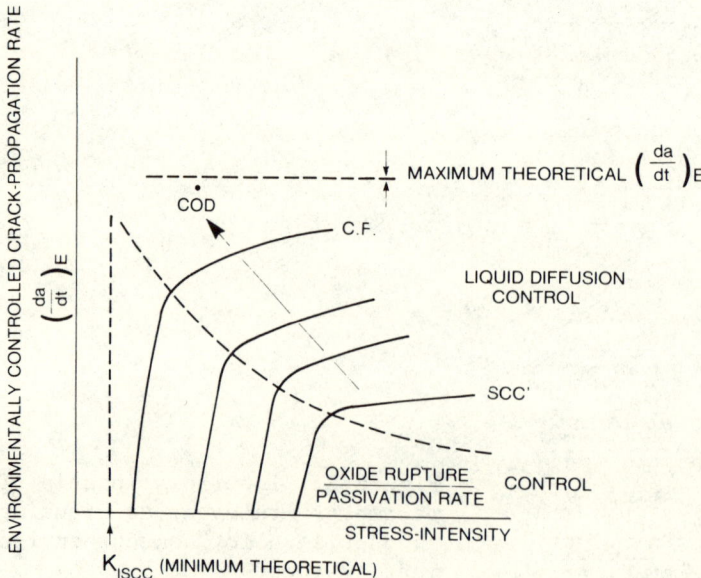

Figure 4. Suggested variation in environme[...] controlled crack propa[...] tion rate $(da/dt)_E$, wi[...] stress intensity for various crack tip defo[...] tion rates, COD. Note the suggested rate con[...] trolling parameters an[...] the fact that these re[...] lationships should be bounded by a maximum theoretical $(da/dt)_E$ a[...] a minimum theoretical K_{Iscc}.

Experimental Technique

The alloy/environment systems investigated were aluminum 7% magnesium (60% cold reduction followed by a 150°C for 4 hour stabilization treatment) in 0•5M Na_2So_4 or 1M NaCl at 25°C and 304 austenitic stainless steel (furnace sensitized at 600°C for 24 hours) in water containing 1.5 ppm O_2 at 98°C. Stress-corrosion and corrosion fatigue are possible in both of these systems[8,9] with changes in crack morphology from transgranular to intergranular depending on the stressing frequency and R value; in addition, neither system offers complications in data analysis due to chemical blunting of the crack tip, at least under the experimental conditions investigated.

Details of the corrosion fatigue and associated electrochemical experiments are given elsewhere.[8-14] In brief, however, the corrosion-fatigue data on the aluminum alloy[8] were obtained on .063" thick centre cracked sheet, whilst that on the stainless steel[11,13] was obtained on 1" thick compact tension specimens; in both cases the environment was confined to the crack region and was continuously renewed from an adjacent reservoir. Crack following under 'zero to tension' loading on the sheet specimens was accomplished by surface optical observation, whilst that on the CTS specimen was achieved by the use of a 10A D.C. potential drop technique. Reference data was obtained in dry argon.

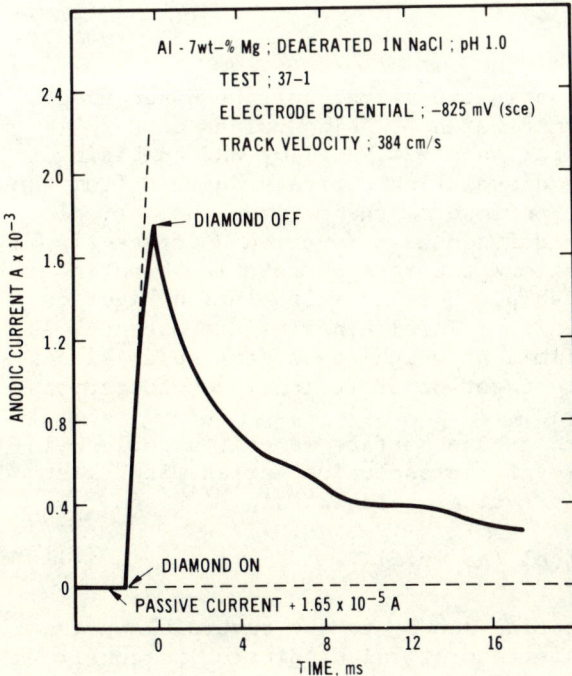

Al - 7wt-% Mg ; DEAERATED 1N NaCl ; pH 1.0
TEST ; 37-1
ELECTRODE POTENTIAL ; -825 mV (sce)
TRACK VELOCITY ; 384 cm/s

DIAMOND OFF

DIAMOND ON

PASSIVE CURRENT + 1.65 x 10⁻⁵ A

ANODIC CURRENT A x 10⁻³

TIME, ms

Figure 5. Observed current-transient on Al-7% Mg when scratched with a diamond stylus.[10] Note the increase in dissolution rate upon mechanical rupture of the protective surface oxide, followed by the decrease when the thermodynamically stable oxide reforms.

Experiments to determine the bare surface reaction rates expected at the strained crack tip used transient straining techniques on potentiostatically controlled electrodes of compositions corresponding to that expected at the tips of intergranular or transgranular cracks. Rupture of the surface oxide to rapidly expose the metal substrate was accomplished by either straining[12] a specimen wire 3.2% at a strain rate of 23 s^{-1} or else by scratching[16] the surface of a rotating disc electrode, with a contact time of 1 ms or less. In both techniques the transient in reaction rates (Figure 5) accompanying the increase in dissolution (and hydrogen-ion reduction) rate as the protective oxide was mechanically ruptured, and the subsequent decrease associated with the protective oxide nucleation and growth, could be analyzed in terms of the reaction rate (or current) <u>density</u>, since the reactive area could be measured or estimated.

Maximum Theoretical Crack Propagation Rate by Slip-Dissolution Model

The maximum environmentally controlled crack-propagation rate according to the slip-dissolution model may be theoretically derived,[6] provided it is assumed that at any instant there is always a small[15] fraction of the crack tip area which is maintained in a bare condition. In this situation the propagation rate, V (cms^{-1}) will be related to the bare-surface dissolution current density, i_a (A cm^{-2}), via Faraday's Laws, by

$$V = \frac{Mi_a}{n\rho F} \tag{1}$$

where ρ is the metal density, n is the valency of the dissolving metal cation and M is the atomic weight. Using values of n = 3, ρ = 2.7 g cm^{-3} and M = 26.98 for an Al-7% Mg alloy and combining Eq. (1) with the bare-surface dissolution currents derived from the transient oxide-rupture experiments, the maximum theoretical crack-propagation rate may be defined as a function of crack-tip electrode potential. This maximum theoretical rate is shown in Fig. 6 as the linear relationship, where the dissolution kinetics are under activation control, i.e., Tafel kinetics are obeyed. As discussed below, there is a limit at which these particular kinetics can no longer apply due to the onset of diffusional considerations.

Under the condition where a bare surface is continuously maintained, the activation enthalpy for dissolution varies with electrode potential according to the theoretical relationship[6,16]

$$\Delta H^* = 10 - 11.5 \ (E-E_o) \ \text{Kcal (gm mole)}^{-1} \tag{2}$$

where E is the electrode potential and E_o is the reversible potential for Al-7% Mg dissolution. This theoretical prediction is confirmed

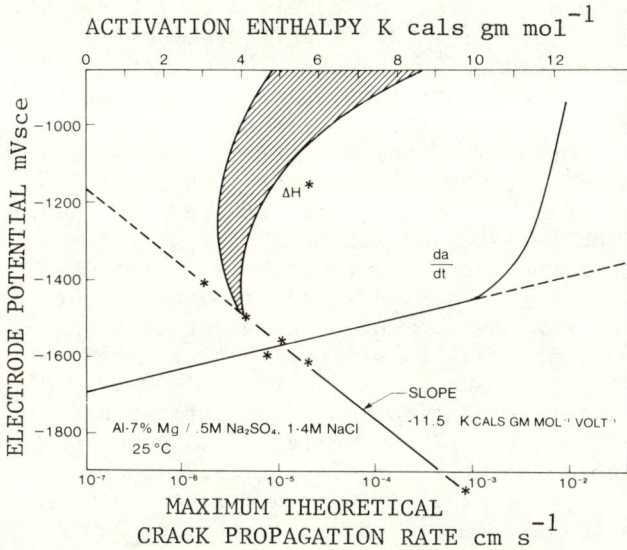

ACTIVATION ENTHALPY K cals gm mol^{-1}

MAXIMUM THEORETICAL
CRACK PROPAGATION RATE cm s^{-1}

Figure 6. Dependence of
maximum theoretical crack
propagation rate on the
crack tip electrode potential for slip-dissolution
theory,[6] and the poten-
tial dependence of the
activation enthalpy
associated with bare sur-
face dissolution of Al-7%
Mg in various dilute en-
vironments at 25°C.[16]

over a limited potential range by the experimental ΔH^* data points
in Fig. 6. Assuming that the concentration of Al^{3+} cations at the
crack tip is 1 M, then the activation enthalpy for dissolution by
Tafel kinetics will be zero at potentials more positive than ∿−1150
mV (SCE). Therefore at the open-circuit potential (∿−800 mV (SCE))
associated with aerated chloride solutions, (and assuming negligible
potential drop down the crack), the crack could advance at a maxi-
um rate equivalent to the removal of an atomic layer every atomic
vibration (∿10^4 cm s^{-1}). It would be extremely unlikely however
that this could ever be achieved since activation control will be
superseded by mass-transport effects associated with either the high
concentration of metal cations at the crack tip or the transport
of solvating water molecules down the crack. This change in rate
controlling reaction for bare surface dissolution is substantiated
by the fact that the ΔH^*/potential relationship in Fig. 6 no longer
obeys the activation law in Eq. 2 at more positive potentials but
deviates to ΔH^* values more symptomatic of liquid diffusion or
oxide precipitation processes on the electrode surface.

Using arguments[6] based on the flux of free water molecules to
the crack tip and estimating the limiting concentrations of these
species necessary to solvate the metal atom before an oxide forms,
yields a crack velocity in stagnant solutions of between 2 x 10^{-3}
and 8 x 10^{-3} cm s^{-1} as a limiting value for crack advancement under
bare surface dissolution.

The maximum theoretical crack propagation rate/electrode poten-

tial relationship (Fig. 6) predicts that the observed environmentally
controlled component of crack advance in Al 7% Mg/aqueous environment
system at 25°C should never exceed the values for activation-
controlled kinetics, regardless of the stressing mode (i.e., static
or cyclic) or stressing frequency in various sulphate and chloride
solutions. This is, in fact, the case; the potential dependence of
the environmental component of the propagation rate in $0.5M$ Na_2SO_4
(pH, 2.0) is illustrated in Fig. 7, together with the maximum theoretical rates for activation controlled dissolution. (The environmentally-controlled component of the crack growth rate is derived by
subtracting the $(da/dN)_{\Delta K}$ value obtained in argon from that observed
in solution, and converting this environmental increment to a time-
based (da/dt) $(-da/dN \times frequency)$.) It is clear that, at the
higher ΔK values and at potentials more negative than -1400 mV (SCE)
the activation-controlled slip-dissolution model correctly predicts
the observed crack propagation rate. Furthermore, it is seen in
Fig. 8 that the maximum theoretical rate at -1400 mV SCE is not
exceeded, even at high loading frequencies; by contrast, at -900 mV
SCE when the crack propagation rate at the bared crack tip should
be under liquid diffusion control, the propagation rate continuously
increases with loading frequency and the associated pumping action
of the crack sides.

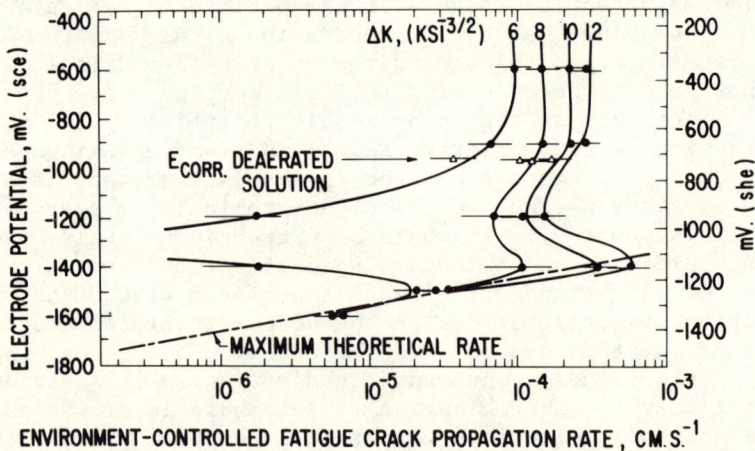

ENVIRONMENT-CONTROLLED FATIGUE CRACK PROPAGATION RATE, $CM.S.^{-1}$

Figure 7.   Variation of environment-controlled component of crack
            propagation with potential for aluminum 7% magnesium
            H36 at various ΔK values.[8] $0.5M$ Na_2SO_4, pH2.0, 11.6 Hz,
            Mode I crack orientation. Also shown is the maximum
            theoretical rate derived from the activation-controlled
            bare surface dissolution rate.[10,16]

     Thus it is concluded that the maximum corrosion-fatigue crack
propagation rate can, under certain potential conditions, be accurately predicted on the premise that a bare surface is maintained at

the crack tip and that the enhancement in crack propagation rate is faradaically equivalent to activation-controlled dissolution.   At lower ΔK values or loading frequencies it is unlikely that a bare surface can be maintained due to the lower oxide rupture rate at the crack tip; under these conditions, therefore, the controlling crack tip processes should change from those associated with activation control and diffusion to passivation controlled reactions.   The variations in activation enthalpy and reaction order over the crack propagation rate/potential range may be predicted[6] from the results of simulated crack tip reactions during transient straining tests (Fig. 9). These predictions are confirmed by the observation (Fig. 10) during corrosion-fatigue of Al-7% Mg[8] of a smooth transition in activation enthalpy from a low value associated with diffusional limitations at high ΔK values to higher values symptomatic of passivation control at lower ΔK values; similar effects are observed under static loading conditions,[17] the inference being that the ΔH*/ΔK/ν relationships form a spectrum of behaviour in the context of Figure 4.

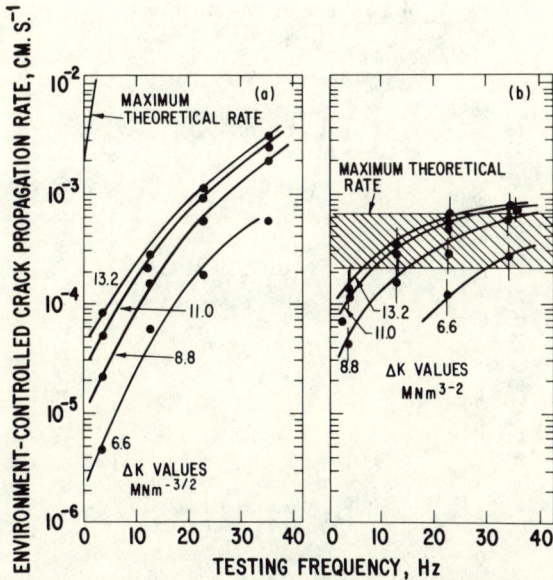

Figure 8. $(da/dt)_E$ versus frequency characteristics at various ΔK
values for aluminum 7% magnesium H36.[8] 20°C, 0•5M
Na_2SO_4 Mode I, pH2.0.   (a) -900 mV sce, (b) -1400 mV SCE
Also shown are the maximum theoretical rates under bare
surface activation or diffusion control.

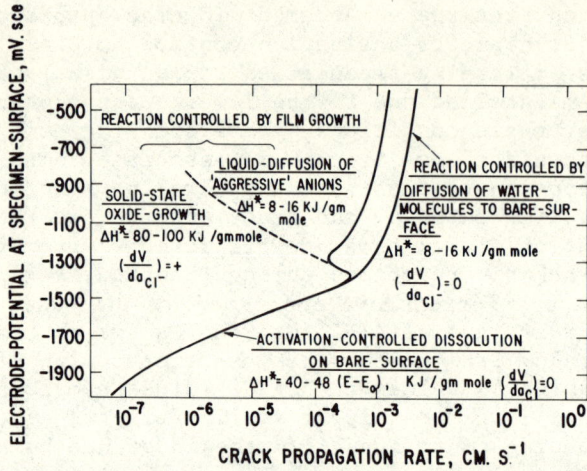

Figure 9. Theoretical prediction of the maximum theoretical crack propagation rates for the slip-dissolution model, derived from transient electrochemical observations on Al-7% Mg in chloride and sulphate solutions in the concentration range 1N to 4N.[8]

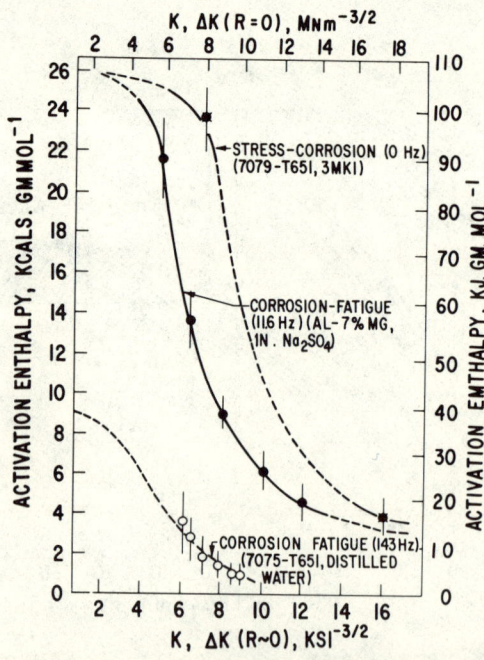

Figure 10. Relationship between activation enthalpy for $(da/dt)_E$ and the ΔK value for various aluminum alloys in aqueous environments and under various loading rate conditions (refs. 8, 19, 21).

The expected spectrum behaviour in Fig. 4 for both intergranular and transgranular crack propagation below the maximum theoretical rate is examined in the next section for the 304 stainless steel/water system, with ultimately the definition of the requirements for 'no' environmentally controlled crack propagation under a variety of loading conditions.

Oxide-Rupture Rate/Passivation Rate Control of Environmentally-Controlled Crack-Propagation

The maximum theoretical crack propagation rate corresponding to the situation where the oxide rupture rate is sufficient to maintain a bared surface at the crack tip is relatively independent of local metallurgical composition (at least for the solid solutions in 304 stainless steel[12] and aluminum-magnesium[16] alloys). As the oxide rupture rate is lowered, however, associated with lower K_{max} or loading frequency conditions, a bare surface is not maintained at the crack tip and the amount of metal penetration becomes governed by the local passivation kinetics and oxide rupture rates; in these cases the local alloy composition is expected to play an important role in the penetration kinetics, especially in the case of the FeCrNi alloys where the chromium and nickel concentrations will vary adjacent to a sensitized grain boundary where $(FeCr)_{23}C_6$ carbides have precipitated.

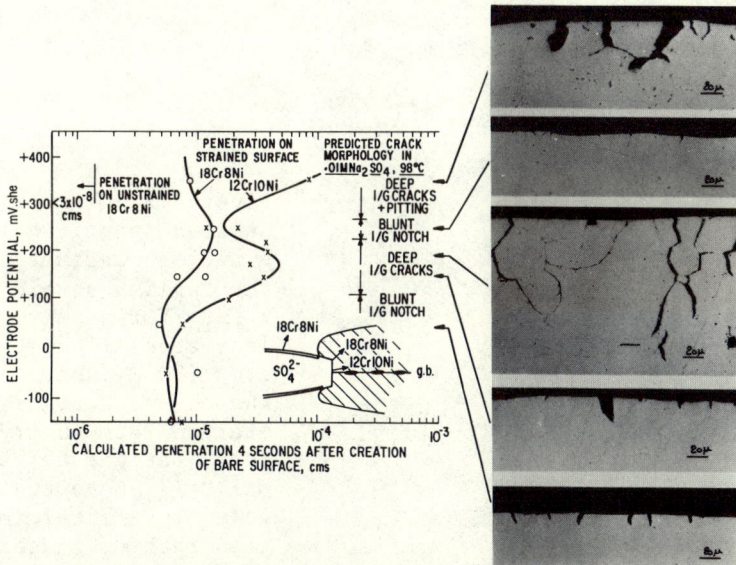

Figure 11. Variation of the calculated crack penetration four seconds after an oxide rupture event at the crack tip, with electrode potential. Note the agreement between predicted and observed morphologies.

Thus in the specific case of 304 stainless steel the crack morphology during environmental cracking may be predicted[12] from the current/time transients following the rapid straining of alloys of composition expected at the tip of an advancing crack. For instance, the relative penetrations in the transgranular and intergranular directions may be determined from the charge passed following rupture of the surface oxide of electrodes of composition Fe 18% Cr 8% Ni or Fe 12% Cr 10% Ni respectively (assuming that the average composition of the chromium depleted zone adjacent to a sensitized grain boundary is Fe 12% Cr 10% Ni[18]). The validity of this experimental approach to crack morphology prediction is shown in Fig. 11, which confirms that the deep intergranular cracking in sensitized 304 stainless steel in dilute environments at 98°C is to be expected, and is observed, in only a narrow potential region, and that outside this region the material degradation is either by pitting or by shallow blunt intergranular notches.

Expanding on this approach for morphology prediction from current-density/time transients at modelled crack tips, the variation in crack growth rate with stressing frequency may be predicted. In Fig. 12 the environmentally controlled crack growth rates are predicted for both intergranular and transgranular cracks as a function of crack tip strain rate, dimensionalized by $\dot{K}/K_{min}$. The crack growth rate, V, has been calculated from

$$V = \frac{M}{n\rho F} \cdot \frac{Q}{t} \tag{3}$$

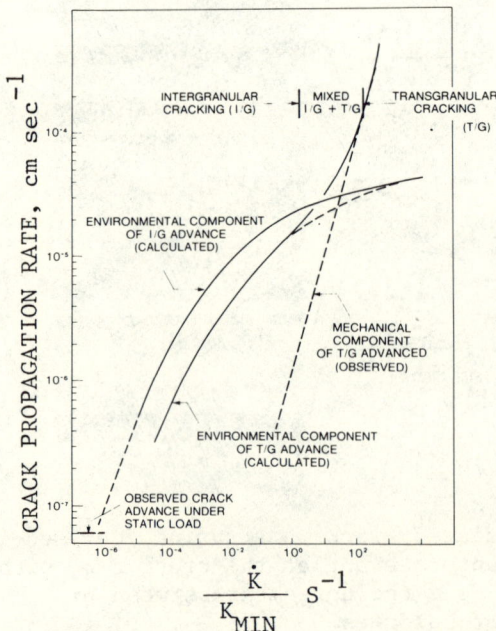

Figure 12. Variation in the calculated intergranular and transgranular penetration rates with the applied strain rate parameter $\dot{K}/K_{min}$ for sensitized 304 stainless steel in .01N Na_2SO_4, at 98°C and +150 mV she. Also shown is the observed transgranular penetration rate in dry argon due to 'dry' fatigue for a K_{max} of 25 Ksi.$^{-3/2}$ Note that at low $\dot{K}/K_{min}$ values the intergranular penetration rate dominates, but this is superseded by the total transgranular penetration rate at higher $\dot{K}/K_{min}$ conditions.

where Q is the charge density (C cm^{-2}) passed following oxide rupture of either an Fe 18% Cr 8% Ni or an Fe 12% Cr 10% Ni electrode and t is the time corresponding to the oxide rupture periodicity at a crack tip.

The value of t, used in the computation of Fig. 12, is obtained from an estimate of the oxide fracture strain (~.003) and the local applied strain rate, $\dot{K}/K_{min}$. In interpreting the data in Fig. 12, it is assumed that the crack will penetrate down that vector where the crack propagation rate is highest. Thus, it is seen that at low $\dot{K}/K_{min}$ values (i.e., low ΔK and/or low loading frequencies and high R values) the crack morphology should be intergranular, whereas at high $\dot{K}/K_{min}$ values the total transgranular penetration due to the environmental component and the mechanical fatigue component observed in argon (at a K_{max} value of 25 Ksi$^{-3/2}$) exceeds the intergranular penetration, and the crack morphology should change to totally transgranular. These predictions on the strain rate effect on crack propagation rate and morphology are upheld, in part, by the data in Fig. 13(a) and (b). For the corrosion-fatigue and stress-corrosion of sensitized 304 stainless steel in $H_2O/1.5$ ppm O_2 at 98°C, the crack propagation rate increases with $\dot{K}/K_{min}$, regardless of whether $\dot{K}/K_{min}$ (= $\Delta K/t \cdot K_{min}$) is defined by changes in ΔK, R value or loading time, with a corresponding gradual change from intergranular morphology at low crack tip strain rates to transgranular at high strain rates.

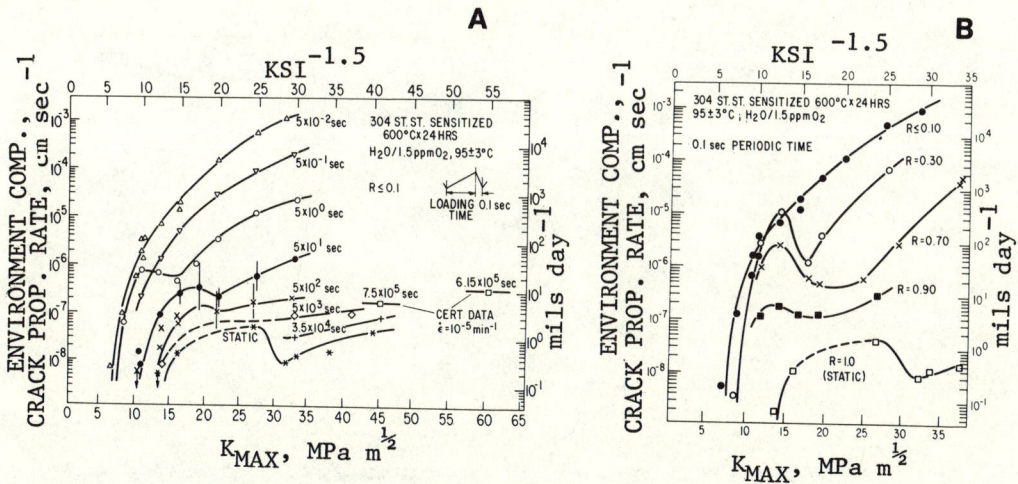

Figure 13. Variation of the environmentally-controlled component of crack propagation rate (da/dt)$_E$ with maximum stress intensity for the 304 stainless steel/H_2O system at 98°C under both static and cyclic loading conditions (Refs. 11,13). (a) Effect of loading time under cyclic loading at R = .1. (b) Effect of R value for static and cyclic loading at 10 Hz loading frequency.

The data in both Figures 13 (a) and (b) may be normalized in terms of the fundamental controlling parameters which are varying in these particular tests; oxide rupture and the ease of solution flow to the crack tip (categorized by the change in crack tip opening displacement ΔCTOD) and the subsequent oxide rupture rate, dimensionalized by $\dot{K}/K_{min}$. (Note that for the fixed alloy/environment system the other fundamental parameter, passivation rate, remains constant.) In Figure 14(a) it is seen that a given environmentally-controlled crack propagation rate is only achieved, at a variety of R values and loading rates, upon the satisfaction of two criteria; (i) the achievement of a critical amount of deformation at the crack tip ΔCTOD, and (ii) that this deformation is achieved at a given rate, $\dot{K}/K_{min}$. Thus, a crack may be propagating at a given velocity because of the satisfaction of these two criteria, but if either are contravened, for instance, the $\dot{K}/K_{min}$ criterion due to stress relaxation, then the crack may slow down or arrest. Such a beginning of an insight into these crack propagation criteria offers a possible explanation of such (experimentally annoying) phenomena as the inability to reproduce crack propagation data following periodic equipment shut-downs, even though the stressing conditions are nominally reproduced.

A   B

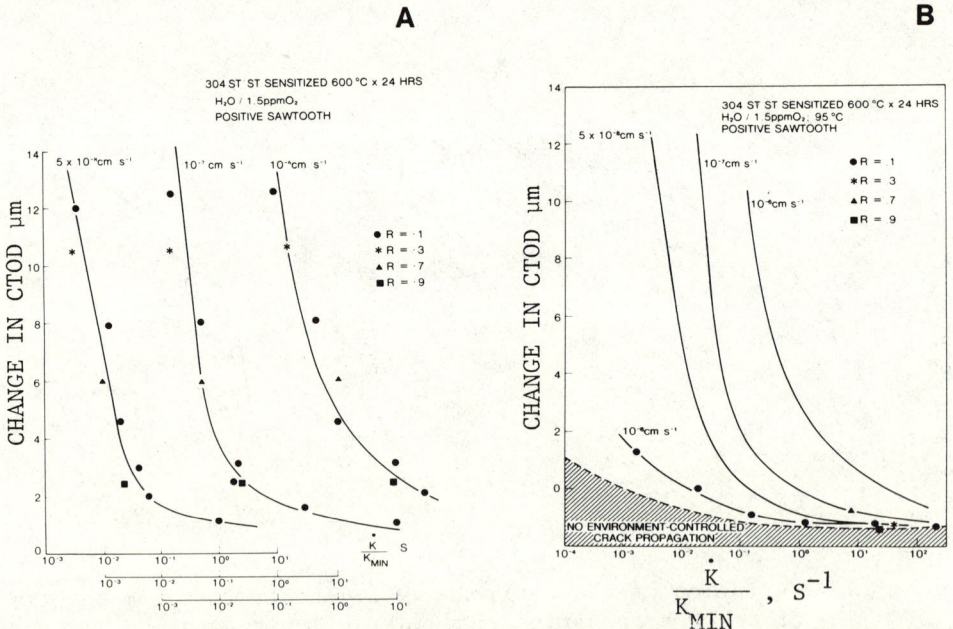

Figure 14.   (a) The crack tip deformation (ΔCTOD) and deformation rate ($\dot{K}/K_{min}$) criteria for environmentally controlled crack propagation in the 304 stainless steel/H_2O system at 98°C at specific propagation rates. Note the wide range of R values which are used in these evaluations.

It is seen in Fig. 14(b) that by applying these criteria to lower crack propagation rates, which practically could be equated with 'no' crack growth that the condition of 'K_{Iscc}' will be dependent on not only $\Delta CTOD$, but also on $\dot{K}/K_{min}$, especially at the lower applied strain rates. Thus, it is to be expected that K_{Iscc} (defined as the K_{max} condition where $da/dt = 10^{-8}$ cm s^{-1}) will vary with R and loading frequency with a minimum value which corresponds to a $\Delta CTOD$ which (presumably) relates to the primary fundamental requirement that the oxide at the crack tip <u>first</u> be ruptured before <u>any</u> environmental effect can be observed. This predicted effect is seen in Fig. 16a and 16b for stainless steel in water at 98°C,[11,13] and aluminum alloys[8,19] in dilute environments at 25°C. It is interesting to note that the minimum K_{Iscc} value is defined when the ratio K_{max}/E is in a relatively narrow band which is common to a wide range of alloy/environment systems[3,20] (presumably implying a similar fracture strain for the thin oxides in these systems) and that this corresponds to a K_{max} of 5-6 $ksi^{-3/2}$ for an $R \leq .1$; thus, this oxide rupture criterion supersedes any other criterion and the thermodynamic predictions of a minimum K_{Iscc} of $\sim .1$ $ksi^{-3/2}$ based on the <u>assumption</u> that the oxide is permanently ruptured, are never realized.

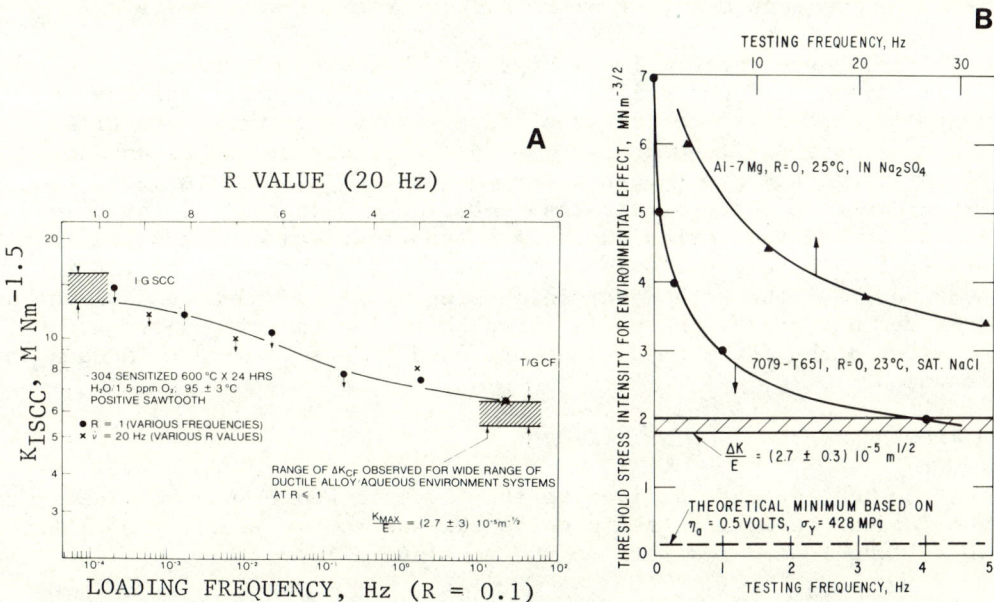

Figure 15. Variation of K_{Iscc} (defined as the K_{max} value above which environmental effects are observed on (fatigue) crack propagation) with loading frequency and R. (a) for 304 stainless steel in H_2O at 98°C. (b) for various aluminum alloys in Na_2SO_4 or NaCl at 25°C.[8,19] Note that the minimum theoretical K_{Iscc} value based on thermodynamic reasoning using the assumption that a bared surface is alwasy maintained, is not achieved.

Conclusions

It is concluded that as well as the spectrum of mechanisms of environmentally controlled cracking under static load there is a spectrum of rate determining steps between stress-corrosion and corrosion-fatigue. For the ductile alloys in aqueous environments, it has been shown that the rate determining steps in the environment-controlled component of crack propagation $(da/dt)_E$ may be the oxide rupture rate, liquid diffusion rate and passivation rate. Since the former two parameters are affected by dynamic straining, the $(da/dt)_E$ value and (through the interaction of oxide rupture periodicity and passivation rate) the crack morphology will change gradually with such operational parameters as R, K_{max} and loading frequency. The maximum $(da/dt)_E$ value can be correctly predicted under stressing conditions which maintain a bare surface at the crack tip, through the faradaic equivalence of dissolution rate and crack propagation rate. The $(da/dt)_E$ values at stressing conditions which do not maintain a bared crack tip can be uniquely described at a given electrode potential and passivation rate, by the mechanical criterion that a combination of a critical change in crack tip displacement together with a subsequent critical oxide rupture rate be achieved.

From these factors it is apparent that simple superposition prediction models of corrosion-fatigue will underestimate the environmentally controlled component of fatigue crack growth in ductile alloy/aqueous environment systems, and that environmental enhancement of fatigue crack growth can occur when the K_{max} value is below the K_{Iscc} observed under static load conditions. It follows that, in systems where the corrosion-fatigue behaviour can be predicted from static load stress-corrosion tests (e.g., high strength steels in hydrogen[2]), the rate determining steps in these advancement mechanisms are not one of those discussed above for the slip-dissolution model, and the advancement is controlled by a strain-rate independent reaction.

Acknowledgments

Acknowledgment is given to the Electric Power Research Institute who funded the work on stainless steel and to the Ministry of Defence (U.K.) who funded the work on the aluminum-alloy.

References

1. R.N. Parkins, British Corr. J. 7, 15, 1972.
2. R.P. Wei and J.D. Landes, Mats. Res. and Standards, July 25, 1969.
3. M.O. Speidel, M.S. Blackburn, T.R. Beck, and J.E. Feeney, "Corrosion Fatigue-Chemistry Mechanics and Microstructure," University of Connecticut, Storrs, June 1971 (Eds. O. Devereaux, A. McEvily, R.W. Staehle; Pubs. NACE Houston 1972) p. 324.

4.   I.M. Austen and E.F. Walker, "Mechanisms of Environment Sensi-
      tive Cracking," University of Surrey U.K., April 1977 (Eds.
      R.E. Swann, F.P. Ford, A.R.C. Westwood; Pubs. Metals Society)
      p. 334.
5.   F.P. Ford, Chapter "Stress-Corrosion Cracking" in Advances in
      Corrosion Science-I (Ed. R.N. Parkins; Pub. Applied Science
      Publishers).
6.   F.P. Ford, Ref. 4, p. 125; Metal Science, July, 326, 1978.
7.   F.P. Ford, Third International Conference on Mechanical Behav-
      iour of Materials, University of Cambridge, August 1979 (Eds.
      K.J. Miller and R.F. Smith; Pub. Permagon Press) Volume 2,
      p. 431.
8.   F.P. Ford, Corrosion 35, 281, 1979.
9.   F.P. Ford and M.J. Povich, Corrosion 35, 569, 1979.
10. F.P. Ford, G.T. Burstein and T.P. Hoar, J. Electrochemical Soc.
      127, 1325, 1980.
11. F.P. Ford and M. Silverman, Paper 181 Corrosion-80, Conference
      Chicago, March 1980.
12. F.P. Ford and M. Silverman, "Mechanisms of Environmentally-
      Enhanced Cracking," First Semiannual Report of EPRI Contract
      RP1332-1, June 1979.
13. F.P. Ford and M. Silverman, ibid, Second Semiannual Report,
      December 1979.
14. T.P. Hoar and F.P. Ford, J. Electrochemical Soc. 120, 1013,
      1973.
15. D.A. Vermilyea and R.B. Diegle, Corrosion 32, 25, 1976.
16. F.P. Ford, PhD Thesis, Cambridge University, 1973.
17. M.O. Speidel, Proc. of Conf. "Theory of Stress-Corrosion Crack-
      ing in Alloys," Ericeria Portugal, p. 289, 1971 (Ed. J.C.
      Scully; Pub. NATO).
18. R. Rao, "Microstructural and Microchemical Studies in Sensitized
      Stainless Steel," Micon-78, Houston April 1978; to be published
      in ASTM STP 672.
19. M.O. Speidel, Proc. Int. Congress Met. Corrosion 1974, pp. 439-
      442 (Eds. Satro and Mario; Pub. NACE).
20. J.D. Harrison, Metal Construction and Br. Weld. J. 93, 1970.
21. R.P. Wei, Int. J. Fracture Mech. 4, 159, 1968.

SURFACE REACTIONS AND FATIGUE CRACK GROWTH

R. P. Wei and G. W. Simmons

Lehigh University
Bethlehem, PA 18015

INTRODUCTION

Recent fracture mechanics and surface chemistry studies of environment assisted crack growth in gaseous environments[1-4] have shown that crack growth may be controlled in some systems by the rate of surface reactions and in others by the rate of transport of the aggressive environment to the crack tip. Based on considerations of surface reactions and gas transport, a model for surface reaction and transport controlled fatigue crack growth in single component gaseous environments was proposed and experimentally verified.[5-8] This model is able to account for the influences of gas pressure and cyclic load frequency on the rate of fatigue crack growth. In practice, however, there is usually more than one gas in a given environment. The various component gases can compete for surface reaction sites and therefore alter the fatigue crack growth response. The effect of this competition needs to be considered. In this paper, the development of the model for surface reaction and transport controlled fatigue crack growth in single component gaseous environments is briefly reviewed. Extension of the same considerations to fatigue crack growth in a binary gas mixture, in which one of the components acts as an inhibitor, is described. Quantitative application of this model to the consideration of oxygen on fatigue crack growth in humid air is discussed.

MODELING OF FATIGUE CRACK GROWTH IN ONE COMPONENT GAS

Modeling of environment assisted fatigue crack growth in single-component gaseous environments was based on the proposition that the rate of crack growth in an aggressive environment, $(da/dN)_e$, is composed of the sum of three components.[5,8,10]

$$(da/dN)_e = (da/dN)_r + (da/dN)_{cf} + (da/dN)_{scc} \qquad (1)$$

$(da/dN)_r$ is the rate of fatigue crack growth in an inert environment and, therefore, represents the contribution of "pure" (mechanical) fatigue. This component is essentially independent of frequency at temperatures where creep is not important. $(da/dN)_{scc}$ is the contribution by sustained-load crack growth (that is, stress corrosion cracking) at K levels above K_{Iscc}.[11] The contribution by sustained-load crack growth, i.e., the $(da/dN)_{scc}$ term, has been examined in some detail previously,[11,12] and appears to be adequately accounted for by the superposition model.[11] $(da/dN)_{cf}$ represents a cycle-dependent contribution requiring synergistic interaction of fatigue and environmental attack, and was not considered by Wei and Landes.[11] This model,[5] therefore, was directed specifically to the cycle-dependent term, and was limited to establishing a formal framework for estimating frequency and pressure dependence of $(da/dN)_{cf}$ in single-component gaseous environments. (Examinations of the influences of other loading variables (such as, load ratio) are in progress.)

In the model,[5] environmental enhancement of fatigue crack growth is assumed to result from embrittlement by hydrogen that is produced by the reaction of hydrogenous gases (e.g., water vapor) with the freshly produced fatigue crack surfaces. More specifically, $(da/dN)_{cf}$ is assumed to be proportional to the amount of hydrogen produced by the surface reaction during each cycle, which is proportional in turn to the "effective" crack area[1] produced during the prior loading cycles and to the extent of surface reaction. The time available for reaction is assumed to be equal to one-half of the fatigue cycle.[5] Transport of gaseous environments to the crack tip is assumed to be by Knudsen flow[13] at low pressures. The rates of hydrogen diffusion and embrittlement were assumed to be much faster than those of gas transport and surface reaction, and, therefore, did not need to be considered.

The governing differential equations for flow and surface reactions, and the relationship for $(da/dN)_{cf}$ are as follows:[5]

---

[1] In the original derivation,[5] it was assumed that only the area produced by the previous cycle of crack growth contributed to embrittlement. It is recognized that a greater portion of the crack surface remained active, particularly for slow surface reactions and at high frequencies, and can contribute hydrogen to the embrittlement process.[8] To provide consistency, while retaining the form of the original model, an "effective" crack area or "effective" crack increment Δa^* is defined here.

$$\frac{dp}{dt} = -\frac{SN_o RT}{V}\frac{d\theta}{dt} + \frac{F}{V}(p_o - p) \tag{2}$$

$$\frac{d\theta}{dt} = k_c p f(\theta) = k_c p(1 - \theta) \tag{3}$$

$$(da/dN)_{cf} \propto \theta \cdot \Delta a*^{1/} \tag{4}$$

The terms in the equations are as follows:[5]

> $F = 8.72 \times 10^2 \; \beta(\sigma_{ys}/E)^2 \; (T/M)^{1/2} \; B\ell$ (in m^3/s) = Knudsen flow parameter that depends on dimension and shape of the capillary, molecular weight (M) of the gas and temperature (T). The specific form of this expression reflects an attempt to account for constriction in flow by the real crack, where ℓ is a selected distance (of the order 10^{-6} m) from the crack tip used in defining a crack opening and β is an empirical quantity to be determined from the crack growth data.[5,13]

> k_c = reaction rate constant; $(Pa-s)^{-1}$.

> N_o = density of surface sites; molecules $(atoms)-m^{-2}$.

> p = pressure of gas at the crack tip; Pa.

> p_o = pressure of gas in the surrounding environment; Pa.

> R = gas constant = 1.38×10^{-23} $Pa-m^3-molecules^{-1}$.

> S = area of "effective" crack surface per cycle = $\alpha \; (2B\Delta a*)$, where $\Delta a*$ = "effective" increment per cycle, B = specimen thickness, and α = empirical constant for surface roughness and crack geometry.[5]

> T = absolute temperature.

> V = control volume at the crack tip, *i.e.*, volume associated with the distance ℓ.

> θ = fractional surface coverage or extent of reaction of surface per unit area.

From eqn. (2), it can be seen that the rate of change of pressure at the crack tip depends on the decrease in pressure produced by reaction of the environment with the active ("effective") crack surface and on the increase in pressure from the influx of gas from the external environment. The form $f(\theta) = 1 - \theta$ in eqn. (3) incorporates the assumption that the surface reaction is first-order in relation to available surface sites.[4-7]

Solutions for eqns. (2) and (3), with $f(\theta) = 1 - \theta$, were obtained for two limiting cases, for $0 < \theta < 1$, and were used in

conjunction with eqn. (4) to estimate the influences of key environmental and loading variables on the cycle-dependent component of fatigue crack growth rate, $i.e.$, on $(da/dN)_{cf}$.[5]

Case I: Transport controlled.

$$\theta \approx \frac{F}{SN_o RT} P_o t \quad \text{for} \quad \frac{SN_o RTk_c}{F} >> 1 \tag{5}$$

Case II: Surface reaction controlled.

$$\theta \approx 1 - \exp(-k_c p_o t) \quad \text{for} \quad \frac{SN_o RTk_c}{F} << 1 \tag{6}$$

In the transport controlled case, because of the rapid reactions of
the environment with the freshly created crack surfaces (high k_c) and
the limited rate of supply of the environment to the crack tip, significant attenuation of gas pressure takes place at the crack tip.
The extent of surface reaction (θ) during one cycle is controlled by
the rate of transport of the aggressive environment to the crack tip,
and thus varies linearly with time (see eqn. (5)). For the surface
reaction controlled case, the reaction rates are sufficiently slow
so that the gas pressure at the crack tip is essentially equal to the
external pressure. The extent of reaction, for $f(\theta) = 1 - \theta$, becomes
an exponential function of time, eqn. (6).

As a modification to the original model,[5,8] the "effective"
crack increment (Δa^*) is now chosen, along with ℓ in the expression
of F, to be equal to the growth increment per cycle at saturation
($i.e.$, for $\theta \approx 1$, $\Delta a^* = \ell = (da/dN)_{e,s} \cdot 1$). By taking t equal to $\tau/2$
or $1/2f$,[5,8] the following expressions can be obtained from eqns. (4),
(5) and (6):

Case I: Transport controlled.

$$(p_o/2f)_s \approx \frac{SN_o RT}{F} = \left[4.36 \times 10^2 \frac{\beta}{\alpha} \frac{\sigma_{ys}^2}{N_o RTE^2}(T/M)^{1/2}\right]^{-1} \tag{7}$$

$$\frac{(da/dN)_{cf}}{(da/dN)_{e,s}} \propto 4.36 \times 10^2 \frac{\beta}{\alpha} \frac{\sigma_{ys}^2}{N_o RTE^2}(T/M)^{1/2} \frac{P_o}{2f} \tag{8}$$

$$\frac{(da/dN)_{cf}}{(da/dN)_{cf,s}} = \frac{(p_o/2f)}{(p_o/2f)_s} \tag{9}$$

Case II: Surface reaction controlled.

$$\frac{(da/dN)_{cf}}{(da/dN)_{e,s}} \propto 1 - \exp(-k_c p_o/2f) \tag{10}$$

$$\frac{(da/dN)_{cf}}{(da/dN)_{cf,s}} = 1 - \exp(-k_c p_o/2f) \tag{11}$$

Subscript s is used to denote the corresponding values at saturation. The modification affects only the form of the expressions for the surface reaction controlled case, and provides a physically more acceptable model. In their present form, the model appears to be in excellent agreement with the experimental observations[2] .[5-9,14] An example of this correlation is illustrated in Fig. 1 for transport controlled crack growth in a 2219–T851 aluminum alloy.[6]

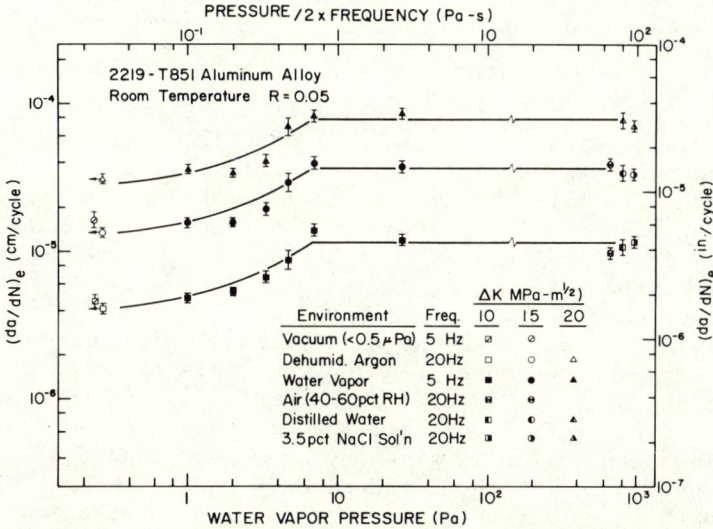

Fig. 1.   Influence of water vapor pressure (or pressure/2×frequency) on fatigue crack growth rates in 2219–T851 aluminum alloy at room temperature. Solid lines represent model predictions.[5,6]

---

[2] There is a discrepancy of 10^2 in exposure ($p_o/2f$) between the model predictions and the experimental data on AISI 4340 steel,[9] suggesting that capillary condensation of water vapor may have taken place. With capillary condensation, the effective exposure can be much higher than that given by $p_o/2f$.

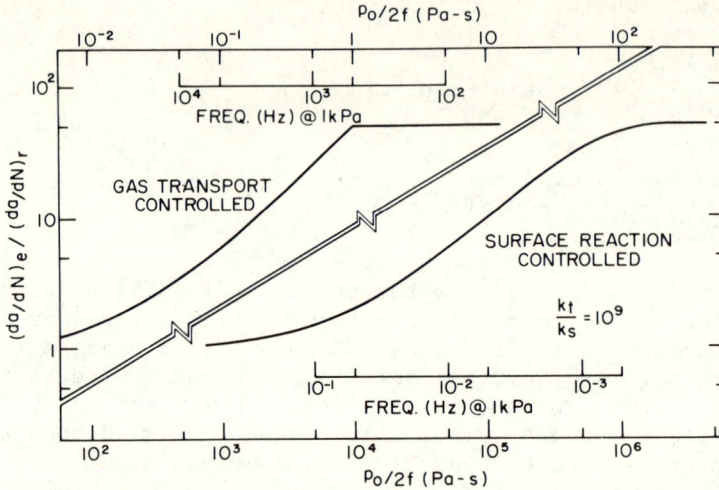

Fig. 2.   Schematic illustration and comparison of gas transport and
surface reaction controlled fatigue crack growth.

The engineering significance and implications of the model may
be discussed in relation to eqns. (7) and (11), and considered
through the schematic illustration shown in Fig. 2. The model provides a formalism for "predicting" environmentally assisted fatigue
crack growth response in relation to gas pressure and cyclic load
frequency. Specifically, eqn. (7) gives, for the transport controlled case, the value of $p_o/2f$ at which "saturation" in environmental effect can be expected in terms of the properties of the alloy, molecular weight of the aggressive gas, and temperature. Fatigue crack growth response is given, for this case, as an explicit
function of $p_o/2f$ below saturation by eqns. (8) and (9). For the
surface reaction controlled case, the dependence of fatigue crack
growth response on $p_o/2f$ is given by eqns. (10) and (11). It should
be noted that eqns. (8) to (11) deal only with relative rates. The
actual growth rates depend on the interactions of the embrittling
specie (hydrogen) with the alloy, which are not adequately understood at this time.

As further illustration, two hypothetical cases -- one representing gas transport and the other surface reaction control -- are
shown in Fig. 2. Crack growth response curves are given in terms of
the ratio $(da/dN)_e/(da/dN)_r$, as functions of $p_o/2f$, which can be
readily obtained from eqns. (1), (8) and (11). In addition, frequency scales are shown for an external pressure (p_o) of 1 kPa. For
this illustration, $(da/dN)_{e,s}/(da/dN)_r = 50$, and $k_c = 2 \times 10^3$ and
2×10^{-6} (Pa-s)$^{-1}$ are used. Although the curves differ somewhat in
detail from experimental data, they do serve to illustrate fatigue

crack growth response for a high strength steel exposed to hydrogen
sulfide (transport controlled) and water vapor (reaction controlled).[7]
It can be seen that the range of exposures ($p_o/2f$) or frequencies
over which apparent frequency dependence may be observed differ by
about 6 orders of magnitude for the two cases. The apparent independence of frequency or pressure, or $p_o/2f$, does not imply absence
of environmental effect. Neither does testing at high frequencies in
itself ensure absence of environmental effects. Most importantly,
each material-environment system must be considered individually.
Generalizations and extrapolations on the basis of limited data can
be very misleading and should be avoided.

MODELING OF FATIGUE CRACK GROWTH IN BINARY GAS MIXTURE

Experimental evidence obtained to date[1-4,6-9] clearly indicates
that the rate of environment assisted crack growth is dependent on
the rate of hydrogen production at the crack surfaces. The model for
fatigue crack growth developed earlier[5,8] may be readily extended to
the consideration of crack growth in gas mixtures. For simplicity,
the case of a binary mixture is considered. One of the component
gases is taken to be an inhibitor (that is, a gas which will react
with the clean metal surface but will not enhance crack growth). As
such, the results may be used, for example, to examine the influence
of oxygen (an inhibitor) on fatigue crack growth in humid air, where
water vapor is the aggressive component.[14-16]

It is assumed that (i) both gases are strongly adsorbed on the
clean metal surface, (ii) chemical adsorption of either gas at a
given surface site precludes further adsorption at that site, (iii)
the ratio of partial pressures of the gases at the crack tip is essentially the same as that of the surrounding environment, and (iv)
no capillary condensation of either gas occurs at the crack tip. In
line with the model of Weir et al.,[5] the cycle-dependent component
of crack growth rate in the gas mixture, $(da/dN)_{cf,m}$, is assumed to
be proportional to the extent of surface reaction with the aggressive gas during one loading cycle, or to θ_a. If one now assumes,
for simplicity, that the kinetics of surface reaction for both gases
are first order with respect to pressure and to available surface
sites, then the rate equations for the surface reactions may be
written as follows:

$$\frac{d\theta_a}{dt} = k_a p_a (1 - \theta) \qquad\qquad (12)$$

$$\frac{d\theta_i}{dt} = k_i p_i (1 - \theta) \qquad\qquad (13)$$

where the subscripts a and i denote the aggressive and inhibitor
gases respectively. The quantities k_a, p_a, k_i and p_i are the
reaction rate constants and partial pressures of the gases at the
crack tip, respectively. The coverages θ_a and θ_i denote the fraction
of surface that has reacted with the aggressive and inhibitor gases,
respectively. Coverage for both gases is denoted by θ, where $\theta =$
$\theta_a + \theta_i$ and $0 \leq \theta \leq 1$.

Eqns. (12) and (13) may be solved straightforwardly to obtain
the extent of reaction of a fresh surface with each gas after being
exposed to the gas mixture for a time t.

$$\theta_a = \frac{k_a p_a}{k_a p_a + k_i p_i}\{1 - \exp[-(k_a p_a + k_i p_i)t]\} \qquad (14)$$

$$\theta_i = \frac{k_i p_i}{k_a p_a + k_i p_i}\{1 - \exp[-(k_a p_a + k_i p_i)t]\} \qquad (15)$$

It follows that the fraction of surface that has reacted with the
aggressive gas (θ_a) in relation to the total surface coverage (θ)
may be given by eqn. (16).

$$\frac{\theta_a}{\theta} = \frac{\theta_a}{\theta_a + \theta_i} = \left(1 + \frac{k_i p_i}{k_a p_a}\right)^{-1} \qquad (16)$$

If the total pressure of the gas mixture at the crack tip ($p_m =$
$p_a + p_i$) and cyclic-load frequency, or $p_m/2f$, are such to produce
"complete" surface reaction or "saturation" coverage ($\theta \approx 1$) during
one cycle, then the fraction of the crack surface that has reacted
with the aggressive gas is given by eqn. (17).

$$\theta_a = \left(1 + \frac{k_i p_i}{k_a p_a}\right)^{-1} \qquad ; \quad (\theta \approx 1) \qquad (17)$$

Since $(da/dN)_{cf} \propto \theta \cdot \Delta a*$ (see eqn. (4)), the cycle-dependent
component of fatigue crack growth rate in the mixed gases,
$(da/dN)_{cf,m}$, can be obtained from the cycle-dependent rate in a pure
gas at saturation, $(da/dN)_{cf,p,s}$, and eqn. (17).

$$\frac{(da/dN)_{cf,m}}{(da/dN)_{cf,p,s}} = \left(1 + \frac{k_i p_i}{k_a p_a}\right)^{-1} = \left(1 + \frac{k_i p_{io}}{k_a p_{ao}}\right)^{-1} \qquad (18)$$

$$(\text{for } \theta \approx 1)$$

The subscript o is used to denote pressures in the external environment, and reflects the assumption that the partial pressure ratio at
the crack tip is the same as that in the external environment. It
should be noted that only the competition between two gases has been
modeled here, and eqn. (18) is expected to apply equally well to
both transport and surface reaction controlled crack growth. Schematic illustration of eqn. (18) is shown in Fig. 3 as $(da/dN)_{e,m}/$
$(da/dN)_r$ for different ratios of $(da/dN)_{e,s}/(da/dN)_r$; the ratios
reflect different degrees of severity of environmental effects in
the pure gas. A ratio of k_i/k_a of 0.14, which approximates the ratio
between the reaction rate constants for oxygen and water vapor over
a 2219-T851 aluminum alloy,[6,17] is used. The upper scale for p_a in
Fig. 3 is computed for p_i = 152 torr (or 20.2 kPa), and therefore
corresponds to the partial pressure of water vapor in humid (moist)
air.

Assuming that the surface reaction data on 2219-T851 aluminum
alloys[6,17] may be used for other high-strength aluminum alloys, comparisons of model predictions with data reported by Hartman et al.[15]
and Feeney, McMillan and Wei[16] on 7075-T6 aluminum alloy, and Bradshaw and Wheeler[14] on DTD 5070A aluminum alloy are made and are
shown in Figs. 4 and 5, respectively. For the 7075-T6 aluminum alloy, data in water and salt water[16] are assumed to represent the
"saturation" level in pure water vapor (θ = 1).[6] Excellent agreement is observed at the higher K levels and higher crack growth
rates. At the lower K levels and lower crack growth rates, the data
tend to follow those of pure water vapor, and may reflect partial
capillary condensation during a portion of each loading-unloading

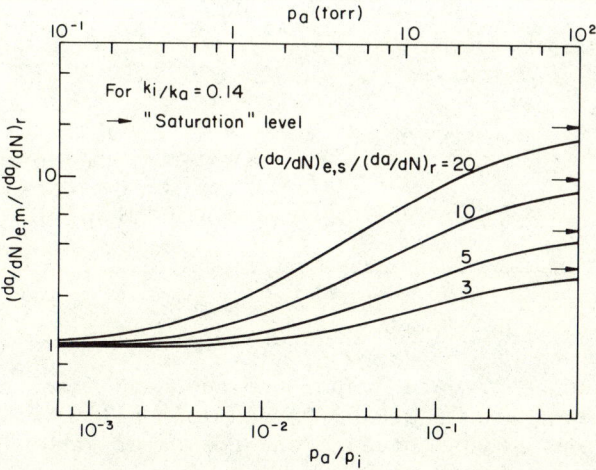

Fig. 3.   Schematic illustration of the influence of inhibitor gas
          on fatigue crack growth in a binary gas mixture. (In the
          upper scale, p_a corresponds to p_i = 152 torr or 20.2 kPa.)

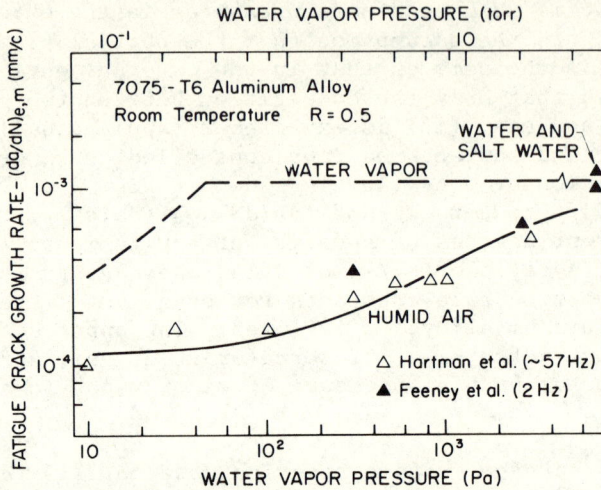

Fig. 4. Influence of water vapor pressure on fatigue crack growth
         in 7075-T6 aluminum alloy in humid air.[15,16] Dashed and
         solid lines represent model predictions from eqns. (7),
         (8) and (18).

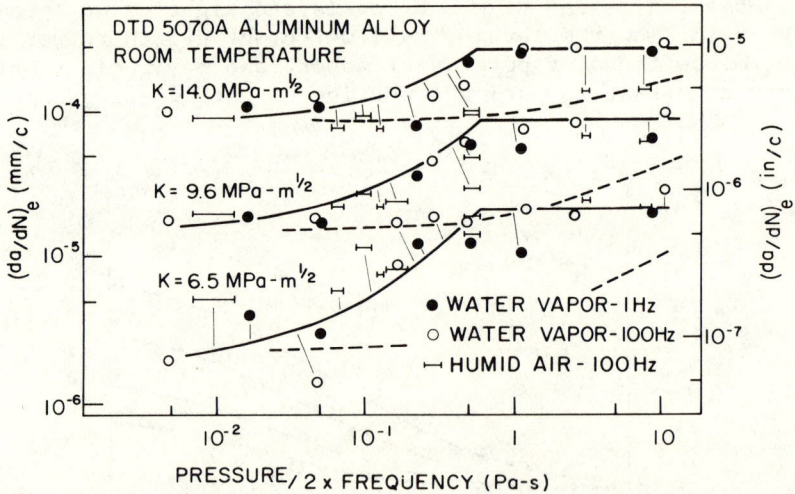

Fig. 5. Influence of water vapor pressure and frequency on fatigue
         crack growth in DTD 5070A aluminum alloy.[14] Solid lines
         represent model predictions for water vapor, and dashed
         lines for humid air.

cycle or some other factor. Further studies are needed to clarify
this discrepancy.

DISCUSSION

On the basis of available experimental results, it is seen that the models reasonably account for environmentally assisted fatigue crack growth response in single-component gaseous environments and in binary gas mixtures, in which one of the components acts as an inhibitor. A similar consideration of environmentally assisted crack growth in binary gas mixtures, under sustained loading, has also been made and provided good agreement with experimental observations.[18] Thus, the models provide a formalism for the planning of experiments, and for the interpretation and extrapolation of data for engineering applications.

In deriving these models and in their application, only simple surface reactions have been considered. Many reactions that are of interest (for example, reactions of steels with water vapor or hydrogen sulfide), however, are more complex. Reactions of different components to form new complexes are also possible. Incorporation of these complexities into future modeling efforts must be considered. Extension of these ideas to the consideration of fatigue crack growth in mixtures of H_2S and CO has been made.[19,20] Preliminary results indicate that modeling of the more complex environments would be feasible.

SUMMARY

Based on considerations of surface reactions and gas transport, models of environmentally assisted fatigue crack growth in single-component gaseous environments and in binary gas mixtures (in which one of the components acts as an inhibitor) have been developed. Good agreement with experimental data has been observed. The models, therefore, provide a formalized framework for experimental design, and for the interpretation and extrapolation of data for engineering applications. Further research will be needed to refine and extend this work to the consideration of more complex environments, and to incorporate the influences of metallurgical variables.

ACKNOWLEDGEMENT

Partial support of this work by the Office of Naval Research under Contract N00014-75-C-0543, NR 036-097 is gratefully acknowledged.

REFERENCES

1. G. W. Simmons, P. S. Pao, and R. P. Wei, Met. Trans. A 9A:1147 (1978).

2. M. Lu, P. S. Pao, N. H. Chan, K. Klier, and R. P. Wei, in:
    "Hydrogen in Metals", Suppl. to Trans. Japan Inst. Metals
    21:449 (1980).

3. N. H. Chan, K. Klier, and R. P. Wei, in: "Hydrogen in Metals",
    Suppl. to Trans. Japan Inst. Metals 21:305 (1980).

4. M. Lu, P. S. Pao, T. W. Weir, G. W. Simmons, and R. P. Wei,
    Rate Controlling Processes for Crack Growth in Hydrogel Sul-
    fide for an AISI 4340 Steel, submitted for publication to
    Met. Trans. A (1980).

5. T. W. Weir, G. W. Simmons, R. G. Hart, and R. P. Wei, Scripta
    Met. 14:357 (1980).

6. R. P. Wei, P. S. Pao, R. G. Hart, T. W. Weir, and G. W. Simmons,
    Met. Trans. A 11A:151 (1980).

7. R. L. Brazill, G. W. Simmons, and R. P. Wei, J. Engr. Matls. &
    Tech., Trans. ASME 101:199 (1979).

8. R. P. Wei and G. W. Simmons, Recent Progress in Understanding
    Environment Assisted Fatigue Crack Growth, Int. J. Fract.
    (to be published) (1980).

9. P. S. Pao, W. Wei, and R. P. Wei, in: "Environment Sensitive
    Fracture of Engineering Materials", Z. A. Foroulis, ed., TMS-
    AIME, New York:565 (1979).

10. R. P. Wei, in: ASTM STP 675 "Fatigue Mechanisms", J. T. Fong,
    ed., Am. Soc. Testing Matls., Philadelphia:816 (1979).

11. J. D. Landes & R. P. Wei, J. Engr. Matls. & Tech., 95:2 (1973).

12. G. A. Miller, S. J. Hudak, and R. P. Wei, J. Testing & Eval.,
    ASTM 1:524 (1973).

13. S. Dushman, "Scientific Foundations of Vacuum Techniques", 2nd
    ed., J. M. Lafferty, ed., Wiley:88 (1962).

14. F. J. Bradshaw and C. Wheeler, RAE Tech. Rept. No. 68041, Eng-
    land (1968).

15. A. Hartman, F. J. Jacobs, A. Nederveen, and R. DeRijk, NLR Tech.
    Note No. M.2182, The Netherlands (1967).

16. J. A. Feeney, J. C. McMillan, and R. P. Wei, Met. Trans. 1:1741
    (1970).

17. R. G. Hart, G. W. Simmons, and R. P. Wei, Studies of the Reac-
    tion of Oxygen and Water Vapor With 2219-T851 Aluminum Alloy,
    submitted for publication to J. Electrochem. (1980).

18. R. P. Wei, Rate Controlling Processes and Crack Growth Response,
    prepared for the 3rd Intl. Conf. on Effect of Hydrogen on
    Behavior of Materials, Jackson Lake Lodge, Wyoming, Aug. 26-
    31, 1980.

19. R. P. Wei and G. W. Simmons, Fracture Mechanics and Surface
    Chemistry Studies of Steels for Coal Gasification Systems,
    Final Tech. Rept., Contract No. Ex-76-S-01-2527, DOE-Fossil
    Energy Research, Lehigh University, Bethlehem, PA (1980).

20. R. L. Brazill, G. W. Simmons, and R. P. Wei, unpublished re-
    sults, Lehigh University, Bethlehem, PA (1980).

DETERMINATION OF PREFRACTURE DAMAGE AND FAILURE PREDICTION

IN CORROSION-FATIGUED AL-2024-T4 BY X-RAY DIFFRACTION METHODS

T. Takemoto,* S. Weissmann* and I. R. Kramer**

*College of Engineering, Rutgers University,
Piscataway, NJ
**D. W. Taylor Naval Ship R&D Center, Annapolis, MD

ABSTRACT

X-ray double crystal diffractometry and reflection topography were employed to examine the dislocation structure induced in corrosion-fatigued AL-2024-T4 alloys. The alloys were cycled in tension-tension, R = 0.1, at the maximum stress levels of S_m = 241, 276 and 310 MPa in 3.5 pct NaCl solution. The rocking curve measurements of the grains, carried out as a function of fatigue cycles and also as a function of depth distance from the surface, showed that the grains at the surface work-hardened much faster than the grains in the bulk. By determining the build-up of the excess dislocation densities in surface and bulk, using $CrK\alpha_1$ and $MoK\alpha_1$, respectively, the accrued prefracture damage could be determined and the onset of catastrophic failure predicted.

INTRODUCTION

Recent studies[1,2,3] of cycled aluminum crystals and AL-2024 alloys have shown the importance of the increased plastic resistance of the surface layer to deformation and have also demonstrated the special dependence of the dislocation structure in the bulk on the deformation behavior of the surface layer. It was shown that in aluminum alloys cycled in inert media the accrued damage could be determined and the remaining life could be predicted, provided the gradual build-up of the excess dislocation density in the bulk could be measured. This objective was accomplished by carrying out x-ray rocking curve measurements of the grains located in the bulk of the material using MoKα radiation which penetrated beyond the work hardened surface layer. The results of these studies suggested at once the exploration of the behavior of these alloys

when fatigued in hostile environment. The present investigation
may be viewed, therefore, as a logical extension of the aforementioned fatigue studies[3] to corrosion fatigue.

EXPERIMENTAL

    Corrosion fatigue (CF) tests of AL-2024-T4 alloys with average
grain size of 25 µm were carried out in tension-tension at three
applied stress levels with R = 0.1; the latter being defined as the
ratio of minimum stress/maximum stress. The maximum stress levels
were 241 MPa, 276 MPa and 310 MPa and corresponded to values of 87,
100 and 112 percent of the static yield stress in inert medium,
respectively. The corrosive medium was a 3.5 percent NaCℓ solution
which, confined in a non-corrosive plastic container, was kept
circulating around the test specimen. Special grips were made of
MC nylon and the pins were made of stainless steel, covered with
poly-imide, to prevent generation of electrical current between
grips and specimen during testing. Prior to CF testing the specimens
were electrolytically polished and to prevent oxidation and contami-
nation effects a surface of at least 100 µm was removed. Electro-
polishing was also applied for the in-depth analysis of deformation
(depth profile analysis) when surface layers were removed increment-
ally. A solution of 20 percent perchloric acid and 80 percent
methanol was used at 20 volts and -20°C. The non-destructive method
of analysis of the defect structure induced by CF was based on the
principle of x-ray double crystal diffractometry.[4,5,6] According
to this method the polycrystalline specimen is irradiated with a
crystal-monochromated x-ray beam, and each reflecting grain is
considered to function independently as the test crystal of a double-
crystal diffractometer. Depending on the perfection of the grains,
the specimen is rotated in intervals of seconds or minutes of arc,
and the spot reflections, recorded along the Debye arcs of a
cylindrical film for each discrete specimen rotation, are separated
by film shifts. The automated and programmed multiple-exposure
technique gives rise to an array of spots for each reflecting grain,
such as are shown in Fig. 1. The spot arrays, with their intensity
dependence on the angular position, represent rocking curves. On
the basis of rocking curve parameters, such as the width β at half
of the intensity maximum, the excess dislocation density (accumulation of dislocations of the same sign) of the reflecting grain
can be calculated.[7] If a resolvable substructure exists within the
grain, the rocking curve exhibits a multimodal intensity distribution such as that shown by the grain reflection A of Fig. 1. From
the angle, subtending successive peaks of the rocking curve, the
excess dislocation density between subgrains can be determined,
while from the spread of the subpeak curve, the excess dislocation
density within the subgrain lattice can be obtained. Furthermore,
by taking reflection (Berg-Barrett) topographs and performing a
spatial tracing of the reflections to the spot reflections of the
rocking curve, the analyzed rocking curve can be correlated to the

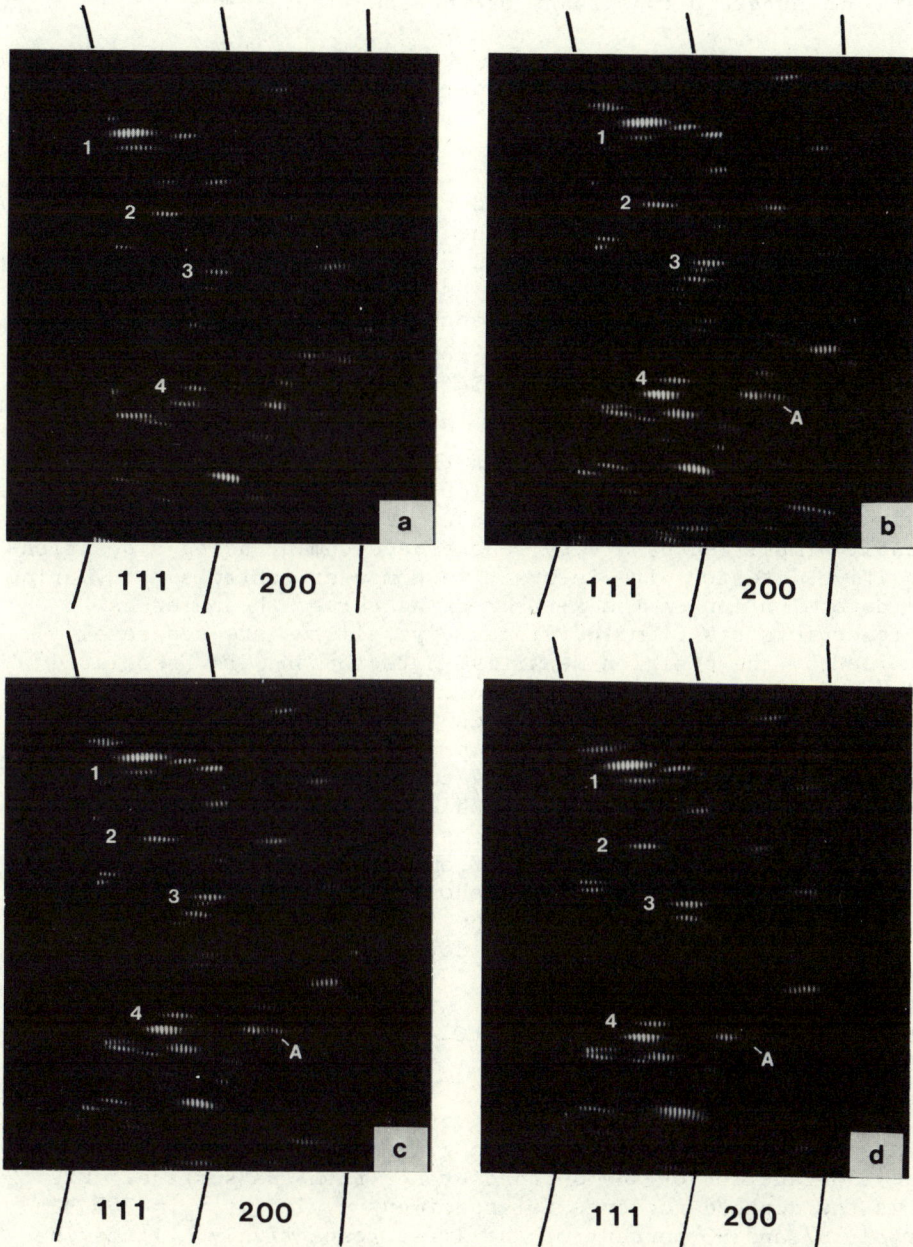

Fig. 1. Dependence of x-ray rocking curves of grain reflections on fatigue cycles. S_{max} = 267 MPa. Detail of (111) and (200) reflections, CrKα$_1$ radiation, specimen rotation = 3 minutes of arc (a) N = 0 cycles, (b) N = 2,500, (c) N = 8,000, (d) N = 11,100.

grain and subgrain topography of the specimen.[4,5,6]

An important feature in the present CF study was the introduction of a precision specimen holder which permitted precise repositioning of the specimen after CF testing or after removal of surface layers for depth profile analysis. If the applied stress did not exceed substantially the static yield stress, excessive grain rotation did not occur, and consequently, due to the precise repositioning of the specimen, the identical grain reflection could be retained. Thus, the identical grains could be analyzed for their induced lattice defects not only as a function of CF cycling but also as a function of depth distance from the surface. Typically, the grains marked 1, 2, 3 and 4 of Fig. 1 may serve to illustrate the identity of grain reflections with increased cycling and the effect of cycling on the spread of the rocking curves.

It should be remembered that the success of the rocking curve analysis depends on the sampling of a large grain population. The tedious and limiting task of data collection and analysis was greatly simplified by a very recent development using a position-sensitive detector with interactive computer controls.[8,9] During the determination of the x-ray rocking curve, the numerous microscopic spots from individual crystallites are separately recorded by the position-sensitive detector and its associated multichannel analyzer at each increment of sample rotation. An on-line mini-computer simultaneously collects these data and applies any necessary corrections. This process is then automatically repeated through the full rocking curve range. At the end of each experiment the computer is available for the rocking curve and analysis of the whole aggregate, as well as each grain. The separation of the contributions from surface and bulk grains is accomplished by appropriate threshold control of the noise level. It is believed that because of the speed and efficiency of data collection and analysis this new development may have technological viability.

DEPTH DISTRIBUTION OF EXCESS DISLOCATION DENSITY INDUCED BY CORROSION FATIGUE

X-ray rocking curve studies were carried out to obtain the excess dislocation density of the grains as a function of CF cycles and as a function of the depth distance from the surface. Fig. 2 shows the depth dependence for specimens cycled at S_m = 267 MPa for 31, 67 and 99 percent of the life, respectively. It may be seen that this alloy, just like the alloy cycled in air[1,2,3] exhibited the largest $\bar{\beta}$ values at the surface. The $\bar{\beta}$ values declined to a depth distance of about 50 μm from the surface and subsequently retained a plateau value throughout the bulk of the specimen for each fraction of the life. However, unlike the specimens cycled in air, the depth profile curve did not exhibit

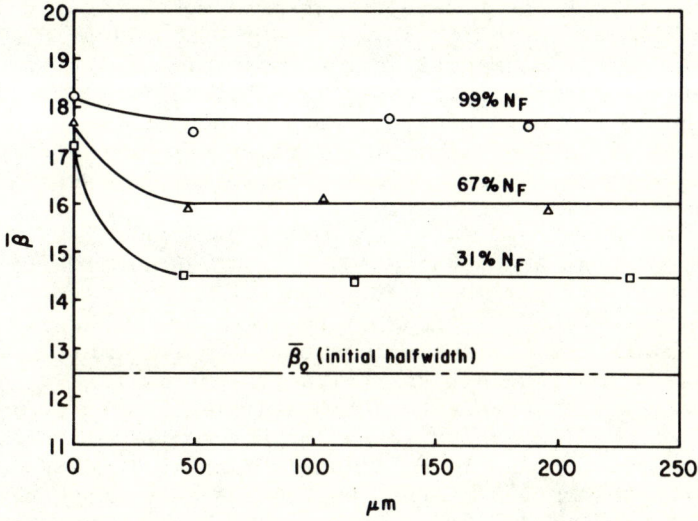

Fig. 2.   Dependence of the average rocking curve half width $\bar{\beta}$ on
depth distance from surface for different fractions of
corrosion fatigue lives, N_F. Sm = 276 MPa, R = 0.1, $CrK\alpha_1$.

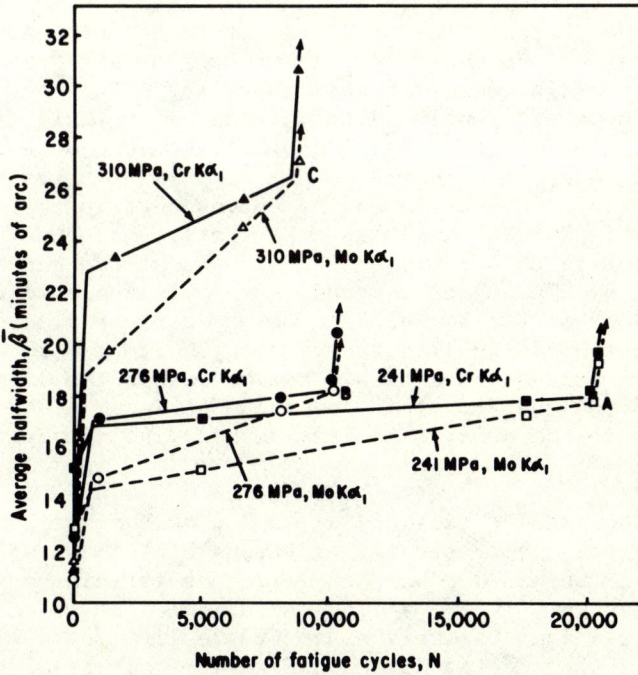

Fig. 3. $\bar{\beta}$ versus N at various stress levels.

a minimum and its shape resembled those of monotonically deformed specimens. Of great significance for both the alloys cycled in inert and in corrosive media, is the aspect that with cycling the $\bar{\beta}$ values at the surface approached rapidly saturation value while those in the bulk, pertaining to the plateaus of the curve, increased continuously during the life. It is this characteristic feature of the latter which leads to the determination of the accrued damage by the nondestructive x-ray method.

DETERMINATION OF ACCRUED DAMAGE AND FAILURE PREDICTION

Fig. 3 shows the relationship between the average rocking curve half width, $\bar{\beta}$, and the number of fatigue cycle, N, for various maximum stress amplitudes. As indicated in this figure, the solid lines of the curves pertain to the measurements with $CrK\alpha_1$ radiation, while the dotted lines refer to the measurements performed with $MoK\alpha_1$ radiation. It should be borne in mind that for chromium radiation the absorption of the specimen is such that only surface grains can be analyzed. By contrast, the short wave length of the molybdenum radiation permits also analysis of the grains exceeding the 50 µm depth distance from the surface (see Fig. 2). It will be seen from Fig. 3 that for the stress level of 241 MPa the $\bar{\beta}$ value rose rapidly during the first 5 percent of the life. This rise was much more pronounced for the surface grains ($CrK\alpha_1$ radiation), in agreement with the depth profile study of Fig. 2. From then until N = 20,000, when catastrophic failure set in at point A, corresponding to the critical value of $\bar{\beta}* = 18$ minutes of arc, the $\bar{\beta}$ values of the surface grains were virtually constant. The $\bar{\beta}$ values of the bulk grains, however, continued to increase. It is important to note that the two curves ($CrK\alpha_1$ and $MoK\alpha_1$) converged at A.

Focusing one's attention on the curves pertaining to the higher stress level of 276 MPa one notes a perfectly analogous behavior. Again a sharp rise of $\bar{\beta}$ early in the life and convergence of the two curves at B. The $\bar{\beta}*$ corresponding to B is the same as that of A, namely 18' of arc. The increase of the stress level from 241 to 276 MPa decreased the life from N = 20,000 to N = 11,100. Increasing the stress level to 310 MPa decreased the life to N = 8,500. Note, that for this low cycle fatigue life $\bar{\beta}*$, corresponding to the point C of Fig. 3, has risen to 26.4 min. of arc.

Fig. 4 shows the ratios of $\bar{\beta}_{Mo}$ to $\bar{\beta}_{Cr}$ as a function of the fraction of fatigue life at various stress levels. A straight line relationship is obtained. When the branches pertaining to $MoK\alpha_1$ and $CrK\alpha_1$ converge, $\bar{\beta}_{Mo}/\bar{\beta}_{Cr} = 1.00$ and critical $\bar{\beta}*$ is reached. Thus the accrued damage at any stage of the life, as well as the fatigue failure, can be predicted. Note that the predictions of failure for the stress levels 241, 267 and 310 MPa corresponded to 94, 95 and 98 percent of the actual life, respectively, when the

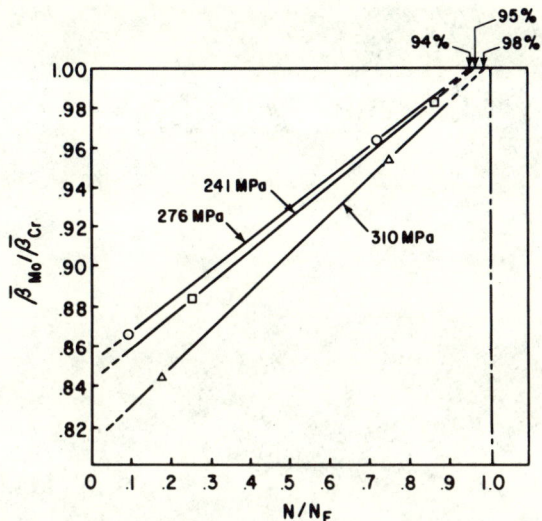

Fig. 4.   Ratio of $\bar{\beta}_{MoK\alpha1}$ to $\bar{\beta}_{CrK\alpha1}$ as a function of N/N_F at various
stress levels.

specimens were cycled to failure. It should be pointed out that
compared to the alloys cycled in air the corrosive environment had
the deleterious effect of reducing the life of the specimens by a
factor of about thirty.

     The light micrographs of Fig. 5 show the importance of the
criticality of $\bar{\beta}*$. The micrograph of Fig. 5a was taken from a
specimen fatigued at a stress level of 241 MPa for N = 20,000
corresponding to point A in Fig. 3, while Fig. 5b was from a specimen
cycled at 267 MPa corresponding to point B in Fig. 3. It can be
seen that at this stage of fatigue a large crack propagated
catastrophically in the direction perpendicular to the tensile axis
(crack propagation stage II).

     Fig. 6 may serve to demonstrate the capability of molybdenum
radiation to penetrate beyond the work hardened surface layer and
thereby making the analysis of the defect structure induced in the
bulk grains possible. The lower, steepest curve was constructed
from the data points of Fig. 2 and were collected from the plateau
values of the bulk grains as analyzed by chromium radiation after
electrochemical removal of the surface layer. The curve pertaining
to the MoKα_1, on the other hand, was obtained with the work hardened
surface layer intact. It is evident that the less steep slope of
the latter curve was due to the fact that in the analysis employing
molybdenum radiation some work hardened surface grains were also
included. Nevertheless, the resulting slope is sufficiently steep

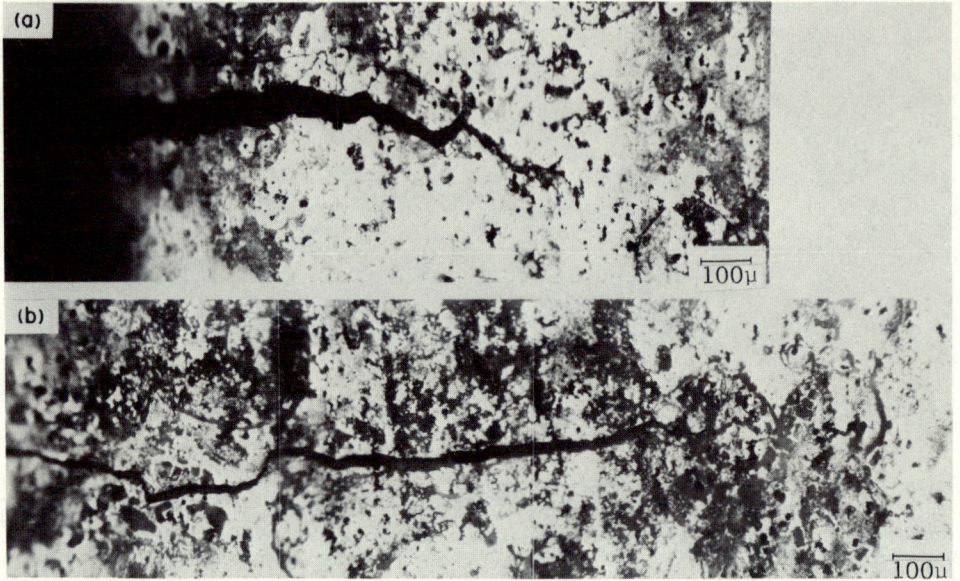

Fig. 5. Micrographs showing cracks (a) CF at Sm = 241 MPA for
N = 20,000 corresponding to point A in Fig. 3, (b) CF
at Sm = 267 MPa for N = 11,100 corresponding to point
B in Fig. 3.

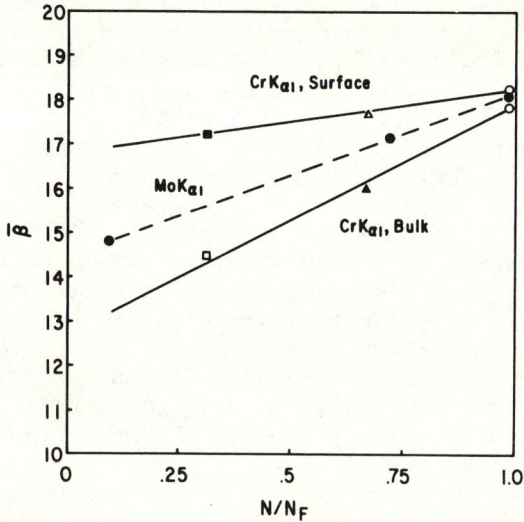

Fig. 6.   Surface and bulk dependence of $\bar{\beta}$ on N/N_F and $CrK\alpha_1$
          radiation used.

to determine the critical $\bar{\beta}*$ unequivocally. This curve can thus
be used as a calibration standard to gage the accrued damage at
any stage of the life.

It should be pointed out that it is possible to make the curve
pertaining to the $MoK\alpha_1$ radiation entirely representative to the
bulk grains by subtracting the contribution of the surface grains.
This can be accomplished by employing a multiple film technique
in recording the rocking curves of the grain reflections and
interposing an absorbing screening material of appropriate thickness,
viz. copper foil, between the films. Due to the larger absorption
of the x-ray beams reflected from the bulk grains their contribution
will appear entirely suppressed in the screened, second film and
only the rocking curves of the strongly reflecting surface grains
will be recorded. By comparing first and second film and eliminating the rocking curves of the surface grains from the total grain
population analyzed in the first film, a steeper slope will be
obtained for the $MoK\alpha_1$ curve.

DISCUSSION

In analogy to the results of the aluminum alloys cycled in
air, this corrosion fatigue study has again demonstrated the
importance of the interdependence in the development of the excess
dislocation density of surface layer and bulk. The initial rapid
rise of $\bar{\beta}$ of the surface grains (Figs. 2 and 3) is indicative of

rapid work hardening of the surface layer with cycling. Since on subsequent cycling the surface grains did not exhibit a substantial increase of $\bar{\beta}$, the plateau branch of this curve (Fig. 3) is interpreted to reflect physically the blocking by the surface layer to prevent relaxation of excess dislocations, accumulated in the bulk. By contrast, the initial rise of $\bar{\beta}$ of the bulk grains, which was much less than that of the surface grains, increased steadily with cycling (Figs. 2 and 3). When the excess dislocation density reaches a critical value $\rho*$ such that the stress field associated with the accumulation or pileup of dislocations exceeds the local fracture strength of the surface layer, catastrophic crack propagation takes place. Experimentally, the determination of $\rho*$ corresponds to the determination of $\bar{\beta}*$ and in accordance with the above concept $\bar{\beta}*$ will lie on the point of intersection of the $\bar{\beta}$ curves of the surface and bulk grains. Thus, points A, B and C of Fig. 3 reflect graphically the local, physical breakdown of the barrier effect of the surface layer for the respective, applied stress levels.

When the applied maximum stress level was 310 MPa, the deformation process should be regarded to fall into the regime of low cycle CF where propagation of subcritical cracks takes up the largest portion of the life. For this case it is interesting to note in Fig. 3 that not only were the ascents of the $\bar{\beta}$ values much higher but $\bar{\beta}* = 26.2$ minutes of arc had set in before the $\bar{\beta}$ values of the surface reached saturation value. It will also be seen from Fig. 3 that steeper $\bar{\beta}$ slopes for the bulk grains lead to a reduction in fatigue life. Compared to cycling in inert media, corrosion fatigue resulted in a thirty-fold decrease of life for the aluminum alloys. Since the generation of the dislocations in the bulk is controlled by the surface layer, the latter extending in this case to about 50 μm (see Fig. 2), the rapid generation of the dislocations in the surface layer leads to a corresponding rapid generation of dislocations in the interior. Apparently when the excess dislocation density in the bulk approaches that of the surface layer fatigue failure will occur.

ACKNOWLEDGMENT

The support of this work by the Office of Naval Research, Arlington, VA and Naval Ship R&D Center, Annapolis, MD is deeply appreciated.

REFERENCES

1. R. N. Pangborn, S. Weissmann, Fatigue failure prediction by x-ray double crystal diffractometry and topography, in: Strength of Metals and Alloys, Pergamon Press, edited by P. Haasen et al., pp. 1279-84 (1979).

2. R. N. Pangborn, S. Weissmann and I. R. Kramer, Determination
    of prefracture fatigue damage, Ship Materials Engineering Dept.
    R&D Report, DTNSRDC-80/006 (1980)
3. I. R. Kramer, R. N. Pangborn and S. Weissmann, Dislocation
    distribution in plastically deformed metals, This volume.
4. S. Weissmann, Method for the study of lattice inhomogeneities
    combining x-ray microscopy and diffraction analysis, J. Appl.
    Phys., 27:389-395 (1956).
5. S. Weissmann, Substructure Characteristics of fine-grained
    metals and alloys disclosed by x-ray reflection microscopy and
    diffraction analysis, Trans. ASM, 52:599-614 (1960).
6. S. Weissmann, Analysis and topography of lattice defects in
    powder diffraction patterns, Nat. Bur. Stand. Special Public.,
    567:411-431 (1980).
7. P. B. Hirsch, Mosaic structure, in: Progress Met. Phys.,
    6:283, Pergamon Press, New York (1956).
8. S. Weissmann, R. Yazici, T. Takemoto, T. Tsakalakos and
    I. R. Kramer, X-ray analysis of accrued damage in stress
    corrosion and corrosion fatigue, ONR Technical Report, A078511
    (1979).
9. R. Yazici, W. Mayo and S. Weissmann, Computer controlled double
    crystal diffractometer with position sensitive detector for
    analysis of defect structures in polycrystalline materials,
    Abstract B4 of Summer Meeting of Am. Cryst. Assoc., Ser. 3,
    p. 11, Vol. 8, No. 1, ISSN-0569-4221 (1980).

ΔK THRESHOLDS IN TITANIUM ALLOYS - THE ROLE OF

MICROSTRUCTURE, TEMPERATURE AND ENVIRONMENT

C.J. Beevers and C.M. Ward-Close[*]

Department of Physical Metallurgy and
Science of Materials, the University
P.O. Box 363, Edgbaston, Birmingham
B15 2TT, U.K.
[*]Materials Department, Royal Aircraft
Establishment, Farnborough, Hants, U.K.

INTRODUCTION

The conditions under which a crack will extend when subjected
to repeated cyclic loading can have substantial implications in
the choice of material and loadings for particular components.
Probably of equal interest are the conditions under which a crack
arrests or does not propagate. This concept of a non-propagating
crack was reinforced by fatigue crack growth measurements at rather
low growth rates $\sim 10^{-8}$ mm/cycle. In this region the log da/dN v
log ΔK curves tend to run asymptotically to a lower bound ΔK level.
This ΔK value is referred to as a threshold ΔK, ΔK_{th}. In this
paper the fatigue crack growth at low ΔK's and the ΔK_{th} values are
reviewed for a range of titaniums and titanium alloys. The influence of grain size/microstructure, temperature and environment on
the ΔK_{th} values are considered.

RESULTS

Fatigue Crack Growth α-Titanium

Single edge notched specimens 140 mm x 35 mm x 5 mm of Ti 115,

Ti 130 and Ti 155* were pin loaded in tension-tension at room temperature with a test frequency of $\sim$ 100 Hz[1]. Two grain sized
materials ($\sim$ 0.04 and 0.2 mm) were tested over the load ratio R
range 0.07 to 0.7. The new ΔK_{th} values from these tests are presented in Table I. The fatigue crack growth over the range 10^{-4}
mm/cycle to 10^{-7} mm/cycle was an order of magnitude slower in the
coarser grained material. The growth curves also exhibited an R
dependence with lower growth rates and a tendency for higher ΔK_{th}
values with lower R values.

In a recent study of fatigue crack growth in Ti 155[2] it was
shown that both grain orientation and environment can influence
the mode of crack extension and the rate of crack growth. Single
edge notch specimens were tested at R = 0.35 at a frequency of
$\sim$ 130 Hz in vacuum (2 x 10^{-6} torr) laboratory and analer water and
3.5% sodium chloride solution. A comparison of the fatigue crack
growth response in air, analer water and sodium chloride solution
(brine) is presented in Figure 1. The brine solution resulted in
higher crack growth rates, the scatter in the results is related
to the coarse grain size, $\sim$ 0.25 mm. A detailed crystallographic
study revealed the formation of cleavage like facets on basal
planes, striations and furrows depending upon the orientation of
the grain presented to the crack front. In Figure 2 the furrow
mode of crack growth is illustrated, the micrograph was taken from
the fatigue fracture surface of a specimen tested in laboratory
air, $\Delta K = 20$ $MNm^{-3/2}$. The formation of cleavage like facets and
irregular or mixed mode failure is illustrated in Figure 3, the
micrograph was taken from the fatigue fracture surface of a specimen tested in analer water, $\Delta K \sim 8$ $MNm^{-3/2}$. The macroscopic fracture plane in both laboratory air and analer water contained many
out of plane features. From an examination of over 200 grains the
influence of grain orientation and environment on the mode of
crack growth was established. These observations can be presented
in the form of fracture maps as illustrated in Figure 4 and 5.
The major effect of making the environment more active, that is
changing from an inert vacuum to a 3½% sodium chloride solution
was to increase the orientation range over which cleavage like
facets formed on the basal planes.

* α-Titanium	115	130	155
Oxygen content wt.%	0.095	0.16	0.34

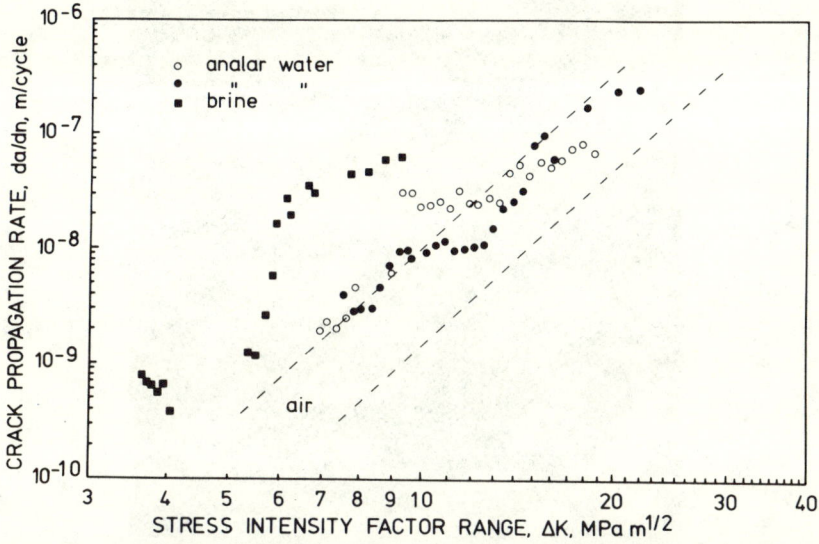

Fig. 1.   Fatigue crack growth data for α-titanium tested in
          laboratory air, analar water and brine.

Fig. 2.   SEM micrograph of the fatigue fracture surface of
          α-titanium showing mainly furrowing with some facets.
          Laboratory air ΔK = 20 MNm$^{-3/2}$.   X 40

Fig. 4.   Fatigue fracture maps for α-titanium tested in vacuum
          ΔK 10–25 MNm$^{-3/2}$.

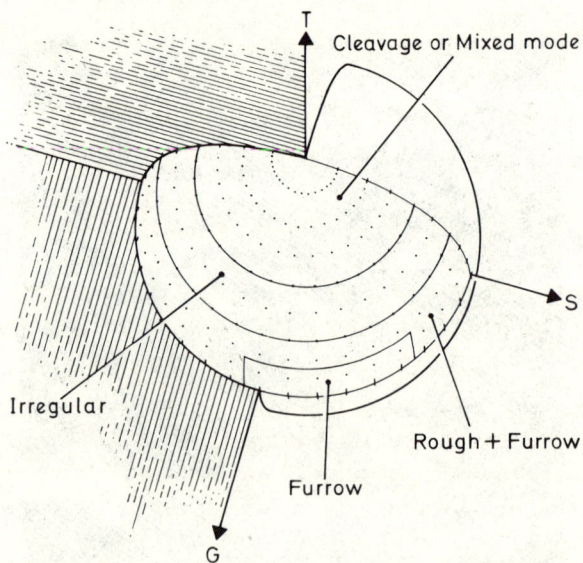

Fig. 3.   SEM micrograph of the fatigue fracture surface of
          α-titanium showing cleavage like facets and mixed modes
          of separation.   Analar water ΔK ∿ 8 MNm$^{-3/2}$.   X 80

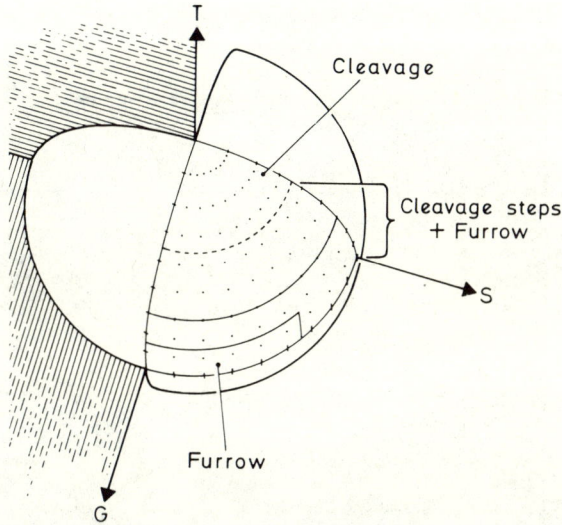

Fig. 5.   Fatigue fracture maps for α-titanium tested in 3.5% NaCl solution ΔK ≃ 3.5 → 9 MNm$^{-3/2}$.

Fatigue Crack Growth in Ti-6Al-4V

Single edge notch specimens 200 mm x 50 mm x 10 mm were pin loaded and tested in tension-tension at room temperature with a frequency of ∿ 100 Hz[3,4]. The specimens were tested in a range of microstructural conditions as received, transformed β, martensitic and β-annealed, the sizes of the α and β structural elements are presented in Table I. Tests on the same specimen geometry were also carried out at 110°C, 257°C and 386°C [5]. The fatigue crack growth curves are presented in Figure 6. The results for the 387°C test were limited by major crack bifocation at a ΔK of ∿ 8 MNm$^{-3/2}$. Specimens were held at a range of K_{max} values for up to 48 hours and no crack extension at a rate greater than 3 x 10^{-7} mm/sec could be detected. However, on the resumption of cyclic loading the crack growth rate accelerated for approximately 0.2 mm before returning to its projected value. A marked feature of fracture surface morphology in these elevated temperature tests was the formation of facets spreading over many adjacent α-grains.

TABLE I

(a)

Alloy	Yield Stress MNm^{-2}	ΔK_{th} $MNm^{-3/2}$	R Ratio	α Grain Size $\times 10^{-6}$ m	β Grain Size $\times 10^{-6}$ m	Temperature $^{\circ}C$	Environment
Ti 115	260	< 5	0.07	35	–	$20\,^{\circ}C$	Laboratory Air
Ti 115	260	< 5	0.35	35	–	$20\,^{\circ}C$	Laboratory Air
Ti 115	260	< 3	0.70	35	–	$20\,^{\circ}C$	Laboratory Air
Ti 130	430	< 5	0.07	40	–	$20\,^{\circ}C$	Laboratory Air
Ti 130	430	< 5	0.35	40	–	$20\,^{\circ}C$	Laboratory Air
Ti 130	430	< 3	0.70	40	–	$20\,^{\circ}C$	Laboratory Air
Ti 130	430	≈ 8	0.35	40	–	$20\,^{\circ}C$	Vacuum < 10^{-5} torr
Ti 155	630	< 5	0.07	20	–	$20\,^{\circ}C$	Laboratory Air
Ti 155	630	< 5	0.35	20	–	$20\,^{\circ}C$	Laboratory Air
Ti 155	630	< 5	0.70	20	–	$20\,^{\circ}C$	Laboratory Air

(b)

Alloy	Yield Stress MNm^{-2}	ΔK_{th} MNm$^{-3/2}$	R Ratio	α Grain Size x 10^{-6} m	β Grain Size x 10^{-6} m	Temperature °C	Environment
Ti 115	260	< 5	0.07	230	-	20°	Laboratory Air
Ti 155	580	> 6	0.07	210	-	20°	Laboratory Air
Ti 6Al 4V	860	≈ 3	0.35	3	0.8	20°	Laboratory Air
Ti 6Al 4V	750	> 5	0.35	8	9*	20°	Laboratory Air
Ti 6Al 4V	1,000	< 3	0.35	35	2	20°	Laboratory Air
Ti 6Al 4V	1,000	> 5	0.12	3.5	2	20°	Laboratory Air
Ti 6Al 4V	860	≈ 8	0.35	3	0.8	20	Vacuum < 10^{-5} torr
Ti 6Al 4V	860	≈ 3	0.35	3	0.8	110°	Laboratory Air
Ti 6Al 4V	860	< 3	0.35	3	0.8	257	Laboratory Air

(c)

Alloy	Yield Stress MNm^{-2}	ΔK_{th} MNm$^{-3/2}$	R Ratio	α Grain Size x 10^{-6} m	β Grain Size x 10^{-6} m	Temperature $^{\circ}$C	Environment
Ti 6Al 4V	860	~ 2.5	0.35	3	0.8	386	Laboratory Air
Ti 6Al 4V	860	~ 8	0.35	3	0.8	350	Vacuum < 10^{-5} torr
Ti 6Al 5Zr 0.5Mo (A)	840	8	0.12	100^{+}	600	25°	Laboratory Air
Ti 6Al 5Zr 0.5Mo (A)	840	6	0.35	100^{+}	600	25	Laboratory Air
Ti 6Al 5Zr 0.5Mo (A)	840	5	0.61	100^{+}	600	25	Laboratory Air
Ti 6Al 5Zr 0.5Mo (A)	840	13	0.12	100^{+}	600	25	Vacuum
Ti 6Al 5Zr 0.5Mo (A)	840	9	0.35	100^{+}	600	25	Vacuum
Ti 6Al 5Zr 0.5Mo (A)	840	7.5	0.61	100^{+}	600	25	Vacuum

(d)

Alloy	Yield Stress MNm^{-2}	ΔK_{th} MNm$^{-3/2}$	R Ratio	α Grain Size x 10^{-6} m	β Grain Size x 10^{-6} m	Temperature °C	Environment
Ti 6Al 5Zr 0.5Mo (B)	820	11.5	0.12	210[+]	730	25	Laboratory Air
Ti 6Al 5Zr 0.5Mo (B)	820	8	0.35	210[+]	730	25	Laboratory Air
Ti 6Al 5Zr 0.5Mo (B)	820	5	0.61	210[+]	730	25	Laboratory Air

* Widmanstatten colony size.

+ α colony size.

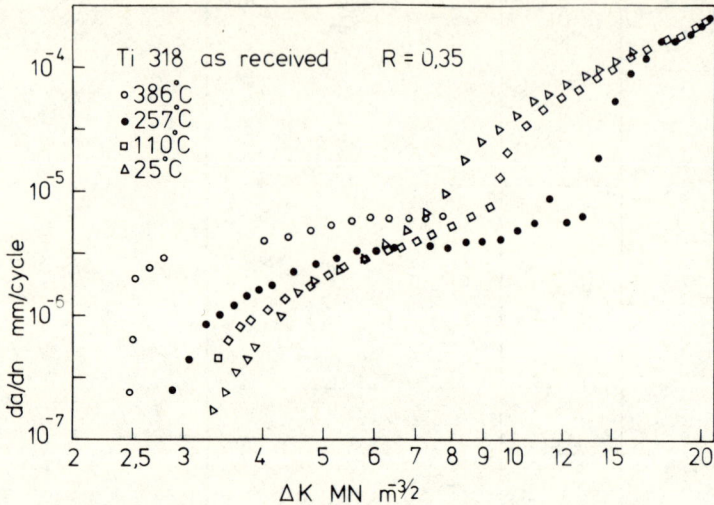

Fig. 6.   The log da/dN - log ΔK curves for specimens tested
at 21°C, 110°C, 257°C and 386°C in air at R = 0.35.

Fatigue Crack Growth in Ti-6Al-5Zr-0.5Mo

Tests on pin loaded specimens 200 mm x 50 mm x 10 mm were
carried out in tension-tension at 25°C in laboratory air and
vacuum at a frequency of 120-150 Hz[6]. The A series material
was solution treated at 1050°C for 30 minutes, oil quenched and
aged 24 hours at 580°C and then air cooled. The B series specimens had the same manual heat treatment as the A series but were
subjected to slower cooling rates from the quenching and ageing
temperatures. The ΔK threshold values obtained from these specimens for growth rates of 10^{-8} mm/cycle are presented in Table I.
The ΔK_{th} values show a strong R dependence in both air and vacuum.
The coarser and more aligned α colonies resulted in higher ΔK_{th}
values particularly at low R values. In the A series tests
load-COD curves were obtained and the effective ΔK, ΔK_{th}^{i} was
obtained

$$\Delta K_{th}^{i} = K_{max} - K_{op} \qquad [1]$$

where K_{max} is the maximum stress intensity and K_{op} the stress intensity for a fully opened crack. The values obtained for ΔK_{th}^i were
5.3, 5.0 and 5.1 $\text{MNm}^{-3/2}$ for R values of 0.1, 0.35 and 0.61
respectively.

In both air and vacuum the fracture surfaces were macro-
scopically irregular with substantial out of plane features in all
specimens. Examples of the fatigue fracture morphology are presented in Figures 7 and 8. In Figure 9 surface battering resulting
from load transfer across the fracture planes is evident.

Short Fatigue Crack "Thresholds" in α-Titanium

This section of the paper refers to tests on parallel sided
specimens with gauge lengths of 14 mm and diameters of 3.5 mm. [(7)]
The specimens were tested at a frequency of $\sim$ 140 Hz with R = -1.
Conventional S-N curves up to 10^7 cycles were established for two
series of specimens, non pre-cracked and pre-cracked. The following treatment was employed to introduce cracks into the specimens
prior to fatigue testing. A hydriding treatment introduced many
transgranular brittle hydride platelets with dimensions of the
order of the grain size ($\sim$ 0.25 mm). A 2% plastic strain at - 196°C
resulted in the failure of the precipitates but not the specimens.
The hydrogen was removed by a 2 hour high temperature (500°C) vacuum
anneal. Thus, the effect of preexisting cracks on the fatigue limit
could be studied. The size of preexisting cracks was investigated
by examining metallographic sections of the fatigue specimens and
the results are presented in Table II. The density of cracks was
$\sim$ 50 cm^{-2} and their orientations varied from 90° to 45° to the
tensile axis. An example of pre-existing cracks associated with
the final fatigue crack is presented in Figure 10.

The presence of the pre-cracks lowered the fatigue limit
(Table II). To obtain a value of ΔK_{th} at which the cracks would
not propagate the following procedure was followed

$$\Delta K_{th} = \Delta \sigma_L (\pi \alpha a)^{\frac{1}{2}}$$ [2]

ΔK_{th} is the threshold ΔK, $\Delta \sigma_L$ fatigue limit, α the geometrical
factor was taken as 1.05 (a/W $\sim$ 0.1). The results in Table II
indicate a value of ΔK_{th} varying from 2 to 3 $\text{MNm}^{-3/2}$. If $\Delta \sigma_L$ had
been represented by the tensile part of the cycle only then
ΔK_{th} values would have been 1 to 1.5 $\text{MNm}^{-3/2}$.

ΔK_{th}^i is referred to later in the discussion as the microstructural/
environmental component of ΔK_{th}.

Table II

Alloy	Fatigue Limit MNm^{-2}		Length of surface cracks 'a' mm	ΔK_{th} MNm$^{-3/2}$	Length of internal cracks 2a mm	ΔK_{th}
	Precracked	Non Precracked				
Ti 115	90	140	0.15	2.0	0.4	2.3
Ti 130	160	200	0.05	2.0	0.2	2.9
Ti 160	160	300	0.05	2.0	0.2	2.9

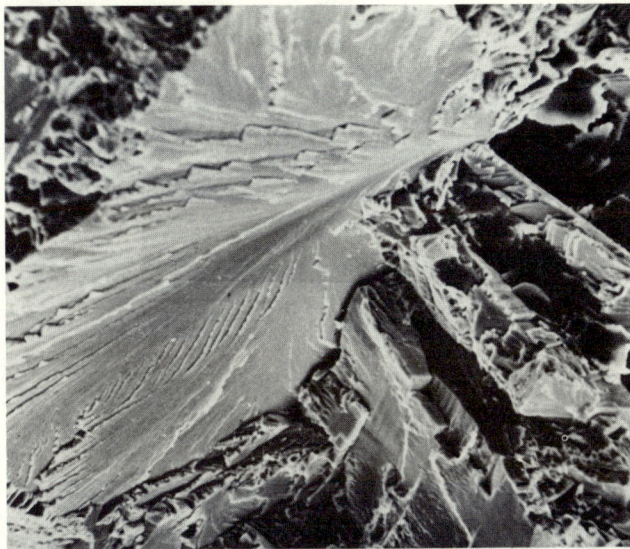

Fig. 7.   SEM micrograph of the fatigue fracture surface of a
           Ti 6Al 5Zr 0.5Mo (A) alloy tested in vacuum at R =
           0.61 showing facet formation.   X 190

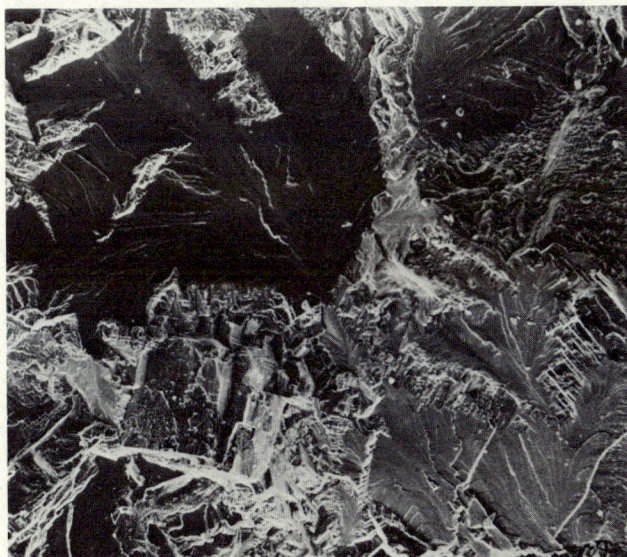

Fig. 8.   SEM micrograph of the fatigue fracture surface of a
           Ti 6Al 5Zr 0.5Mo alloy tested in air at R = 0.35
           showing facets and out of plane features.   X110

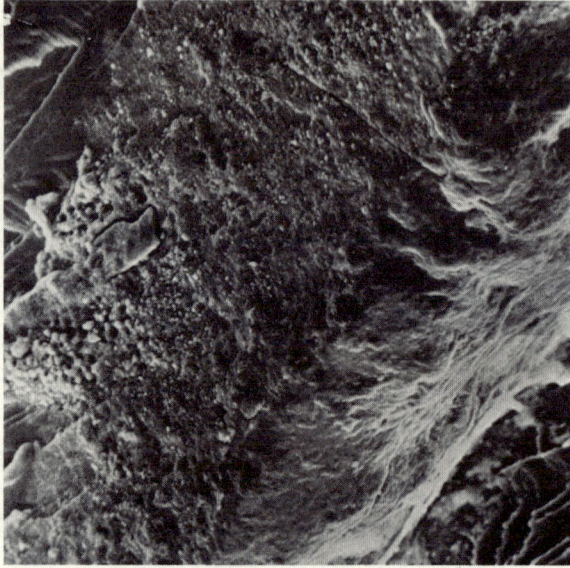

Fig. 9.   SEM micrograph of the fatigue fracture surface of a
          Ti 6Al 5Zr 0.5Mo(A) alloy tested at R = 0.1 near
          threshold showing surface bettering.   X 900

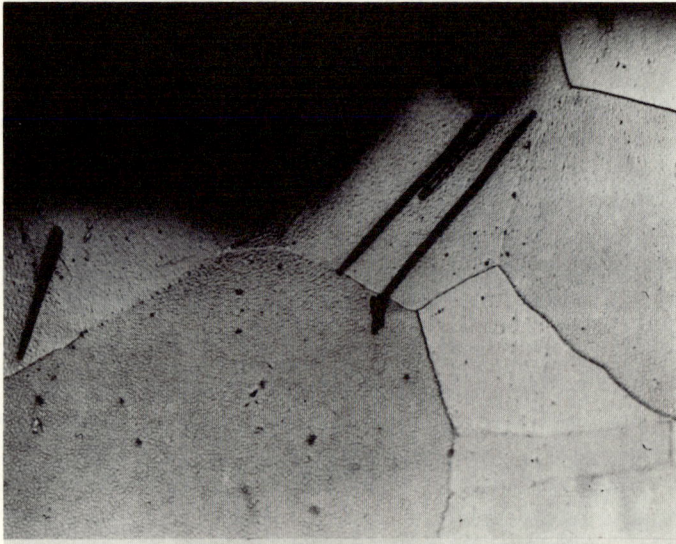

Fig. 10. Pre-cracks lying on planes parallel to part of the
          final fatigue fracture path in Ti 160.   X 425

DISCUSSION

The factors which may influence fatigue crack growth and ΔK threshold include Youngs Modulus, cycical yield strength, grain size and other microstructural elements, mean stress, stress history, residual stress, mode of crack tip opening, environment and test temperature.

The results in this paper refer to a range of titanium alloys tested with nominally Mode I crack face opening. Thus, the major potential variables influencing fatigue crack growth and ΔK threshold are load ratio, test temperature, yield strength, microstructure and environment.

An examination of the results in Table I indicates no overall systematic variation of ΔK_{th} with <u>yield stress</u> but there will be further consideration of this point later in the discussion.

The effect of <u>test temperature</u> $20^{o}C$ to $380^{o}C$ on fatigue crack growth rates and ΔK_{th} is illustrated in Figure 6 and results presented in Table I. The change in ΔK_{th} was small but showed a movement towards lower values with increasing test temperature. The major effect of elevated test temperatures was the sharpening of the log da/dN v ΔK curves and the marked increase in facet formation.

There was a general trend with increased severity of <u>environment</u> to faster crack growth rates and lower ΔK_{th} values (Figure 1) and Table I. The tests in an inert environment such as a vacuum of 10^{-5} torr resulted in ΔK_{th} values of 8 $MNm^{-3/2}$ for α-titanium and α-β titanium (Ti 6Al 4V) at $20^{o}C$ and $350^{o}C$. In the Ti 6Al 5Zr 0.5Mo alloy the ΔK_{th} values in vacuum were higher than 8 $MNm^{-3/2}$ and showed a marked R dependence (13-7 $MNm^{-3/2}$ R = 0.1 to 0.6). The effect of analer water, and 3.5% sodium chloride solutions was to increase the extent of basal plane facet formation. This was achieved by facets forming in grains with their basal poles making increased angles to the direction of the major tensile stress axis (Figures 4 and 5).

The role of <u>microstructure</u> in determining fatigue crack growth rates and ΔK_{th} is well documented in Table I. The trend is for ΔK_{th} to increase with grain size in the Ti 115 and Ti 155, with α grain size in the Ti 6Al 4V alloy and with α colony size in the Ti 6Al 5Zr 0.5Mo alloy. The highest ΔK_{th} values were obtained in the Ti 6Al 5Zr 0.5Mo alloy with the coarsest microstructure and the higher yield strengths.

The influence of <u>load ratio R</u> on ΔK_{th} was as expected with increasing ΔK_{th} as R decreased. The ΔK_{th}^i values of ~ 5 MNm$^{-3/2}$ for the B series Ti 6Al 5Zr 0.5Mo alloys showed that in this alloy the R dependence of ΔK_{th} measured was directly attributable to varying amounts of crack closure.

A recent paper[8] on fatigue crack growth in α-titanium which reported the results of optical and compliance measurements came to the following conclusions. Replica and optical studies of surface cracks showed that crack closure was caused by deviations of the crack trajectory associated with a transgranular mode of crack growth. As the crack faces approached one another over the reducing half of the load cycle small amounts of in-plane shear produced points of contact which tended to wedge the crack open and prevent the stress intensity from falling to that associated with the minimum load. Thus, the wedging open of the crack reduces the range of crack tip opening at the crack tip. The mismatch of the fracture planes and the subsequent wedging open of the crack faces could be expected to occur as a result of small amounts of Mode II opening.

The present results indicate that the increased ΔK_{th} has two components ΔK_{th}^i and ΔK_{th}^c where

$$\Delta K_{th} \text{(measured)} = \Delta K_{th}^i + \Delta K_{th}^c \qquad [3]$$

ΔK_{th}^i intrinsic threshold is related to material and environmental parameters. ΔK_{th}^c is the contribution to ΔK_{th} resulting from 'closure' and a limitation of crack tip opening displacement.

The factors influencing ΔK_{th}^i will now be considered. At R = 0.35 in vacuum the ΔK_{th} is ~ 8 MNm$^{-3/2}$ for α-titanium and mill annealed and β transformed Ti 6Al 4V at room temperature and mill annealed Ti 6Al 4V at 350°C. For Ti 6Al 4V in the tempered martensitic microstructure the ΔK_{th} at room temperature is lowered to ~ 6 MNm$^{-3/2}$ and in the near α Ti 6Al 5Zr 0.5Mo it is raised to ~ 9 MNm$^{-3/2}$. It would thus appear that there is an environmental independent contribution to ΔK_{th}^i which increases with increasing microstructural element size. This may relate to the arrest condition, namely the break up of the crack front. In the finer microstructures such as the martensites the small plastic zone sizes (5–10 μm) can maintain crack extension across individual platelets and thus may also be assisted by interphase and interface separation. In coarser microstructures the ligaments size

created by non-uniform crack extension increases and requires
higher stress intensities for their failure and maintenance of a
coherent crack front.

The introduction of an environment such as water vapor leads
to hydrogen and other assisted modes of crack growth. In all
alloys the ΔK thresholds are decreased by the presence of water
vapor. In the fine α-grained Ti 6Al 4V alloy the ΔK_{th} values are
markedly reduced as a consequence of an increase in the extent of
facet formation [3]. The results in Figures 4 and 5 reinforce
this observation. The holds at maximum load during the elevated
temperature tests showed no evidence of sustained load crack extension. The lowering of the ΔK_{th} through ΔK_{th}^i may thus be
thought of in terms of a synergistic interaction between the environment and the mechanical damage at the crack tip due to reversed plasticity.

The ΔK_{th}^c contribution to ΔK_{th} can now be considered. The
non-closure of the crack faces can limit the range of crack tip
opening and hence the rate of crack extension. Non-closure results from the mis-match of the crack faces and is encouraged by
coarse microstructural features. In the Ti 6Al 5Zr 0.5Mo alloy
the R dependence could be directly related to the ΔK_{th}^c component
with a ΔK_{th} of ∿ 5 $MNm^{-3/2}$ for all R ratios. In other metals and
alloys[9,10] the vacuum ΔK_{th} tends to be insensitive to R ratio.
The vacuum ΔK_{th} values of the Ti 6Al 5Zr 0.5Mo alloy varied from
7.5 to 13 $MNm^{-3/2}$ (R 0.6 to 0.1); thus dependence can be related
directly to the facet formation and out of plane character of the
fracture surface formed during fatigue crack growth in vacuum.

The threshold or arrest condition for a fatigue crack can be
thought of in terms of a lower bend value of crack tip opening
displacement range ΔCTOD

$$\Delta CTOD \simeq \frac{(\Delta K_{th}^i)^2}{4\sigma_y'E} \qquad [4]$$

where σ_y' is cyclical flow stress. For the titanium alloys being
considered ΔK_{th}^i should increase with $\sqrt{\sigma_y'}$. From Table I the ΔK_{th}
values for the α and near α titanium alloys at high R ratio
($\Delta K_{th} \rightarrow \Delta K_{th}^i$) are presented in Table III. The results indicate

TABLE   III

Alloy	Yield Stress MNm^{-2}	ΔK_{th} $\mathrm{MNm}^{-3/2}$
Ti 115	260	2.5 – 3
Ti 155	630	3 – 5
Ti 6Al 5Zr 0.5Mo	840	5 – 7

that for a similar microstructural condition (α grain size and colony size) the ΔK_{th}^i value increases with σ_y (σ_y' values are not available) as expected from [4].

How does this observation relate to the well documented observations

$$\sigma_y = \sigma_o + k_y d^{-\frac{1}{2}} \qquad [5]$$

and $\qquad \Delta K_{th} = \Delta K_{th}(o) + kd^{\frac{1}{2}} \qquad [6]$

that is ΔK_{th} increases as the macroscopic yield stress decreases. The yielded material ahead of the crack tip is

$$r_p = \frac{1}{3\pi} \left(\frac{\Delta K}{2\sigma_y} \right)^2 \qquad [7]$$

contained in a zone r_p which is very much smaller than the grain size when $\Delta K = \Delta K_{th}$. Thus, the yield stress of the material in the process zone is independent of the grain size and for other factors constant ΔK_{th}^i should not vary significantly with grain size. There is a small proviso to this namely, the possible influence of ligament size on ΔK_{th}^i varying with grain size. On this basis the grain size influence on ΔK_{th} is mainly contained in the ΔK_{th}^c component. The larger grain sizes producing more irregular fracture faces and increased "non-closure" of the fracture surfaces.

The preceding discussion was based on a long crack configuration where the fracture surface topology imposed itself on the subsequent crack extension process. In the "short crack" case there is little prior history by deformation. The results in Table II were obtained from R = - 1 tests on pre-cracked smooth sided specimens. The results are of a similar magnitude to those for ΔK_{th}^i obtained from Ti 115, Ti 130 and Ti 155 (Table I). These observations lend support to the concept that the "short crack" threshold values will be nearer the values of ΔK_{th}^i than ΔK_{th} obtained from deep crack studies.

CONCLUSIONS

1) The measured ΔK threshold ΔK_{th} can be considered to have two major components ΔK_{th}^i and ΔK_{th}^c

$$\Delta K_{th} \text{(measured)} = \Delta K_{th}^{i} + \Delta K_{th}^{c}$$

The magnitude of ΔK_{th}^{i} is controlled by microstructure and environment through their influence on the crack tip opening displacement at which the crack extension becomes diminishingly small. The magnitude of ΔK_{th}^{c} is controlled by the extent of non-closure of the crack faces and the limiting of the range of crack tip opening displacement.

2) ΔK_{th}^{i} can increase with yield stress providing this increase is achieved through the hardening of a basically coarse microstructure, i.e., solid solution hardening.

3) Factors which enhance the out of plane character of crack extension tend to increase the ΔK_{th}^{c} contribution to ΔK_{th}.

4) The variation of ΔK_{th}^{c} is probably the major contributor to the grain size dependence of ΔK_{th}.

ACKNOWLEDGEMENT

The authors wish to thank M. Hicks and R. Jeal for their permission to include results obtained on the Ti 6Al 5Zr 0.5Mo alloy.

REFERENCES

1. J. L. Robinson and C. J. Beevers, Metal Science Journal, 1973, V1, p. 153.
2. M. Ward-Close and C. J. Beevers. To be published in Met. Trans. 1980.
3. P. E. Irving and C. J. Beevers, Met. Trans., 1974, V5, p. 392.
4. P. E. Irving and C. J. Beevers, Material Science and Engineering, 1974, V14, p. 229.
5. C. J. Beevers and P. E. Irving, 1974, Tweksbury Symposium, Melbourne, Australia, p. 179.
6. M. A. Hicks, Ph.D. Thesis, University of Birmingham, 1980.
7. J. L. Robinson, M. R. Warren and C. J. Beevers, J. Less Common Metals, 1969, V19, p. 161.
8. N. Walker and C. J. Beevers, Fatigue of Engineering Materials and Structures, 1979, V1, p. 135.
9. B. R. Kirby and C. J. Beevers, Fatigue of Engineering Materials and Structures, 1979, V1, p. 203.
10. G. Booth, 1975, Ph.D. Thesis, University of Birmingham.

DISLOCATION DISTRIBUTION IN PLASTICALLY DEFORMED METALS

I. R. Kramer, R. Pangborn and S. Weissmann

David W. Taylor Naval Ship R&D Center, Annapolis, MD
Penn State University, University Park, PA and
Rutgers University, New Brunswick, NJ

ABSTRACT

The distribution of dislocations in the surface layer ρ_s and in the interior ρ_i and its affect on work hardening and fatigue fracture has been described. In fatigue ρ_s is always greater than ρ_i but both increase with the number of stress cycles. The dislocation structure in the interior is, however, unstable without the presence of the surface layer. It is shown that fatigue damage can be measured when a sufficiently penetrating X-ray source is used.

INTRODUCTION

It is often mistakenly assumed that the work hardening is uniform throughout the cross-section of specimens strained axially. This view is often tacidly accepted contrary to the evidence presented in a large number of papers published over the past 20 years. However, when uniform work hardening is not assumed it appears much easier to understand the phenomena involved in fatigue, stress corrosion, creep and the influence of environmental effects on the mechanical behavior of metals. Apparently a too simplistic model leads to difficulties in explaining phenomena that occur in real materials. One of the purposes of this paper is to provide evidence for the important role of the surface layer in plastic deformation and fracture processes. The discovery in 1928 by A. F. Jaffe' (1) that the ductility of rock salt crystals which are normally brittle could be greatly enhanced when they were deformed in water is perhaps the first indication of the possibility of altering the mechanical behavior of solids by influencing the surface. Later Gough (2) reported that the fatigue behavior of various metals

was enhanced when the tests were conducted at reduced pressures.
Rehbinder found in 1928 that suface active agents decreased the
hardness and increased the drilling rate of rocks and minerals.
Later it was reported that the plastic flow properties of metals
also were altered by solutions containing surface active sub-
stances (3). The influence of solid films also was found to have
a very large effect on the mechanical behavior of single and polycrystalline metals. A review of the earlier work conducted prior
to 1960 is given in reference (4). The early explanations for the
influences of the surface on mechanical behavior were usually in
terms of the dissolution of microcracks or in terms of a decrease
in surface energy. The concept of a Griffith crack arose because
of the Jaffe' effect. However, electron microscopic as well as
other investigative techniques failed to reveal the presence of
such cracks. Yet, the concept of the existence of Griffith cracks
as a real entity is still assumed in some quarters.

Starting in 1961, a systematic investigation of surface effects
was initiated. The earlier research was an investigation to determine the effect of chemically removing the surface during plastic
deformation on the mechanical behavior of metals (Al, Cu, Au, Fe,
Mo, and Zn (5-8). It was established that the removal of the surface layers of metals during tensile deformation decreased the
slopes of Stages I, II and decreased the flow stress in Stage III.
The extent of the first two stages increased with increasing rate
of removal of the surface layers. These effects were not small,
for example for Al single crystals at a dissolution rate of 13 μm
per minute, the slope of Stage I decreased by a factor of 2 while
the extent increased 2.5 times. Similarly for polycrystalline
metals the flow stress was decreased with increasing removal rate.
From these latter observations it was clear that grain boundaries
did not eliminate the influence of the surface layer on mechanical
behavior as might have been assumed on intuitive grounds. Other
parameters such as activation energy, activation volume, dislocation
velocity stress exponent and the relaxation rate are known to be
affected by the surface. Of interest is the observation that the
mechanical behavior of gold, which is believed not to have an
oxide layer, was influenced by the surface. Apparently, the changes
in the mechanical behavior are not due only to the removal of an
oxide film but also associated with the formation of a surface
layer that influences the behavior of the entire specimen.

From the changes in the mechanical properties, the activation
energy, activation volume and hardness, it was concluded that the
surface layer formed during plastic deformation was "hard" and
contained a higher percentage of dislocations than the interior (7,9).
This conclusion was based primarily on the changes produced in
the mechanical behavior when the surface layer was removed from

specimens that had been plastically deformed. However, the
conclusion was supported by many subsequent investigators, especially
by the USSR investigators who appear to be heavily engaged in
investigations into this subject. A hard surface might be expected
on the basis that the stress to activate dislocation sources at or
near the surface is much lower than that required for sources in
the interior. However, it could be argued that these dislocations
could leave the specimen because of the attractive image forces as
postulated by Head for a single dislocation (10). This argument
fails to consider the interaction of dislocation on neighboring
slip planes in region of the surface (11) or the possibility that
an inverse pile up array may exist because of the influence of the
stress field of the nearest neighbor dislocations on the lead
dislocations (12). Support for the "hard" surface layer concept
may be found from investigations employing transmission electron
microscopy, etch pits and a birefringent technique. In his 1969
investigation on crystals, S. Kitajama and co-workers using an
etch pit technique reported that the dislocation density was
much larger in the surface layer than in the interior (13). Electron
microscopy evidence for the existence of a high density dislocation
surface layer has been provided by Terent'ev (14), Goritaskii (15),
and Vol'shukov (16). Etch pit measurements on polycrystalline
copper by Vellaikal and Washburn revealed a higher dislocation density at the surface (17) and Kolb and Macherauch made similar
observations on Ni (18). Evidence for the presence of a hard layer
was provided by etch pit measurements of Fe-3%Si and microhardness
measurements on single crystals of Cu (19). Using a birefringent
technique Suzuki (20) and Mendelson (21) reported the existence of a
hard layer on deformed KCl and NaCl crystals, respectively. Contrary
to the evidence presented above, Fourie considered the surface
layer to be "soft" (22). His evidence was based on comparison of
the flow stress of specimens about 50-60 µm thick that had been cut
from a 10 mm thick single crystal of Cu. He reported that the
critical resolved shear stress of the thin specimens taken from the
surface region was lower than that of specimens cut from the interior.

   Various models have been proposed to explain the formation of a
"hard" surface layer. From theoretical and experimental observation
it is evident that the stress required to operate dislocation sources
at a near the surface is lower than that required for internal
sources (17, 23, 24). Alekseev from theoretical considerations
concluded that the rate of multiplication of dislocation by a near
surface source is an order of magnitude greater than the corresponding rate by an internal source (23). Gol'dina (24) postulated that
while the first dislocation produced by the source will emerge because
of the attractive image forces, a "tail dislocation" is left behind
that prevents the emergence of subsequent dislocations that reach
the surface.

X-RAY DIFFRACTION STUDY OF SURFACE LAYER

X-ray diffraction analysis can provide a reliable method for the measurement of the changes in the dislocation density as a function of plastic deformation. The techniques are highly sensitive to the changes in the crystal structure and may be employed for a nondestructive evaluation. Both the double crystal diffractometer and the powder X-ray techniques have been employed in these investigations to determine the changes in the dislocation density-depth profile of specimens subjected to tensile deformation and fatigue. The dislocation density ρ may be related to β the widths of the X-ray line, or spots by (25):

$$\rho = \beta/bT \qquad (1)$$

where T is the sub-grain size and b is the Burger's vector or

$$\rho = \beta^2/9b \qquad (2)$$

Equations 1 and 2 give the lower and upper bounds of the dislocation density, respectively. The dislocation density may be obtained from an analysis of the line profile (26) and the method of integral breadths (27). According to the latter

$$\beta \, \cos \, \theta \; = \; \frac{K\lambda}{L} + 4e \, \sin \, \theta \quad (3)$$

where $e = \Delta d/d$ and d is the interplaner spacing. The first term on the right is related to the particle size while the second term is related to the micro strain $\langle\varepsilon^2\rangle^{1/2}$. The dislocation density associated with the particle size and strain is, respectively

$$\rho_p = \frac{1}{L^2} \qquad (4)$$

$$\rho_\varepsilon = 12\langle\varepsilon^2\rangle^{1/2}/b^2 \qquad (5)$$

where $\langle\varepsilon^2\rangle^{1/2} = e/1.25$ is the root mean square microstrain.

DISLOCATION-DEPTH PROFILE OF UNIDIRECTIONAL STRESSED CRYSTALS

Figure 1a shows the distribution of plastic zones at the surface as studied by Lang X-ray traverse topography in a silicon specimen that was deformed at 800 C with an applied stress of 6.86 MPa. Figure 1b shows the distribution after removal of 125 μm. Not only has the density of the microplastic zones decreased with depth into the crystal but the preferred alignment of the zones has changed. The microplastic zones at the surface were aligned perpendicularly or at 60 deg. to the (110) tensile axis. Topographs taken at intermediate depths showed a systematic change in density and alignment. The absence of the 60 deg. zones in the deeper region

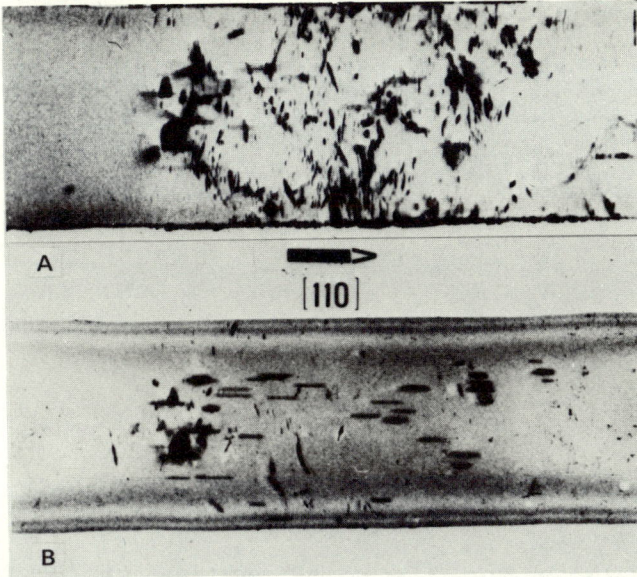

Fig. 1.   Traverse X-ray Topograph of Flat Silicon Specimen Deformed
in Tension at 800^{0}C, Ag, Kα, (200).   (a) Without Surface
Removal, (b) Surface Removal of 125 μm.

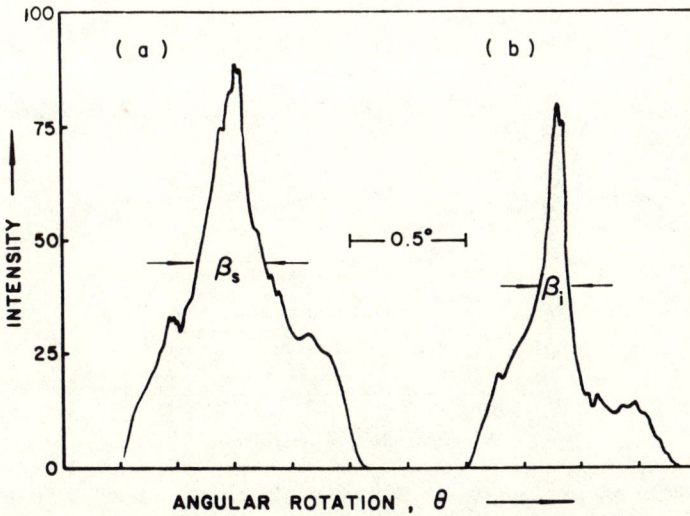

Fig. 2.   X-ray Rocking Curve Profiles of Tensile Deformed Silicon
Single Crystal. ε_p = 10 pct, 650^{0}C, Tensile Axis [110],
(112) Reflection, Cu Kα.   (a) Original Surface, (b)
After Removal of a 100 μm Surface Layer.

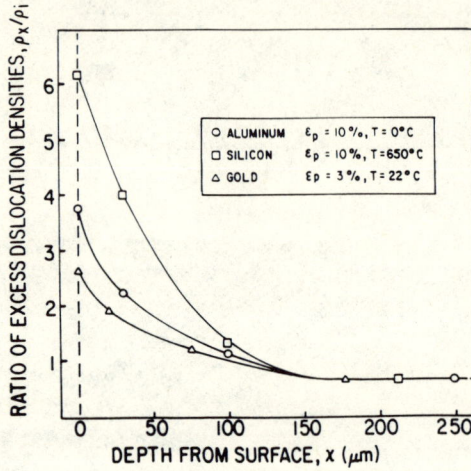

Fig. 3.   Distribution of Excess Dislocations with Depth from the
Crystal Surface.   Tensile Axis and Surface Orientations:
Al, <100> and (100); Si <110> and (112); Au, <123> and
(311).

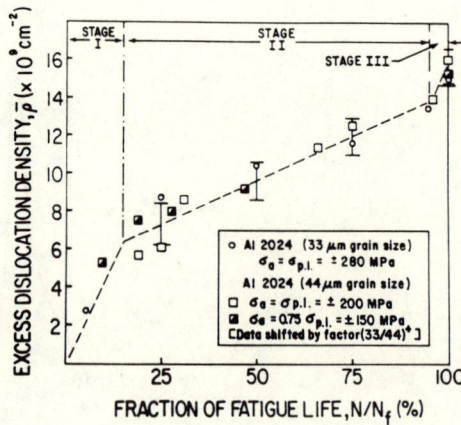

Fig. 4.   Excess Dislocation Density in the Surface for Fatigued
Al 2024 Specimens with G.S 4 and 33 mm, Tested at Various
Stress Amplitudes.   CuKα Radiation.

indicates that many more dislocation reactions are possible at the surface than in the interior of the material (28, 29). The rocking curve profile obtained through the use of the double crystal dif-fractometer for silicon crystals after 10 percent tensile strain is multipeaked and has a half–width β of 935 s of arc, Fig. 2a. A typical value for β for undeformed silicon crystals is about 15 s of arc. Fig. 2b shows the rocking curve after removal of 250 µm. Both the total range and β decreased drastically. Figure 3 shows the dislocation density-depth profile (ρ-x), for silicon a material with a low stacking fault energy; aluminum, a high stacking fault energy material; and gold, which exhibits little or no propensity for the formation of a surface oxide. To prevent possible relax-ation effects in Al the straining and X-ray measurements were carried out at −5 C. The β values given in Fig. 3 and in the other portions of this paper were calculated from Eq. 2. For all three metals there was a decline by factors of 3 to 6 from the high surface layer density ρ_s to a constant value ρ_i at about 100 to 150 µm in the interior.

These results are in agreement with previous investigation (4,7,9,30,31). It is of interest to note that the depth of the sur-face layer (100–150 µm) determined by the X-ray technique is in close agreement with that found by Kramer from surface layer stress measurements.

DISLOCATION DENSITY-DEPTH PROFILE OF FATIGUED SPECIMENS

Figure 4 shows data obtained by analysis of the surface layer, using shallow-penetrating copper radiation (~10 µm) after cycling to various fractions of the fatigue life. The results are plotted for specimens with two average grain sizes differing by about 25 pct in grain diameter, and for tests carried out at two different stress amplitudes. In agreement with Taira, et al, (32, 33), the change in excess dislocation density during the life could be described by a three-stage sequence for all the tests. Rapid increases in the defect concentration occurred early (Stage I) and late (Stage III) in the life. A markedly decreased slope was obtained for the long duration of the second stage, comprising the period from 0.15 to $0.95N_F$. All the data fell within the experimental error band, even for the alternate grain size stock, if a uniform shift factor related to the ratio of grain diameters was applied (34).

The data presented in Fig. 5 (a) were obtained from the rocking curve analysis of Al 2024-T3 specimens cycled for various fatigue life fractions at ±200 MPa, corresponding to the proportional limit. Analogous to the deformation characteristic of monotonically and cyclically deformed single crystals, the ρ-x profile revealed a higher excess dislocation density in the surface layer than in the bulk. Up to about 0.15 pct of the fatigue life (N = 21,000), the

ρ-x profile was similar to that observed after simple tension. Observations made after 5 pct of the fatigue life showed that the dislocation density in the bulk also increased, and a trough appeared on the ρ-x profile in the subsurface region. With further cycling, the excess dislocation density continued to increase in the surface layer and in the bulk.

Figure 6 exhibits the ratio of the excess dislocation density in the bulk to that of the surface as a function of fatigue life for Al 2024 cycled at $\pm$ 200 MPa. At $0.05\ N_F$, ρ_i is only about $0.3\ \rho_s$ The ratio increased with cycling until $0.95\ N_F$, where ρ_i is about $0.7\ \rho_s$.

INSTABILITY OF DEFECT STRUCTURE IN THE INTERIOR

From Fig. 5 it is clear that the ρ_s and ρ_i continues to increase with the amount of fatigue damage. The observation that ρ_i increased in the bulk appears to imply that the damage has occurred throughout the specimen and yet it is known that the fatigue life of metals is completely recovered when the surface layer is removed periodically during the cycling process. The answer to this apparent conflict is that the defect structure in the interior is unstable without the presence of the surface layer. The instability of the interior defect structure is demonstrated in Fig. 5b by specimens that had been cycled 75 and 95 pct of their life and after removal of 400 μm again were cycled at the same stress (200 MPa). The dislocation density determined from the rocking curves employing Cu radiation declined very rapidly during the initial recycling and after about 200 cycles reached a minimum that was approximately the same as that of the original specimen. After 1-2 pct of the life β increased again. When the recycling was continued to 5 pct of the fatigue life the ρ-x profile over the entire cross-section was the same as that of the virgin specimen fatigued 5 pct of the life.

The reversion of the bulk dislocation content back to the virgin state when fatigued specimens were cycled in the absence of the work hardened surface layer explains the extension of the fatigue life by periodic removal of the surface. The observations relative to this instability imply that there is a strong reaction of the dislocation in the interior to the surface layer. At the stress amplitude employed, it is apparent that the dislocation density in the bulk would not have increased during the cycling without the presence of the surface layer. In this sense the increase in the bulk dislocation density is controlled entirely by the surface layer.

The instability observation has important implications relative to the influence of the environment on the work-hardening characteristics. The instability observations lead to the concept that the work-hardening of the interior is dependent on the surface layer. Accordingly, an environment that influences the surface layer also

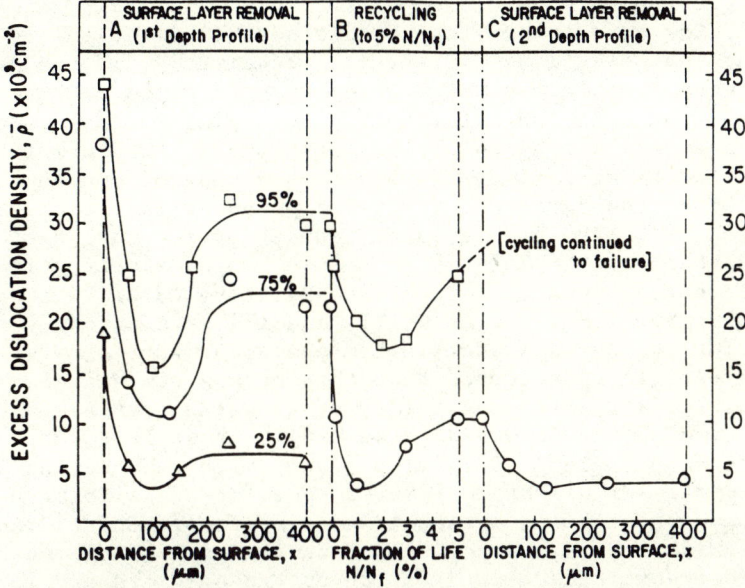

Fig. 5. (A) Excess Dislocation Density-Depth Profile for Al 2024-T3 Specimens Fatigued Various Percentages of Life. (B) Dislocation Density of Surface After Removal of Surface Layer and Cycling. (C) Dislocation Density-Depth Profile After Fatiguing, $N/N_F = 0.75$, Removal of Surface and Cycling $N/N_F = 0.05$.

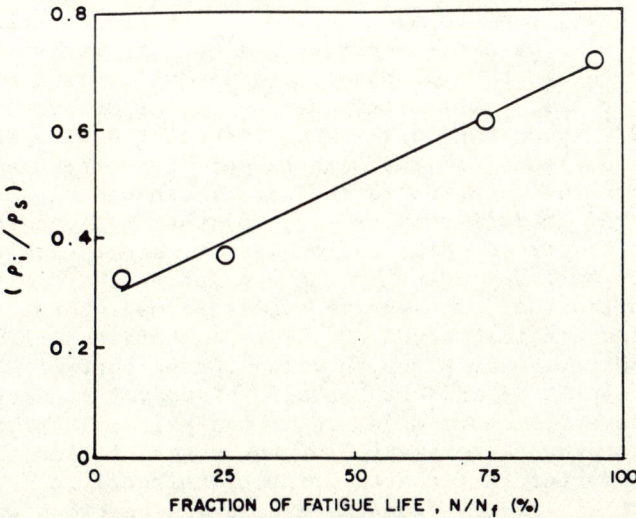

Fig. 6. Ratio of Excess Dislocation Density in the Bulk, ρ_i, to that at the Surface, ρ_s, as a Function of Fatigue Life for Al 2024 Cycled at $\pm$ 200 MPa.

will affect the work hardening of the bulk. For this to occur the stress fields associated with the dislocation array in the surface layer must act over long distances. Conceivably, if the dislocations are arranged in a form similar to an inverse "pile up" the stress may be transmitted over long distances. According to this model the effective stress acting on the dislocations in the interior has in addition to the applied stress a stress component due to the disloca-tion array of the surface layer. The proposal that the stress field associated with the surface layer can act over a long distance to influence the plastic deformation of the bulk material finds support in the investigations of Baranov and Kosykova (35). Whereas it was previously shown that Stages I, II, and III were strongly influenced by removing the surface electrochemically during the deformation (7) these investigators reported that the surface removal also had a remarkable effect on the ductility of tungsten crystals. When no surface removal was involved these single crystals normally failed in a brittle manner along (001) slip planes with a reduction in areas of about 12 pct. With surface removal these crystals fractured with the formation of a neck and a reduction of areas of 83 pct. Apparently, since necking occurs when $d\sigma/d\varepsilon = \sigma$, the removal of the surface decreased the work-hardening of the entire specimen and local instability occurred at a stress lower than the cleavage stress.

ROLE OF DISLOCATION DISTRIBUTION IN SURFACE AND BULK IN FATIGUE

A very interesting and important aspect of fatigue failures as well as other sub-critical crack formation such as stress-corrosion cracking and hydrogen embrittlement is that the cracks form at a relatively low stress level compared to the fracture stress of the material. It has been proposed (30, 31) that such a crack can be formed by the pile-up of dislocation against or within the surface layer. According to this proposal, a crack would form when the surface layer became sufficiently strong to support a critical size "pile-up". When the local stress associated with this accumu-lation of dislocations of like sign exceeds the fracture stress, a crack will form in the affected region. It should be noted that in Fig. 4, and from Taira's results (32, 33) that fatigue failure occurred when the excess dislocation density attained a critical value β^* that was independent of the applied stress amplitude. It was also reported that propagating cracks formed when the surface layer reached a critical strength (31). In this case fracture occurred at the critical strength value independent of the environ-ment, stress amplitude and prior cyclic history. A necessary part of the surface layer fracture model is that a critical barrier strength is required to prevent relaxation of the stress fields. Fracture does not occur prior to the attainment of this critical strength because the stress fields from the excess dislocations would be relaxed by plastic deformation. In fatigue processes where the direction of the dislocation motion changes with the direction of the applied stress it is possible that stress relaxation can also occur

when the dislocation motion is towards the interior. Accordingly,
in cyclic loading processes the build-up of the surface layer,
per se, is a necessary but not sufficient requirement for crack
formation. It appears that the interior must also work harden to
prevent relaxation effects.

DETERMINATION OF FATIGUE DAMAGE

Of interest in Fig. 4 is the observation that all the points
fall on a single curve independent of the stress amplitude. Most
importantly, a propagating crack apparently is initiated, and failure
occurs when a critical value of the excess dislocation density, ρ^*,
is attained. It is evident that if ρ^* can be determined, and the
relationship between fatigue damage and the excess dislocation
density is established, an assessment of the fatigue damage can be
made. It may be seen from Fig. 4, as well as from the data of
Taira et al (32,33) that the slope of Stage II is too low to be used
for accurate and practical determination of fatigue damage. An
examination of the depth profiles shown in Fig. 5 reveals that the
build up of excess dislocation density at the surface layer occurs
much earlier in the life than in the bulk. However, inspection of
the plateau values, established at about 250 μm, discloses that the
average excess dislocation density in the bulk increases steadily
throughout the life. It follows that the fatigue damage can be
determined provided the X-ray beam penetrates sufficiently far into
the materials to sample both the surface layer and bulk. Experimentally, for Al this can be achieved by application of X-ray
radiation with a short wavelength, such as molybdenum Kα radiation.
Figure 7 exhibits a plot which compares the average excess dislocation densities obtained by measurements with copper and molybdenum
radiation. It will be seen that by applying molybdenum radiation, a
sharp incline of the slope is obtained up to the critical value, ρ^*,
which enables one to determine accumulated damage unequivocally. By
contrast, application of copper radiation, which samples only the
defect structure of the quickly saturated surface layer, cannot
accomplish this task, owing to the shallow slope during the
intermediate life fractions.

To test the concept of determining fatigue damage by use of
molybdenum radiation, Al 2024-T3 specimens were fatigued at a series
of stress levels for various number of cycles (Table 1). The
values were obtained after each stress step. Finally, the specimens
were cycled to failure to determine experimentally the remaining
fatigue life. The data presented in Table 1 are typical of the data
obtained to date. After the fourth step, the X-ray measurement
indicated that expended life was 57 pct. Upon cycling at +282 MPa
to failure the expended life was actually 57.5 pct. Included in
Table 1 is a comparison of the fatigue damage as calculated by the
method of Kramer (36) based on a surface layer concept and that of
Miner (37).

The fatigued specimen of Table 1 also was analyzed by a conventional X-ray powder technique using Mo radiation. Using Eq. 3 in the analysis it was noted that the particle size term remained constant, within experimental error, during the fatigue process. The microstrain, however, increased in a linear manner as shown in Fig. 8. An extrapolation of the curve to failure gave a value $\langle \epsilon^2 \rangle^{1/2}$ of 0.008. According to Eq. 3 the particle size was 237A° corresponding to a dislocation density of $1.78 \times 10^{13}/M^2$. At fracture according to Eq. 4, the dislocation density ρ^* was $9.2 \times 10^{13}/M^2$. If it is assumed that the value derived from the double crystal diffractometer analysis is due to the microstrain, the ρ^* at fracture was $9.7 \times 10^{13}/M^2$ (Eq. 3). The value of ρ^* caculated from Eq. 2 was $1.48 \times 10^{12}/M^2$.

Table 1

MEASUREMENT OF FATIGUE DAMAGE – Al2024-T3

Fatigue History	Fatigue Damage, N/N_F		
$\pm$ (MPa)/Cycles, KHz	X-ray	Kramer (36)	Miner (37)
172.4/36	0.22	0.20	0.20
213.7/5.7	0.24	0.31	0.40
248/3.0	0.42	0.49	0.60
282/1.5	0.57	0.63	0.80

Actual expended life at $\pm$282 MPa = 0.575

SUMMARY

Using a double crystal X-ray diffraction technique it was shown:

o For Au, Al and Si strained uniaxially, $\rho_s > \rho_i$

o ρ_s reached a near saturation value after about 20% of fatigue life while ρ_i continued to increase

o ρ_i decreased to the virgin value when the surface layer was removed after fatiguing; implying that the ρ_i is controlled by ρ_s

o Fracture occurred independent of the fatigue stress when the dislocation density reached a critical value ρ^*

o Fatigue damage can be measured provided the X-ray beam penetrates sufficiently far into the metal to sample both the surface layer and the interior.

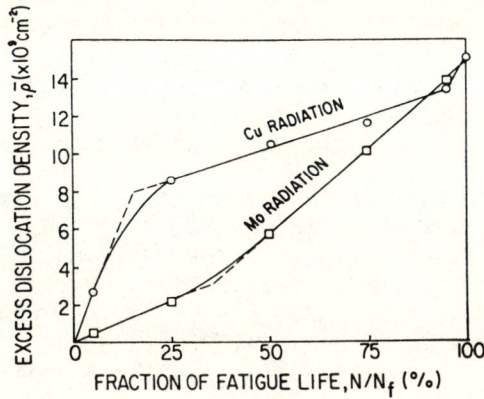

Fig. 7 - Comparison of excess dislocation densities measured by Cu and Mo radiation for Al-2024 fatigued at ±280 MPa.

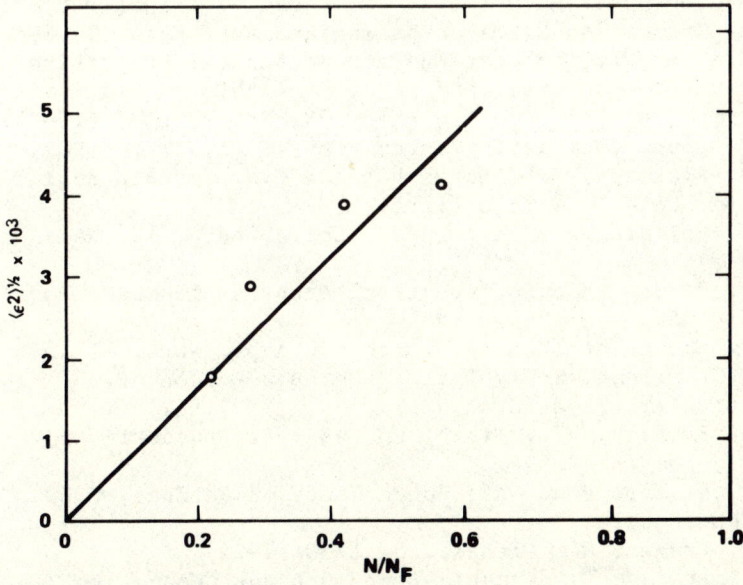

Fig. 8 - Relationship between fatigue damage and microstrain. Al-2024-T3 cycled at various stress amplitudes.

REFERENCES

1. A.F. Joffe Physics of Crystals (1928) McGraw Hill, NY
2. H.G. Gough and D.G. Sopwith, J. Inst. Metals 49, 93 (1932)
3. P.A. Rehbinder, Byull, Akad. Nauk, USSR, Classe, Sci, Mat. Nat.
    Ser. Chem., 639 (1936)
4. I.R. Kramer and L.J. Demer Progress in Material Science, 9,
    No. 3 (1961)
5. I.R. Kramer, Trans. Mat Soc. AIME, 227, 529 (1963)
6. I.R. Kramer, Trans. Met. Soc. AIME, 227, 1003 (1963)
7. I.R. Kramer, Trans. Met. Soc. AIME, 233, No. 8, 1462 (1965)
8. I.R. Kramer and A. Kumer, Scripta Met., No. 4, 205 (1969)
9. I.R. Kramer, Trans. Met. Soc AIME, 230, 991 (1964)
10. A.K. Head, Proc. Phil. Mag. 44, 92 (1953)
11. J.F. Prim and H.G.F. Wilsdorf, Canadian Journal Physics,
    45, 1177 (1967)
12. R. Arsenault, Private communication, 1978
13. S. Kitajama, H. Tanaka, and H. Kaieda, Tran. J.I.M. 10,
    10 (1969)
14. V.F. Terent'ev, Soviet Phys; Doklady, 14, No. 3, 273 (1969)
15. V.M. Goritskii, V.S. Ivanova, L.G. Orlov and V.F. Terent'ev
    Soviet Physics, Doklady, 17, No. 8, 776 (1973)
16. V.I. Vol'shukov and L.G. Orlov, Soviet Physics, Solid
    State, 12, No. 3, 576 (1970)
17. G. Vellaikal and J. Washburn, J. Appl. Phys. 40, 2280 (1969)
18. K. Kolb and E. Macherauch, Phil. Mag. 7, 415 (1962)
19. I.R. Kramer and N. Balusubramanian, Acta Met. 21, 695 (1973)
20. Taira Suzuki, Dislocations and Mechanical Properties of
    Crystals, John Wiley and Sons, Inc., NY, P. 215 (1956)
21. S. Mendelson, J. Applied Phys. 33, No. 7, 2182 (1962)
22. J.T. Fourie, Canadian Journal Phys., 45, 777, (1967)
23. A.A. Alekseev, S.B. Goryachev and B.M. Strunin Soviet Phys.
    Solid State, 16, 2310 (1975)
24. M.G. Gol'diner, Soviet Phys., Solid State, 17, No. 2, 406
    (1975)
25. P.B. Hirsch "Mosaic Structure" Progress in Metal Physics 6,
    236 (1956)
26. B.E. Warren and B.L. Averbach, J. Appl. Phys. 21, 595 (1950)
27. A.J.C. Wilson, X-ray Optics, Methuen London, P. 37 (1940)
28. S. Schafer, Phys. Stat. Sol., 29, 297 (1967)
29. V.N. Erofeev, V.L. Nikitanko and V.B. Osvenskii, Phy. State
    Sol., 35, 79 (1969)
30. I.R. Kramer, Proc. Air Force Conf. on Fatigue, 1969,
    AFFDL-TR-0-144
31. I.R. Kramer, Met. Trans., 5, 1735 (1974)
32. S. Taira and K. Hayashi, Proc. 9th Jap. Congr. on Testing
    Materials, 1 (1966)
33. S. Taira, K. Tanaka and T. Tanabe, Proc. 13th Jap. Congr.
    on Materials Research, 14 (1970)

34.   S. Weissmann, R. Pangborn and I.R. Kramer, Fatigue Mechanisms,
      ASTM STP 675, 163 (1979)
35.   Yu. V. Baranov, E.P. Kostyukova and I. M. Makhmertov Problemy
      Prochnosti, No. 4, April, 483 (1978)
36.   I.R. Kramer, Proc. 2nd Int. Conf. on Mech. Behavior of
      Materials, 812 (1976)
37.   M.A. Miner, J. Appl. Mech. Series E 12, A-159 (1945)

THE EFFECT OF MICROSTRUCTURE ON THE FATIGUE BEHAVIOR OF NI BASE SUPERALLOYS

Stephen D. Antolovich* and N. Jayaraman**

Department of Materials Science and
Metallurgical Engineering
University of Cincinnati
Cincinnati, OH 45221

ABSTRACT

The fatigue crack propagation (FCP) and low cycle fatigue (LCF) behavior of Ni base systems is reviewed. It is seen that for typical Ni base alloys such as Waspaloy, René 95, In 718 and Astroloy the FCP rate decreases with increasing slip planarity, in agreement with previous theoretical suggestions.

The high temperature LCF behavior of alloys such as René 77, René 80 and others is determined by a trade-off between structural coarsening, which is beneficial, and environmental degradation. In René 80 at 871°C and 982°C the trade-off is such that the life increases with decreasing frequency or with imposition of a hold time. Low cycle fatigue data for alloys such as René 77, René 80, Mar M002 and Nimonic 90 are in agreement with an idea which assumes that crack initiation occurs at a unique combination of maximum stress and environmental damage. This idea is used to deduce an expression for the LCF behavior of stable materials in terms of frequency, hold time, temperature, plastic strain range and cyclic stress/strain parameters. The resulting model predicts general trends in high temperature LCF behavior as well as Coffin-Manson LCF exponents.

---

*Professor of Materials Science

**Visiting Assistant Professor of Materials Science

I.   INTRODUCTION

Ni base superalloys are used in jet engine components such as
disks, turbine blades and vanes; as such they are subjected to ex-
tremes in loads, temperatures and environments making resistance to
fatigue a prime design consideration. For example, in a turbine disk
the temperature can vary from 150°C at the hub to 550°C or more at
the rim.   During periods of maximum power application, stresses in
certain locations may be as high as 90% of the yield.   Clearly the
formation and subsequent propagation of a crack is critically important.   In fact implementation of the "retirement for cause" philosophy[1] is based on an ability to characterize crack propagation in
terms of NDI and fracture mechanics methodology.   Improvements in
the fatigue behavior will allow the life to be extended or the
payloads to be increased and improving the fatigue behavior is
a major goal of current research programs[2].

When considering the performance of turbine blades, the low
cycle fatigue (LCF) behavior is a property that is of utmost importance.   The effects of prolonged holds (creep/fatigue) and hostile
environments are even more important in blades than for disks since
the temperature regimes are much higher, ranging from perhaps 600°C
at the base to as high as 1090°C at the tips for some of the more
recent alloy systems.

The first part of this paper deals primarily with the effects of
microstructural variations on the FCP behavior of Ni base alloys
while the second part deals with creep-fatigue-environment interactions at higher temperatures.

II.   THE EFFECT OF MICROSTRUCTURE ON FCP

   A.   Waspaloy at Low Temperature

      1.   Material Heat Treatment and Test Procedure

            Waspaloy is a Ni base alloy having a low lattice mis-
match.   Consequently heat treatments can easily be carried out in
which dislocations shear small particles (~100A°) and loop around
large particles (~1000A°) while maintaining a relatively constant
yield strength[3].   Specimens containing small precipitates will
thus exhibit well-defined planar slip (slip bands) while those containing the large particles will deform by wavy glide or at least
will be less planar.   The degree of slip homogeneity also can be controlled by controlling the grain size with large grains favoring less
homogeneous (planar) deformation.   In order to study the effects of
slip homogeneity on FCP behavior, specimens containing large and
small particles were prepared from specimens that were previously
heat-treated to have two very different grain sizes.   The heat treatments, γ' sizes and grain sizes are shown in Table I.   All specimens

Table 1

Schedule of Heat Treatments with Corresponding
Microstructures

Solution Treatment	Precipitation Treatment	
	875°C 24 hr $\gamma^1 \sim 1000$ A	730°C 6 hr $\gamma^1 \sim 100$ A
1010°C 2 hr ~ASTM GS #9	(A) Fine GS, Large γ^1	(B) Fine GS, Small γ^1
1040°C 2 hr ~ASTM GS #4	(C) Intermediate GS, Large γ^1	(D) Intermediate GS, Small γ^1
1100°C 2 hr ~ASTM GS #3	(E) Coarse GS, Large γ^1	(F) Coarse GS, Small γ^1

were taken from a single heat of material which was within the normal composition limits for Waspaloy.

Testing was done at 25°C using compact specimens as defined by the ASTM[4]. The test frequency was 10Hz and a triangle wave with a load ratio (Pmin/Pmax) of 0.05 was used. Crack length measurements were made using crack propagation gages and occasionally verified by a traveling microscope. Some experiments were also done to study crack retardation in which a single overload was applied. The overload ratio ($\Delta P_{oL}/\Delta P_{cont}$) was 1.5.

2.   Experimental Results and Discussion

The results of the constant amplitude FCP tests are shown in Fig. 1 and summarized in Table II. The overload test results are shown in Fig. 2. From Fig. 1 it is apparent that even in the so-called Paris régime there is a pronounced effect of microstructure and that the more planar or inhomogeneous the slip, the lower the FCP rate. The effect of decreasing the precipitate size from 100A° to 100A° (for constant grain size) was to decrease the FCP rate by a factor of about 5 in tho lower ΔK range of the data shown in Fig. 1. It is not possible to attribute this effect to gross differences in yield behavior as can be seen from Table III where tensile properties are given. The effect of increasing the grain size from ASTM #9 to ASTM #3 is to further decrease the crack growth rate, by a factor of about 2. The data show that the FCP rate can be varied by about one order of magnitude for the parameters

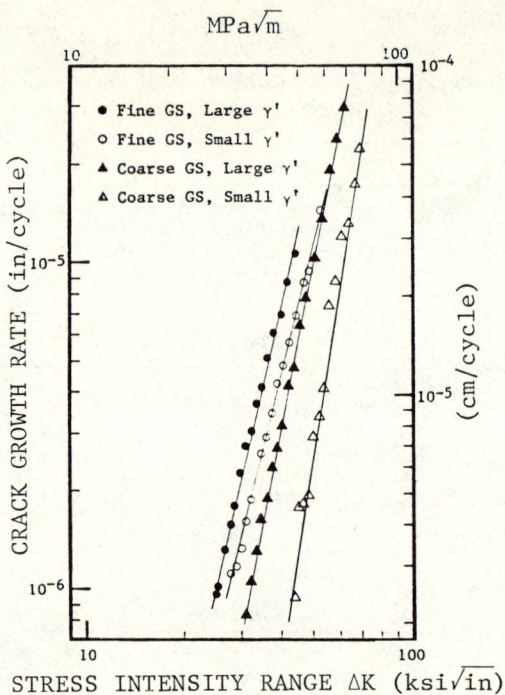

Fig. 1. Composite Plot of Waspaloy FCP Data. All Testing Done at 25°C.

Table II

Paris Law Fatigue Crack Propagation Constants

$$\frac{da}{dN} = R\,(\Delta K)^m$$

Heat Treatment	Stress Intensity Range Limits (ΔK) ksi$\sqrt{\text{in}}$ (MPa $\sqrt{\text{m}}$)	R (when K is in ksi$\sqrt{\text{in}}$)	m	Correlation Coefficient
A	.25.1 - 44.8 (27.6 - 49.2)	1.3×10^{-12}	4.20	0.997
B	28.1 - 52.7 (30.9 - 57.9)	1.8×10^{-12}	3.99	0.993
E	31.4 - 63.3 (34.5 - 69.6)	2.2×10^{-14}	5.09	0.999
F	44.5 - 71.0 (48.9 - 78.0)	1.4×10^{-17}	6.62	0.992

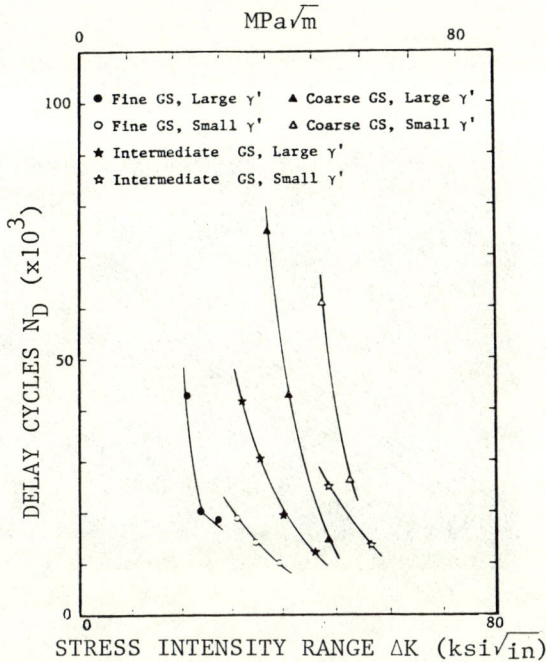

Fig. 2. Waspaloy Tensile Overload Data. Overload Ratio = 1.5
Specimen Thickness = 1/8 in. All Testing Done at 25°C.

used in this study and that the greater the slip planarity the lower
the FCP rate.

Different fracture surface features and deformation modes were
observed for specimens given different heat treatments. As an
example, the fracture surfaces of the most and least fatigue resistant structures are shown in Fig. 3 while the corresponding

Table III
Monotonic Tensile Data

Heat Treat-ment	0.2% Yield Strength Ksi (MPa)	Ultimate Ten-sile Strength Ksi (MPa)	Percent Reduc-tion in Area	True Frac-ture Strain	Strain Harden-ing Exponent
A	113 (778)	229 (1581)	46	0.61	0.32
	110 (760)	227 (1562)	46	0.61	0.32
B	145 (1003)	249 (1715)	39	0.50	0.25
	145 (1003)	249 (1715)	43	0.57	0.25
E	96 (664)	227 (1565)	26	0.30	0.34
	97 (668)	228 (1573)	23	0.27	0.34
F	96 (664)	231 (1593)	52	0.72	0.36
	94 (649)	230 (1587)	54	0.78	0.36

(a)

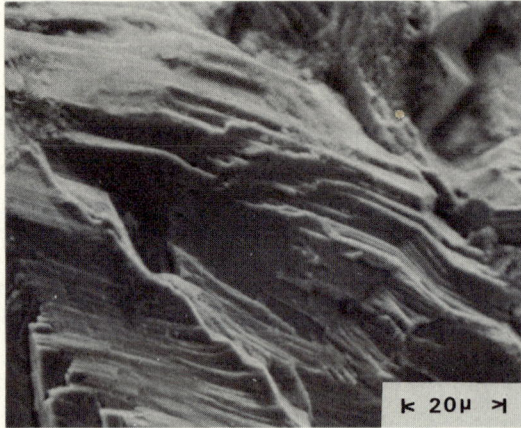

(b)

Fig. 3.   SEM Fractographs of Constant Amplitude FCP Specimens (a)
Fine GS Large γ' ΔK = 42 ksi $\sqrt{in}$ (b) Coarse GS Small γ'
ΔK = 42 ksi $\sqrt{in}$.

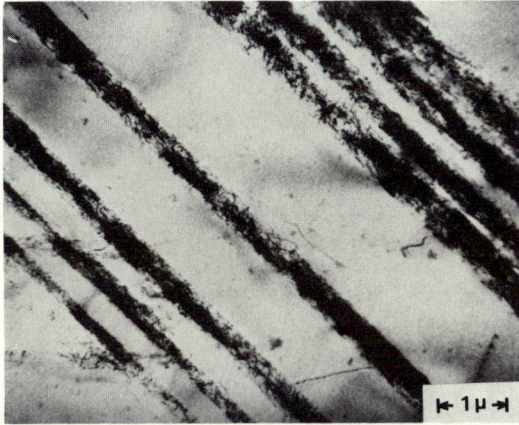

(a)

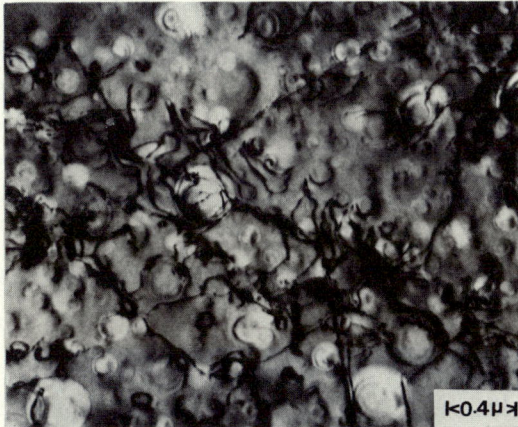

(b)

Fig. 4. Deformation Substructure of Waspaloy Heat Treated so as to Contain (a) Large Grain/Fine γ' and (b) Small Grain/Large γ'. The Transmission Electron Micrographs were Taken from Specimens that were Tested in LCF at 25^0C. In (a) the Plastic Strain Range was 0.94% and (b) it was 0.34%.

deformation substructures are shown in Fig. 4. For the most fatigue resistant microstructure the fracture surface was very crystallo-graphic containing facets that were, in many cases, as large as the grains. The least fatigue resistant structure on the other hand had a fracture surface that was on average more nearly perpendicular to the load axis and which contained indistinct facets. The observa-tions of the substructures by transmission electron microscopy (TEM) were compatible with the fracture surfaces as seen in Fig. 4. The most fatigue resistant material contained well-defined slip bands, while the least fatigue resistant structure deformed by wavy glide. Results similar to these have been observed in other Ni base systems tested at higher temperatures indicating that the beneficial effect of planar glide microstructures is not eliminated by environmental factors, which can themselves be quite important. For example, a review of other results in the literature suggests that promoting slip planarity improves the FCP behavior of such diverse alloys as René 95[5], Astroloy[6] and In 718[7] tested in air in the tempera-ture range of 550°C–650°C. The FCP curves for Astroloy and In 718 are shown in Figs. 5 and 6. In both cases, the improved FCP rates were associated with increases in the grain size, consistent with the results obtained for Waspaloy. For the René 95, both grain size and γ' size were varied. The residual fatigue life is shown as a func-tion of γ' size for two grain sizes in Fig. 7. The results show that increasing slip planarity increases the life, in agreement with the previous results.

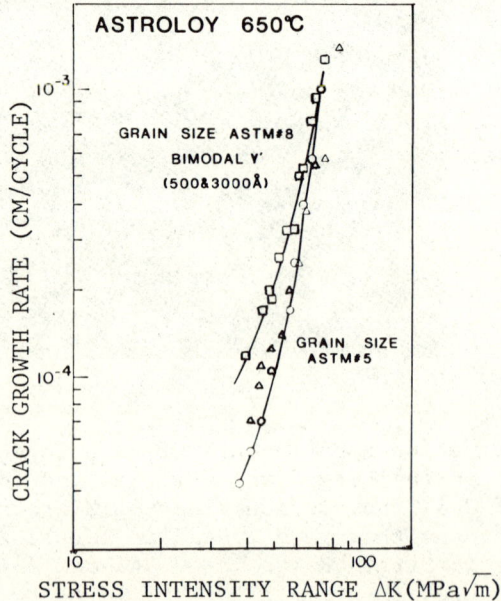

Fig. 5. Effect of Heat Treatment on the Cyclic Crack Growth Rate of P/M Astroloy Tested at 650°C. Frequency = 1.0Hz. [6].

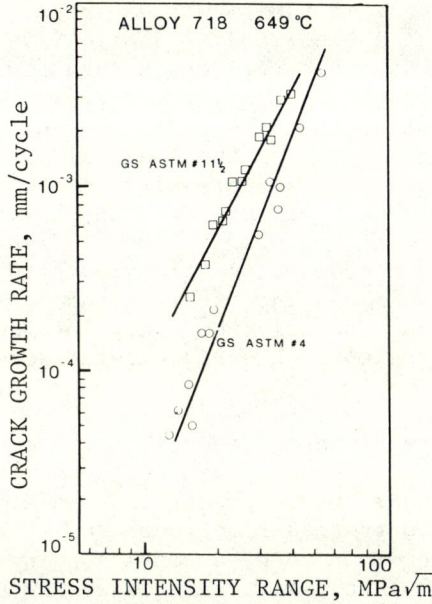

Fig. 6. Fatigue Crack Growth Behavior of Alloy 718 Tested in an Air Environment at 649°C [7].

Reasons for improved FCP behavior with increased slip planarity are not well-established. For example, increased slip reversibility

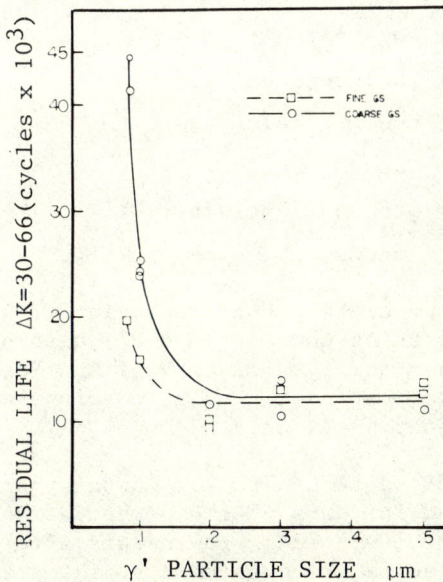

Fig. 7. Effect of γ' Particle Size and Grain Size on Residual Cyclic Life for ΔK = 30 to 66 MPa √m. Testing Done at 538°C [5]

may be advanced as one mechanism wherein plastic deformation is accomplished without an increase in damage[8]. Another possibility is that the increase in the real crack path *decreases* the apparent growth rate even though the local crack growth rate is unchanged[9,10]. Yet another explanation that has been advanced is that as the crack deviates from a path normal to the load axis, there is a reduction in the effective local stress intensity[11]. The efficacy of the latter process would seem to depend on the deviations in crack path being large compared to the plastic zone. Finally for planar glide, the plastic zone is concentrated in narrow slip bands and the probability of these slip bands intersecting and cracking weak regions in the structure (such as grain boundary carbides and intermetallics) is lower than for the more extended plastic zones associated with wavy glide.

The results may be interpreted in terms of a phenomenalogical model in which FCP is viewed as a LCF process ahead of a crack. where the strain is averaged over a process zone[12]. It has been shown previously that this process zone increases with slip planarity[13] which in turn is accompanied by a corresponding decrease in the FCP rate. The results of that model are expressed by the following equation:

$$\frac{da}{dN} = C_1 \left(\frac{C_2}{\varepsilon_f' \, \sigma_{ys}' \, E} \right)^{2m} \cdot \frac{1}{\ell^{\frac{m}{2}-1}} \cdot \Delta K^m \quad \ldots \quad (1)$$

where C_1, C_2 = constants

ε_f' = fatigue ductility

σ_{ys}' = cyclic yield

E = Youngs Modulus

ℓ = process zone size

m = exponent

In addition to predicting a decrease in FCP with increasing ℓ (all other factors being constant) eq (1) also predicts a linear relationship between m and the logarith of the Paris law constant. It is shown elsewhere that such a relationship is observed for the Waspaloy data presented previously[14].

The overload FCP test results for Waspaloy are shown in Fig. 2. The data is presented in terms of the number of delay cycles required for a crack to resume its normal growth rate after application of an overload. The results are even more striking than was the case for the constant load amplitude fatigue; the greater the degree of slip planarity, the greater the retardation. Fractographic features

were similar to those observed previously for constant load amplitude cycling.

In summary, the FCP and overload behavior of Ni base alloys may
be markedly improved by heat treating to increase the slip planarity
(small, shearable precipitates and large grains). The exact heat
treatment would of course depend on factors such as the amount of
γ' and the mismatch in lattice parameter between matrix and precipitate. The results cited above imply that FCP problems in a particular application can be readily eliminated without a costly
replacement or alloy development program.

III. LCF of Ni Base Alloys at High Temperatures

A. Background

At elevated temperatures the problem of what constitutes damage
during fatigue becomes much more complex than at room temperature.
Not only may damage occur as a result of attaining a critical dislocation density and distribution, but the effects of creep and the environment must be considered. A further complicating factor is
that the precipitate structure generally encountered in Ni base
alloys is not stable under stress at elevated temperatures[15,16].
The LCF behavior of some typical Ni base systems is considered in
the following sections.

B. LCF of René 77 and René 80 at Elevated Temperature

1. Microstructures

Blade alloys are generally heat treated so as to con-
tain a duplex γ' morphology consisting of large cuboidal precipitates
which confer good creep resistance and small spherical precipitates
which increase the tensile strength. The precipitate morphology is
shown in Fig. 8 for René 77, a wrought alloy. The carbon content of
these alloys is in the range of 0.15% and the carbon is tied up in
the form of carbides which occur both on grain boundaries and distributed throughout the matrix. Depending on the alloy and temperature
the carbides are of the form MC, M_6C or $M_{23}C_6$ where M refers to the
various metal atoms that may substitute for one another[17]. The
MC carbides are usually blocky and found throughout the material
whereas the M_6C and $M_{23}C_6$ tend to occur on the grain boundaries, frequently associated with grain boundary γ'. The carbide and grain
structure of René 80, a cast blade alloy, is shown in Fig. 9.

2. LCF Test Results for René 80

René 80 was studied extensively at 871°C and 982°C
under conditions of continuous cycling at various rates as well as
with 90 s hold times. The holds were usually applied at maximum
strain although in a few instances the 90 s hold was applied at minimum strain. In some cases, before testing, the specimens were exposed at 982°C for 100h either stress free or at 97 MPa, (~1/3 σ_{ys}).
The results are shown in terms of Coffin-Manson plots in Figs. 10-12.

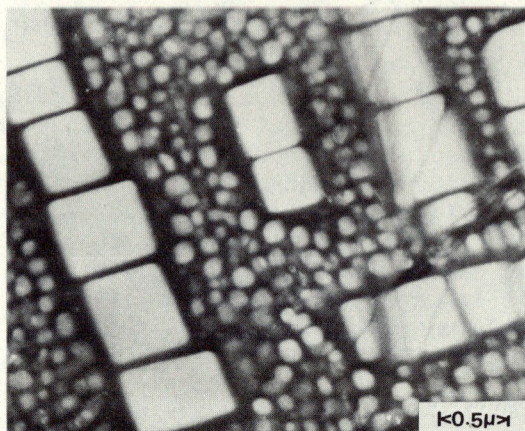

Fig. 8.   As Heat Treated γ' Structure in René 77 [16].

It is noteworthy that when the rate of cycling was low or when a hold
time was imposed, the life *increased* in contrast to the frequent observation that when creep is added to a fatigue cycle the life
decreases. It is also noteworthy that the effect of prior exposure
under stress is to significantly reduce the fatigue life even though
at the low stress levels used in these experiments there was no measurable creep. In some cases the exposed specimens were re-machined
prior to testing and the life was equal to or greater than the life
of the as-heat treated material, indicating that the primary damage
mechanism was surface related[18].

3.   Effects of Cyclic Deformation on Microstructure and
Substructure

Even though the precipitate structure of materials
such as René 77 and René 80 are stable with respect to thermal exposure, cyclic plastic deformation causes rapid structural coarsening.
Typical TEM micrographs are shown in Fig. 13 for René 77 and 80.
These should be contrasted to the initial structure, Fig. 8. Dislocations linking large and small precipitates are shown in Fig. 13a
by dark field TEM. Such dislocations clearly provide a rapid path
for the diffusion associated with such coarsening. The majority of
dislocations were found on the faces of the precipitates, Fig. 13b,
and these have been shown to be near edge in character[18]. Such
dislocations do not constitute damage to any significant degree. Instead they actually *decrease* the energy of the system by taking up
the mismatch between the precipitate and the matrix. Associated with
the coarsened structure is a general decrease in the stress and an

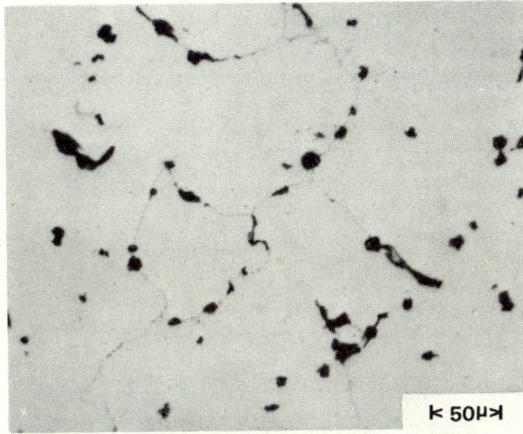

(a)

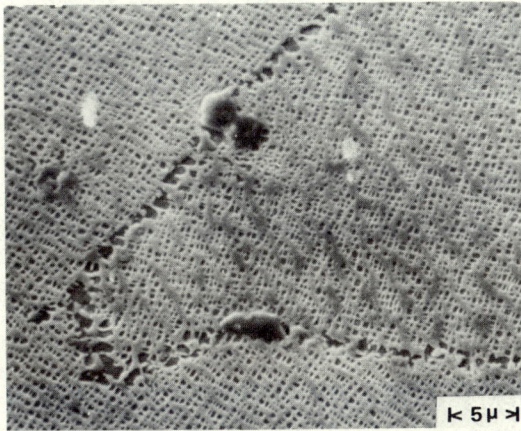

(b)

Fig. 9. Microstructure of as Heat Treated René 80. In (a) the Ir-
regular Nature of the Boundaries and the Carbides is
Revealed Using a Murakamix Etch. In (b) a SEM Micrograph
Shows the Cuboidal Matrix γ', as well as the Discrete
Carbides and γ' in the Boundaries [18].

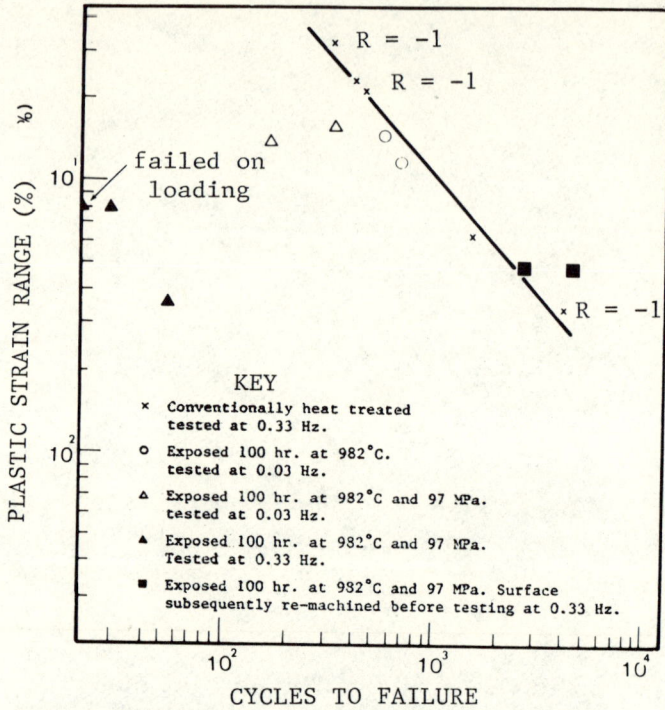

Fig. 10.   Coffin-Manson Plot of LCF Data for René 80 at 871°C.
All Testing Done at R = 0.05 Except as Noted; Includes
Data from Prestressed Specimens (18).

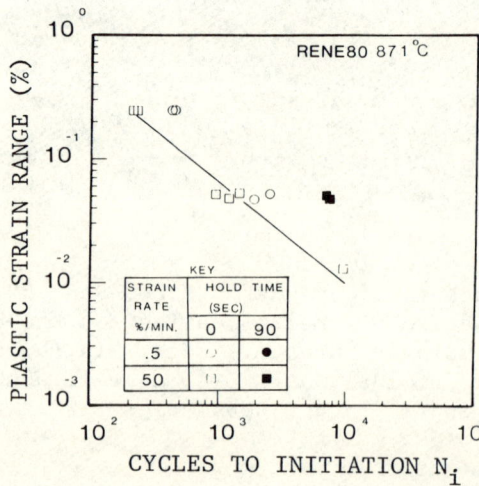

Fig. 11.   Coffin-Manson Behavior of René 80 at 871°C. (25).

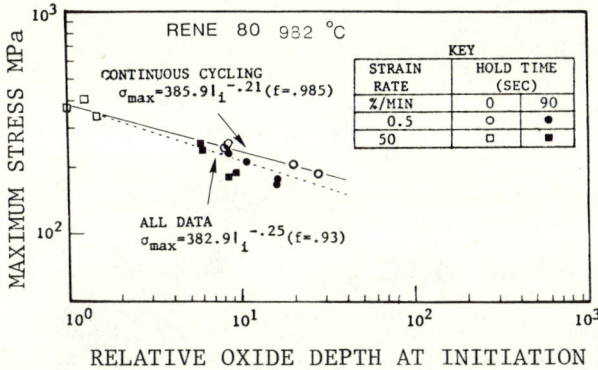

Fig. 12.   Coffin–Manson behavior of René 80 at 982°C. [25].

increase in the apparent ductility[16,18]. Thus the effects of
cyclic deformation that are reflected in the structure should tend
to *increase* the fatigue life with decreasing frequency or increased
hold time and such behavior was in fact observed, Figs. 10-12. The
mechanism of failure must then in some way be associated with environment. This idea is considered in the following section.

    4.   Environmental Damage

        The severe degrading effect of the environment has al-
ready been alluded to for stress exposed René 80 specimens. In
general, failure of such specimens tended to be intergranular, exactly where oxidation is expected to be most severe. Examples of preferential oxidation are seen in Fig. 14 for René 77 and in Fig. 15
for René 80. In Fig. 14, considerable oxidation can be seen along
grain and twin boundaries while in Fig. 15 preferential oxidation is
visible along grain boundaries. Rather than oxide cracking it is
possible that crack formation occurred as a result of oxygen pick up
and subsequent embrittlement of grain boundaries and that macroscopic
oxidation occurred *after* microcracks had formed. Whatever the detailed mechanisms, cracking is fundamentally associated with environment.

    C.   Environmental/Deformation Model of High Temperature LCF

        It appears reasonable to assume that there is a combina-
tion of environmental penetration and stress at which a microcrack
can form[19] that could propagate even in the absence of environmental assistance for the plastic strain range of interest. The exact
mechanism of environmental damage need not be specified. Two possible examples might be formation and cracking of an oxide or embrittlement of boundaries through oxygen accumulation. In either case the
process would be expected to be time and temperature dependent and
to obey diffusional kinetics (which would be possibly modified by
the cyclic deformation). The basic idea can be expressed through the

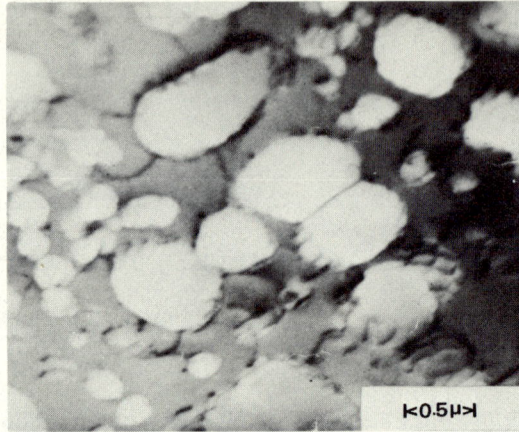

(a)

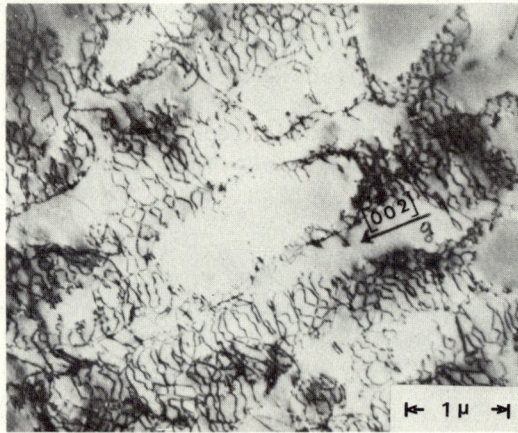

(b)

Fig. 13. Microstructural Coarsening of (a) René 77 Tested in Air
          at 927°C[16].   (b) René 80 Tested in Air at 981°C [18].
          In (a) the Plastic Strain Amplitude was 0.15%, while in
          (b) it was 0.07%.

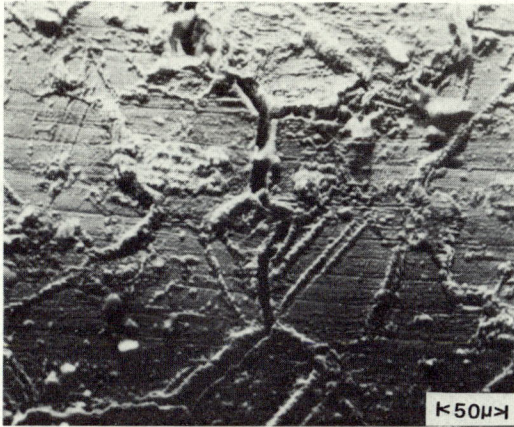

Fig. 14. SEM Surface Micrograph of René 77 Tested at 927°C. The Plastic Strain Amplitude was 0.15% and the Test was Stopped After 50 Cycles (Total Estimated Life ~ 250 Cycles). Note the Oxidation of Grain Boundaries and Twin Boundaries.

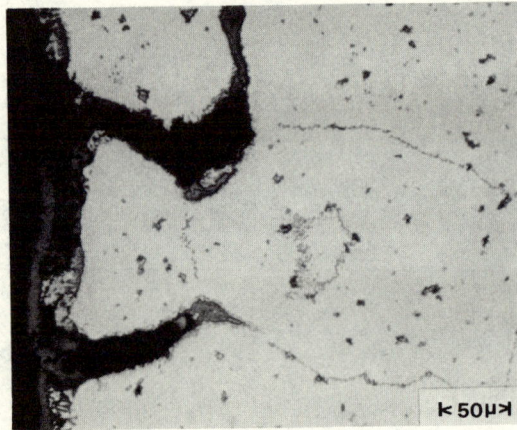

Fig. 15. Grain Boundary Oxidation and Cracking in
          René 80. The Plastic Strain Amplitude
          was 0.07% and the Life was 564 Cycles.
          This Specimen was Held at 982°C for 100
          Hrs. Before Testing at 871°C. [18].

following equation:

$$\sigma_i^{max} \cdot \ell_i^P = Co \qquad \qquad \cdots \text{(2)}$$

where σ_i^{max} = max stress at initiation

ℓi = oxygen penetration at initiation

P, Co = material constants

The oxygen penetration for an initiated crack may be computed assuming that parabolic kinetics are obeyed:

$$\ell i = \alpha \sqrt{Dt_i} \qquad \qquad \cdots \text{(3)}$$

where α = geometric constant

t_i = time to initiation

D = diffusion constant

The applicability of eq. 2 can be examined by taking the time for crack initiation in a given test and comparing it to the shortest crack initiation time for a given set of tests:

$$\frac{\ell_i}{\ell_i^o} = \left(\frac{t_i}{t_i^o}\right)^{\frac{1}{2}} \qquad \qquad \cdots \text{(4)}$$

where ℓ_i^o = initiation crack length for shortest test

t_i^o = time corresponding to ℓ_i^o

For purposes of developing a correlation, ℓ_i^o can be arbitrarily taken as unity. The value of ℓi in arbitrary units is then given by the right hand side of eq. 4. Data corresponding to a set of tests can be used along with calculated values of ℓ_i at initiation as a first order check of these ideas. This has been done for René 80 at 871°C and 982°C as well as for René 77 at 927°C and the results of these correlations are shown in Figs. 16-18. It can be seen that the correlations are quite good, which lends credence to the simple model of environment cracking. A similar correlation has been observed for René 150[20] at elevated temperatures. There is also data available from a recent AGARD study which can be used to further check the model. Materials examined included Nimonic 90[21] and Mar M002[22] tested at various temperatures with different cycle characters. Correlation of the data with the model discussed

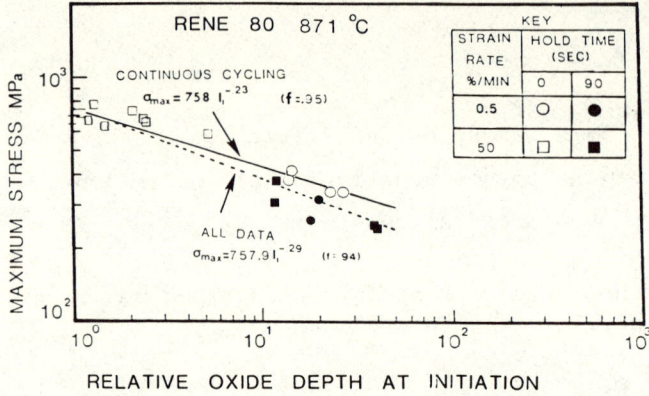

Fig. 16.   Maximum Stress at Initiation as a Funciton of Oxide Depth
          at Initiation for René 80 Tested at 871^{0}C.   The Equation
          of Each Line Along with the Coefficients of Determination
          are Given (26).

previously were quite good as shown in Figs. 19 and 20.   (In these
plots the data are presented in terms of failure rather than initiation since data on initiation were generally not available.)   It is
interesting to note that when different cycle characters are employed
(the cycle character is denoted by the strain range partitioning
(SRP) conventions) different correlations are obtained for each
cycle character.   This implies that the way in which the environment
degrades the material depends on the cycle and that there may be a
mechanistic basis for the SRP methodology, at least for some systems,
in terms of environmental damage.   This is rather different from

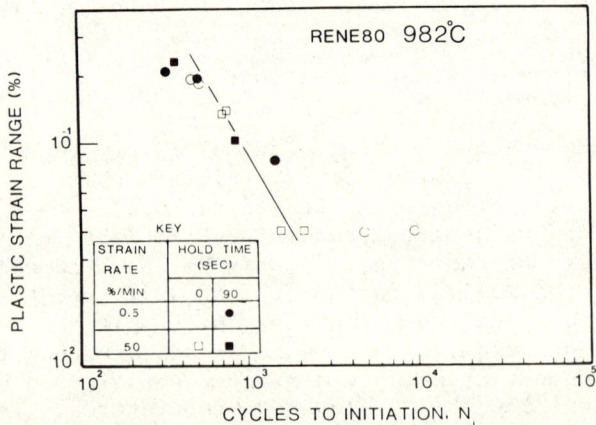

Fig. 17.   Maximum Stress at Initiation as a Function of Oxide Depth
          at Initiation for René 80 Tested at 982^{0}C.   The Equation
          of Each Line Along with the Coefficients of Determination
          are Given (26).

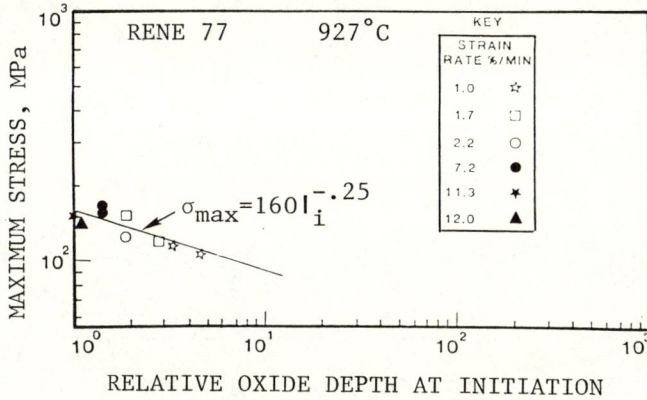

Fig. 18.   Maximum Stress at Initiation as a Function of Oxide Depth
           at Initiation for René 77 Tested at 927^{0}C. The Equation
           of the Line is Given.

the mechanisms usually proposed for SRP, which are based on plasticity[23]. Finally this approach has been applied to the data for
Ti 6242 presented by McEvily[24] as shown in Fig. 21. Again, there
appears to be reasonable agreement especially for the α/β worked
material.

     The results just presented indicate that for many systems, the
LCF behavior depends more on the environment than on accumulation of
deformation debris. It would consequently appear worthwhile for investigators of the "creep/fatigue" phenomenon to consider that the
major damage mechanism might actually be of the form "environment/
fatigue."

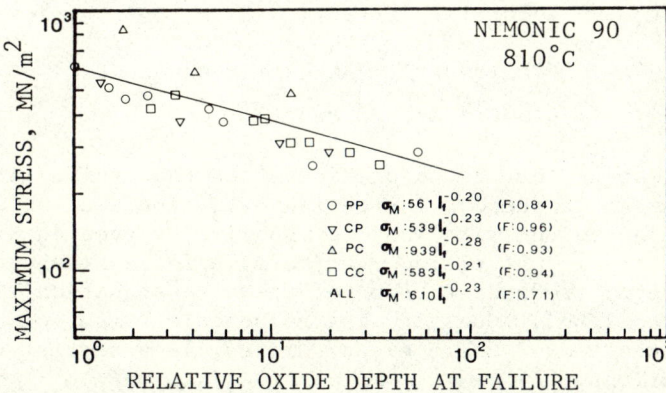

Fig. 19.   Maximum Stress at Half Life as a Function of Oxide Depth
           at Failure for Nimonic 90 Tested at 810^{0}C. The Equations
           of Each Line Along with the Coefficients of Determination
           are Given.

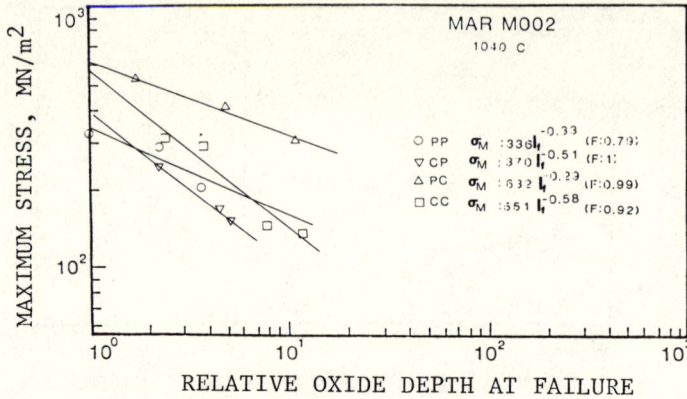

Fig. 20.   Maximum Stress at Half Life as a Function of Oxide Depth
           at Failure for Mar M002 Tested at 1040°C.   The Equation
           of Each Line Along with the Coefficients of Determination
           are Given.

    The simple notions expressed by eqns. 2-4 have been used to
develop an expression for the number of cycles to initiation as a
function of the plastic strain range and other variables[25,26].   In
carrying out the calculation it was assumed that the specimen is
strain cycled about zero mean strain and that the material is stable.
With these assumptions the cycles to crack initiation are given by:

$$Ni = C \ (\nu/1+\nu t_h) \ exp-Q/RT \ \Delta\varepsilon_p^{-8n'} \qquad \qquad \cdots \ (5)$$

where ν = continuous cycling frequency

        t_h = hold time (if any)

        n' = cyclic strain hardening exponent

        $\Delta\varepsilon p$ = plastic strain range

It is interesting that eq. 5 predicts that the cyclic life decreases
with decreasing frequency, increasing hold time and increasing temperature in agreement with what is usually observed in high temperature LCF of stable materials.   Furthermore for a given set of test
conditions eq. 5 reduces to the Coffin-Manson equation with the exponent equal to $1/8n'$.   Substituting reasonable values of n', Coffin-
Manson exponents of $0 \cdot 6 - 0 \cdot 8$ are predicted, in agreement with custom-
ary observations.

    D.   Summary

        High temperature LCF involves factors such as structural
changes, environmental degradation and creep.   In many materials,

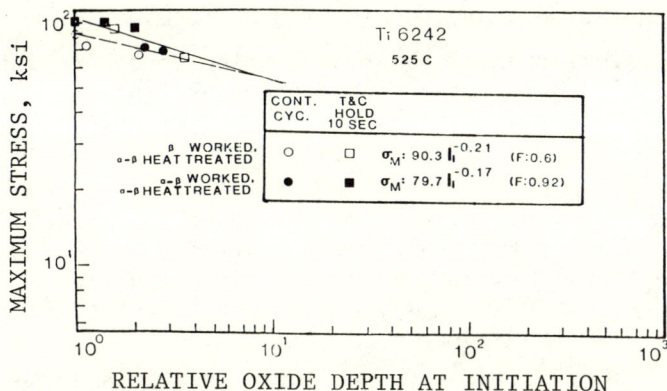

Fig. 21.   Maximum Stress at Initiation as a Function of Oxide Depth
at Initiation for Ti-6242 Tested at 525°C. The Equations
of Each Line Along eiht the Coefficients of Determination
are Given.

especially Ni base alloys at high temperatures, cracking depends on
a trade-off between structural coarsening (beneficial) and environmental degradation. In such cases a phenomenological theory incorporating stress and oxidation appears to describe the process quite
well. The details of the process undoubtedly depend on oxide/metal
properties as well as on the effects that cyclic deformation have on
environmental degradation. Study of dynamic environmental interactions would appear to be important in order to give a better understanding of high temperature LCF in Ni base alloys.

IV.   SUMMARY AND CONCLUSIONS

A.   Crack Propagation in Ni Base Alloys

1.   FCP rates can be changed by simple heat treatments in
such a way that order-of-magnitude improvements in crack growth resistance are obtained.

2.   Factors tending to promote more planar, inhomogeneous
slip also tend to lower the FCP rate. Such factors include decreasing the precipitate size to produce particle shearing as well as
increasing the grain size. Another possible factor would be to decrease the stacking fault energy which would also tend to produce
planar slip.

3.   Improvements in the overload FCP behavior paralleled
those for the constant load amplitude cycling; the greater the slip
planarity, the greater the number of delay cycles.

B.   High Temperature LCF in Ni Base Alloys

   1.   Depending on the system being considered, it is possible for the life to _increase_ with decreasing frequency or with imposition of a hold time.   Such behavior was attributed to structural coarsening in René 80 which was beneficial in as much as it increased the ductility.

   2.   The most severe form of damage was associated with environmental interactions.   Specimens that were exposed under stress prior to testing showed a greatly reduced fatigue life.   The life was restored and even increased when a small surface layer was machined prior to testing.

   3.   The fatigue life was determined by a trade-off between structural coarsening and environmental damage.

   4.   There was a correlation between the maximum stress in a cycle at initiation and the depth of environmental attack.   This correlation can be used to develop a relationship between cycles to initiation and parameters such as cycle frequency, hold time, temperature and cyclic stress/strain properties.

ACKNOWLEDGEMENT

   The authors would like to thank the Air Force Office of Scientific Research (Grant AFOSR 80-0065) and the NASA Lewis Research Center (Grant NSG 3263) for their financial support of much of the work cited in this report.   They especially acknowledge many helpful discussions with Dr. A. Rosenstein of the AFOSR and Dr. M. Hirschberg of NASA who were the Program Managers for these grants.

REFERENCES

1.   T. Nicholas: "Life Predictions for Turbine Engine Components" See this volume.
2.   R. H. Vanstone and D. M. Carlson: "Long Life Engine Disks from Gas Atomized Powder" DARPA Contract #F33615-78-C-5100.
3.   R. Stolz and A. Pineau: "Dislocation Precipitate Interaction and Cyclic Stress Strain Behavior of a γ' Strengthened Superalloy". Mat. Sci. and Engr. 34, 1978, pp. 275-284.
4.   ASTM E399-78a: "Standard Test Method for Fracture Toughness of Metallic Materials".   1979 Annual Book of ASTM Standards ASTM, Philadelphia, Pennsylvania, 1979, pp. 540-561.
5.   J. Bartos and Stephen D. Antolovich: "Effect of Grain Size and γ' Size on FCP in René, 95". Fracture 1977, Vol. 2, pp. 996-1006.
6.   H. F. Merrick and S. Floreen: "The Effect of Microstructure on Elevated Temperature Crack Growth in Ni Base Alloys": Met. Trans., 9A, 1978, pp. 231-233.

7. W. J. Mills and L. A. James: "Effect of Heat Treatment on Elevated Temperature Fatigue Crack Growth Behavior of Two Heats of Alloy 718". ASME Publication 7-WA/PUP-3, 1979.

8. J. Lindigkeit, G. Terlinde, A. Gysler and G. Lütjering: "The Effect of Grain Size on the FCP Behavior of Age Hardened Alloys in Inert and Corrosive Environment" Acta. Met., 27, 1979, pp. 1717-1726.

9. A. Pineau: "Influence of Micromechanics of Cyclic Deformation at Elevated Temperature on Fatigue Behavior". Creep-Fatigue-Environment Interactions, Ed. R. M. Pelloux and N. S. Stoloff, AIME, New York, 1980, pp. 24-45.

10. D. Eylon, J. A. Hall, C. M. Pierce, and D. L. Ruckle: "Microstructure and Mechanical Properties Relationships in the Ti-11 Alloy at Room and Elevated Temperatures" Met. Trans., 7A, 1976, pp. 1817-1826.

11. H. Kitagawa, R. Uyyki and T. Ohira: "Crack-Morphological Aspects in Fracture Mechanics", Engrg. Frac. Mech., 7, 1975, pp. 515-529.

12. Stephen D. Antolovich, A. Saxena and G. R. Chanani: "A Model For Fatigue Crack Propagation", Eng. Fr. Mech., 7, 1975, pp. 649-652.

13. A. Saxena and Stephen D. Antolovich: "Low Cycle Fatigue, Fatigue Crack Propagation and Substructures in a Series of Polycrystalline Cu-Al Alloys". Met. Trans., 6A, 1975, pp. 1809-1828.

14. B. Lawless: "Correlation Between Cyclic Load Response and Fatigue Crack Propagation Mechanisms in the Ni-Base Superalloy Waspaloy" M.S. Thesis, University of Cincinnati, August 1980.

15. J. K. Tien and S. M. Copley: "The Effect of Orientation and Sense of Applied Uniaxial Stress on the Morphology of Coherent γ' Precipitates in Stress Annealed Ni Base Superalloy Crystals". Met. Trans., 2, 1971, pp. 543-553.

16. Stephen D. Antolovich, E. Rosa and A. Pineau: "Low Cycle Fatigue of René 77 at Elevated Temperatures." To appear in Mat. Sci. and Engrg. Feb., 1981.

17. R. F. Decker and C. T. Sims: "The Metallurgy of Ni Base Alloys". The Superalloys, Ed. C. T. Sims and W. C. Hagel, John Wiley and Sons, New York, 1972, pp. 33-77.

18. Stephen D. Antolovich, P. Domas and J. L. Strudel: "Low Cycle Fatigue of René 80 as Affected by Prior Exposure", Met. Trans., 10A, 1979, pp. 1859-1868.

19. Stephen D. Antolovich: "Metallurgical Aspects of High Temperature Fatigue", La Fatigue des Materiaux et des Structures Ed. C. Bathias and J. P. Bailon, Maloine S. A., Paris 1980, pp. 465-496.

20. P. K. Wright and A. F. Anderson: "The Influence of Orientation on the Fatigue of Directionally Solidified Superalloys", Superalloys 1980, Ed. J. K. Tien, S. T. Whodek, H. Morrow III, M. Gell and G. E. Maurer, ASM, Metals Park, Ohio, 1980, pp. 689-698.

21. G. F. Harrison and M. J. Weaver: "The LCF Behavior of Nimonic 90A at Elevated Temperature" AGARD-CP-243, 1978, pp. 6.1-6.19.

22. V.T.A. Antunes and P. Hancock: "SRP of Mar M002 Over the Temperature Range 750°-1040°C. Ibid. pp. 5.1-5.9.

23. S. S. Manson: "The Challenge to Unify Treatment of High Temperature Fatigue - A Partisan Proposal Based on SRP". ASTM STP 520 pp. 744-782.

24. A. J. McEvily: "Creep Fatigue Effects in Ti Alloys". See this volume.

25. Stephen D. Antolovich, R. Baur and S. Liu: "A Mechanistically Based Model for High Temperature LCF of Ni Base Alloys", Superalloys 1980, pp. 605-613.

26. Stephen D. Antolovich, S. Liu and R. Baur: "Low Cycle Fatigue Behavior of René 80 at Elevated Temperatures". Met. Trans., 12A, 1981, pp. 473-481.

CREEP CRACK GROWTH

S. Floreen

Inco Research & Development Center, Inc.
A Unit of Inco Corporate R & D
Sterling Forest, Suffern, NY 10901

INTRODUCTION

Creep has been a subject of considerable practical and theoretical interest. In recent years, however, there has been a shift
in the way creep has been viewed. In earlier times creep was
looked at as a plastic deformation process with the emphasis on
studying plastic flow as a function of time, temperature, stress,
etc. Engineering properties of concern were typically minimum
second stage creep rates or times to reach some amount of deformation. Theoretical studies often involved constructing dislocation
models to account for the deformation behavior.

This type of work is continuing, but particularly within the
last decade creep has been viewed as a fracture problem. This
change in perspective arose in part because of the shortcomings of
considering creep solely as a plastic flow phenomenon. On the
practical side, elevated temperature failures occurred that were
not predicted by conventional creep data. Basic studies showed
extensive microstructural damage could develop during creep. In
addition, investigators studying other elevated temperature problems such as fatigue or corrosion often found that creep crack
growth phenomena were affecting their results.

One convenient way to view elevated temperature crack growth
behavior is through the fracture mechanism maps developed by Ashby
and co-workers.[1] Figure 1 shows an example of one of these maps
for a nickel-base high temperature alloy. In this map the

coordinates are the normalized stress (tensile stress divided by
Young's modulus) and homologous temperature (temperature divided
by melting point) and the dominant fracture mechanisms are plotted
in terms of these parameters.

NIMONIC 80A

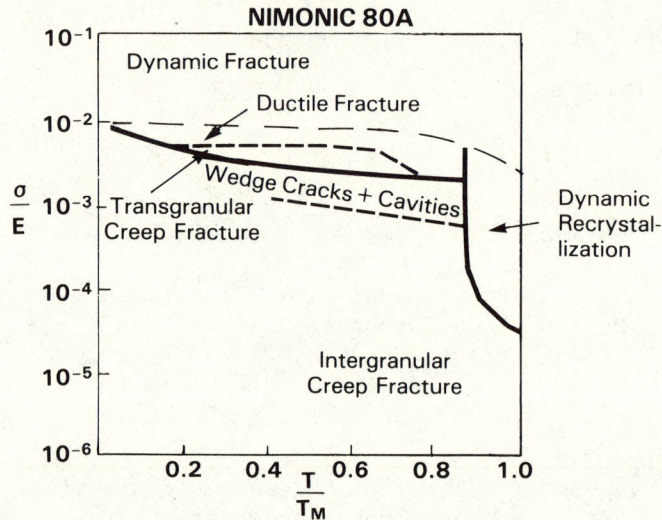

Fig. 1.   Fracture mechanism map for a Ni-base alloy (Ref. 1).

Most creep crack growth studies have been concerned with behavior at intermediate temperatures, at homologous temperatures on
the order of 0.4 to 0.6.   Note in Figure 1 that intergranular crack
growth processes dominate at lower stresses in this temperature
range.   Similar diagrams have been constructed for other materials,
and tests on numerous commercial alloys generally confirms that
intergranular crack growth is a very common, pervasive problem at
these temperatures.

That creep crack growth can be a serious problem is illustrated
by the results in Figure 2.   This figure shows the rupture life
versus initial stress intensity data for a number of nickel-base
superalloys tested at 705°C.   The samples were compact tension
fracture toughness specimens that were fatigue precracked at room
temperature and then dead weight loaded in creep frames.   The

*NIMONIC is a trademark of the Inco family of companies.

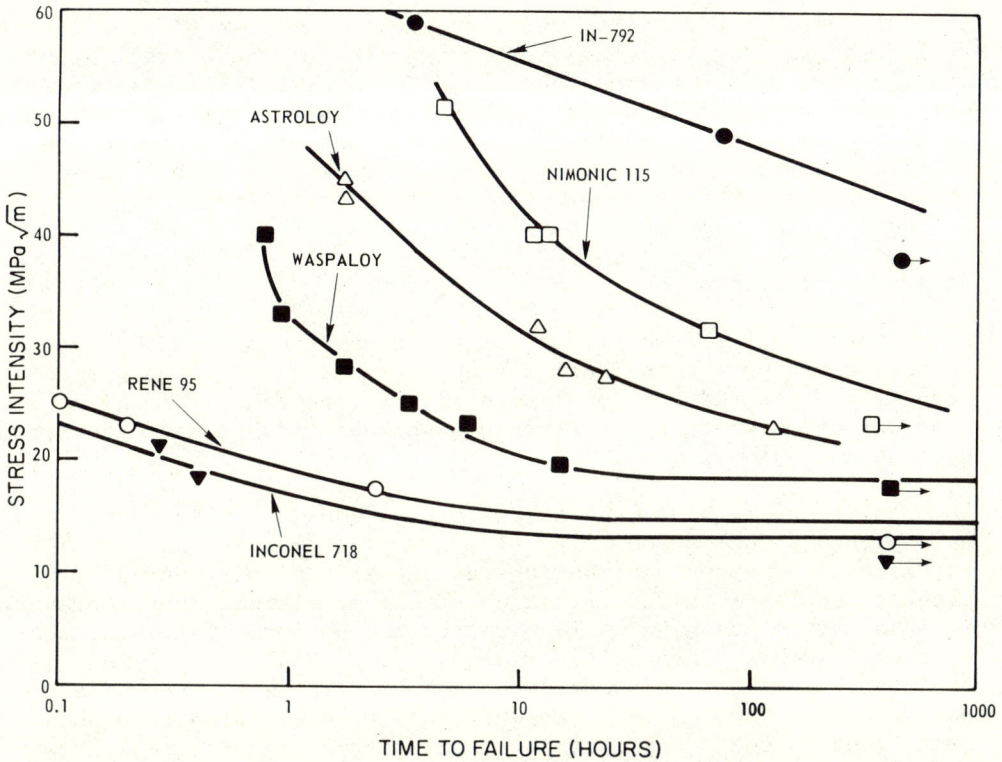

Fig. 2.   Time to failure vs. original stress intensity for Ni-
base superalloys tested at 705°C (Ref. 2).

stress intensity values are based on the lengths of the starting
cracks.   Creep crack growth in these materials was entirely intergranular with no shear lips or changes in specimen thickness in
the slow crack growth region. Measurements of the crack growth
rates in these alloys correlate with time to failure data.   Some
alloys look much better than others, for reasons that are discussed elsewhere.[2]   The point to be made is that creep crack
growth and failure took place at very low stress intensity values
in many of these materials.

        A number of other studies also have shown that creep crack
growth can occur quite readily at intermediate temperatures.   As
might be expected, high strength alloys generally are more prone

*INCONEL is a trademark of the Inco family of companies.   Rene is
a trademark of Teledyne Allvac.

to problems, but even soft, ductile alloys sometimes display little resistance to crack growth. Several reviews of these crack growth studies have been prepared recently,[3-5] and no attempt will be made here to repeat this information in detail. What will be attempted is to very briefly summarize the state of the art in terms of the atomistic models of crack nucleation and growth, the applied mechanics aspects of crack growth, the effects of environment, and the influence of metallurgical parameters on the crack growth behavior.

Atomistic Considerations

Intergranular crack growth often appears to proceed by processes involving the nucleation and growth of cavities on grain boundaries oriented more or less normal to the axis of loading. A number of theoretical treatments have been made to model cavity nucleation and/or growth.

Cavity nucleation generally has been considered in terms of heterogeneous nucleation at particles in the grain boundaries.[6-8] Particle size, shape and spacings are usually significant parameters. Experimental confirmation of the models has been achieved in materials containing well-characterized uniform distribution of particles such as internally oxidized alloys. Also, cavities frequently can be seen associated with inclusions or precipitate particles in other alloys. However, cavity nucleation often has been found to occur continuously with deformation and crack growth, and examinations of the fractured surfaces do not always show obvious nucleation sites. These observations have prompted suggestions that plastic flow may nucleate cavities independently of grain boundary particles.[9-10]

Cavity growth has been looked at in terms of diffusion controlled growth, stress controlled growth, or, more recently, by sequential or combined diffusion plus stress control models. Diffusion control growth is illustrated schematically in Figure 3. H-R growth refers to the classic model developed by Hull and Rimmer[11] and modified by a number of later investigators.[12-17] These models involve the grain boundary diffusion of vacancies to boundaries perpendicular to the stress axis and the growth of cavities on these sites. The models usually predict a threshold stress for cavity growth, a growth rate proportional to the applied stress, and growth kinetics controlled by the rate of grain boundary diffusion. Variations on this theme, such as by Chuang and Rice,[18] are based on surface diffusion instead of grain boundary diffusion.

Also shown in Figure 3 are some notations to indicate that several other diffusion controlled processes may be taking place. These other processes are not generally considered in most of the cavity growth models, but they may affect the crack growth process

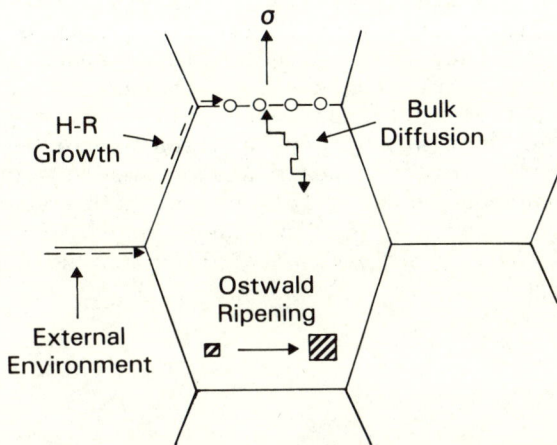

Fig. 3. Diffusion controlled crack growth processes.

and should be mentioned. One such process is Ostwald ripening,
or more generally changes in precipitate morphology during elevated
temperature exposure. Such changes probably can be ignored in
short test times, but during long time service exposures profound
microstructural changes can occur that may drastically alter the
properties.

Two other diffusion processes of concern are bulk diffusion
into the grain boundaries and intergranular diffusion of species
from the external environment. Of particular importance in bulk
diffusion are the possible changes in grain boundary chemistry
that may occur because less soluble elements segregate to the
boundaries. This question will be discussed in more detail later.
The effects of environment on the crack growth behavior will also
be discussed later. For the present we will note that "back of the
envelope" diffusion calculations suggest that the grain boundaries
ahead of a crack tip can be significantly enriched by elements from
the exterior environment during laboratory creep tests or fatigue
tests at slower frequencies. That is, diffusion rates generally
would be faster than crack growth rates.

Stress controlled crack growth models, as exemplified in Figure 4, generally envisage combinations of transgranular slip plus grain boundary shear.[19-26] The latter is of importance to the wedge-type crack growth at grain boundary intersections. Cavity enlargement under stress controlled growth also will take place. It is important to note that sliding and slip are not independent but coupled processes; i.e., the extent of sliding usually is controlled by transgranular slip. Stress controlled models usually predict stress dependencies proportional to the stress exponent from creep experiments on smooth samples. The temperature dependence usually follows the temperature dependence for second stage creep. Also shown in Figure 4, for the sake of completeness, is the notation that applied stresses may change the precipitate morphology.

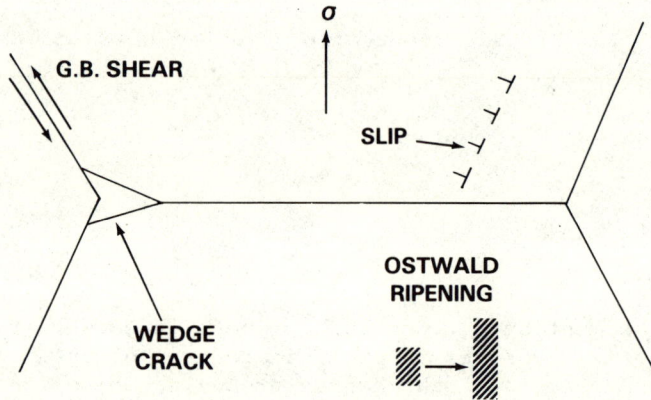

Fig. 4.   Stress controlled crack growth processes.

Comparisons of diffusion control and stress control cavity growth models generally suggest that diffusion control will predominate at very small cavity sizes, while stress control should become dominant at large cavity sizes. The transition cavity size depends upon the specific models but typically is on the order of several microns. Cavities of these dimensions often are observed on fracture surfaces.

This overlap in possible cavity growth controlling mechanisms, and also some of the difficulties associated with the individual models, has prompted consideration in more recent years
to sequential models or coupled models.[21,27-32]   Figure 5 shows
an example of a coupled model in which a zone of diffusion controlled growth lies within a cage of matrix material that is
stress controlled.

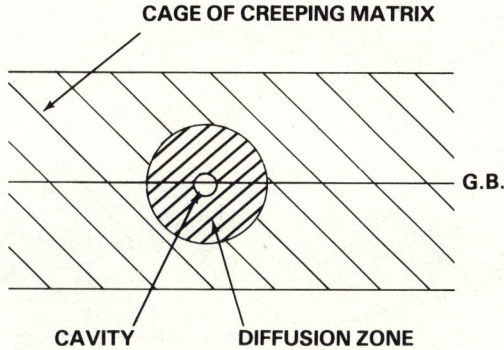

Fig. 5.   Coupled diffusion and stress controlled cavity growth
           (Ref. 21).

     Cavity growth models have worked impressively in specific in-
stances; i.e., the behavior of an individual alloy may agree quite
well with the predictions of a model. At this stage however, none
of the models are generally adequate for all of the results that
have been obtained. Working backwards, in principle one should be
able, based on the stress dependence and temperature dependence
of the crack growth rates and perhaps the cavity morphology, to
broadly distinguish whether the cracking behavior in a material
was diffusion controlled or stress controlled. In some cases
this is true, but in others it is not so easy or unambiguous.
Considerably more needs to be done before atomistic models can be
used with confidence to generally predict material behavior.

Applied Mechanics Considerations

     Crack growth processes imply localized stresses and strains
at the tip of the advancing crack. At elevated temperatures these

parameters will be time dependent. Figure 6 shows in a very
schematic way the principle stresses and strains ahead of a crack
tip at two different times. At time t_1, the stress may be localized at the crack tip, but with stress relaxation the stress will
decay to the curve shown for a later time, t_2. Similarly, the
strain will increase in the crack tip region from t_1 to t_2. Different stress analysis techniques give somewhat different predictions as to the specific shapes of the curves. The point for the
present discussion however, is not the exact shapes of the curves,
but the general implications of time dependent changes in stress
and strain on the crack growth processes.

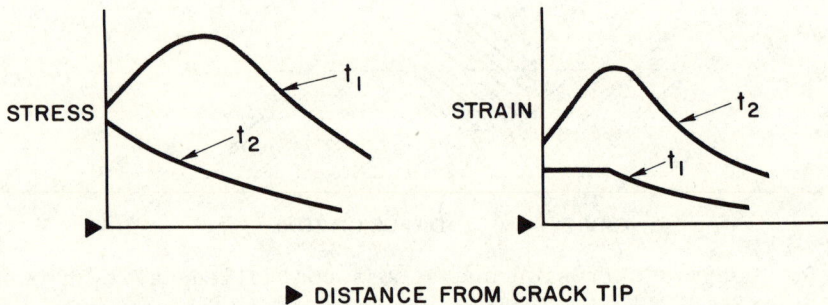

Fig. 6.   Schematic view of stress and strain changes ahead of
          crack tip.

        Returning for a moment to the atomistic considerations, near
a crack tip cavity growth does not take place under a uniform
tensile stress but instead under biaxial or triaxial stress gradients that vary with time. Inclusion of these factors into the
atomistic models usually provides a more realistic description of
cavity growth.[33-39]

        It is also possible to model the crack growth behavior with-
out explicit consideration of the atomistic processes. Thus, for
example, one might predict that crack growth will occur if the
maximum stress at the crack tip stays above some value for a critical time, or that crack growth will occur when the local strain
ahead of the crack tip reaches a critical value. Here again data

for certain alloys have been found to agree nicely with the predictions of such models. Once again, however, these models do
not appear adequate to cover the broad range of results that have
been observed.

One practical consequence of the crack tip stress and strain
changes concerns the question of what parameters should be used to
describe the loading conditions causing crack growth. If crack
growth occurs before little stress relaxation takes place, then
linear elastic fracture mechanics may provide the most appropriate
method to characterize the stress state. At the other extreme, if
the stress is completely relaxed then the net section stress or
the equivalent stress would be appropriate. For intermediate situations post yield fracture mechanics techniques such as the J integral would be implied.

At present there are no guidelines that define how a specific
material should be treated. Some experimental work has been done
however, that provides at least a rough estimate as to how various
classes of materials perform. These studies have been done by
measuring the crack growth behavior of an alloy with different
types of fracture toughness specimens. The results are then compared to see what type of stress parameters most closely correlate
the data from the different specimens.

Figure 7 shows an example of this kind of comparison for tests
on a Cr-Mo-V steel at 565°C. Note in this instance that when plotted versus equivalent stress, the data fall within a reasonably
narrow band, whereas when the same data are plotted versus stress
intensity (K) a much broader scatter band results. In this instance therefore, equivalent stress appears to be the more appropriate yardstick.

The converse situation is shown in Figure 8 for a nickel-base
superalloy. In this case the data are plotted in terms of the
time to failure versus the initial loading condition at the start
of the test. The threshold values for crack growth in the two
specimen geometries agree quite closely when stress intensity is
used, but differs significantly when net section stress is used as
the loading parameter.

To the extent that comparison tests of these kinds have been
made, the results suggest that creep crack growth in high strength
materials, such as nickel-base superalloys, generally is best described by linear elastic stress intensity parameters. Low strength
alloys such as ferritic or austenitic steels seem better described
by J integral, reference stress or net section stress parameters.

Cr – Mo – V STEEL

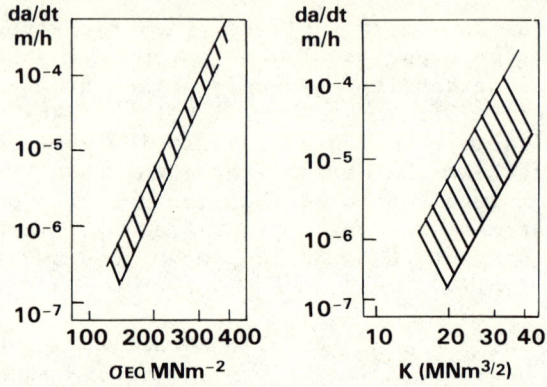

Fig. 7. Crack growth rate data for a Cr–Mo–V steel at 565°C vs.
equivalent stress and vs. stress intensity (Ref. 40).

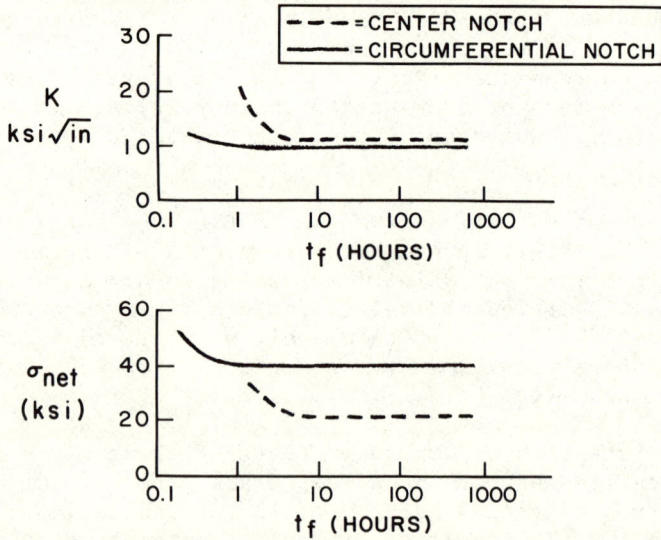

Fig. 8. Time to failure data for INCONEL 718 at 705°C vs. stress
intensity and vs. net section stress (Ref. 41).

Specimen dimensions may also be important variables. The
interplay between material factors such as strength or the tendency for slip to be localized, and engineering factors such as
specimen size, crack length, mode of stressing (e.g., tension vs.
bending) or environment are shown schematically in Figure 9. As
indicated in the figure, linear elastic fracture mechanics techniques become valid at high strength levels and large section
sizes. At ambient temperatures this linear elastic borderline
is defined by the ASTM fracture mechanics criteria. At elevated
temperatures, however, just where this borderline lies is unknown.
Similarly at ambient temperatures one could probably draw a reasonably well-defined border between the post yield fracture mechanics
(FM) area and the reference stress area, but this boundary is
presently undefined at elevated temperature.

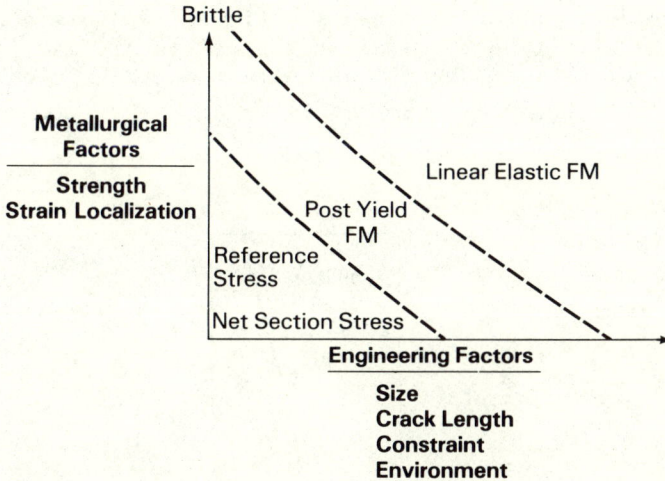

Fig. 9.   Effects of metallurgical and engineering variables in
          fracture mechanics (Ref. 42).

One concern that therefore arises is whether laboratory sized
creep crack growth specimens will predict the behavior of large
components. For the high strength alloys that lie within the
linear elastic field in laboratory sized dimensions, this transla-
tion may not be a problem since the constraints on flow in large
sections presumably would be no worse. With softer alloys however,

the situation is not so clear. In principle it might be possible
for a material that exhibits good crack growth resistance in small
sized laboratory specimens to behave in a brittle fashion when
used in large section sizes. Whether such transitions in behavior
occur is not known, and would be difficult to measure. Considerations of this type may be needed however, if we are to ensure
against failure in large sections.

Environmental Effects

    It has long been known that the environment can significantly
influence the creep behavior. Much of this earlier work has been
reviewed recently[43] and again no attempt will be made here to summarize this work. Several observations can be made. One is that
prior exposures to an environment may drastically alter the subsequent creep behavior. The second is that the creep behavior of
unnotched specimens can be significantly altered by the environment.

    Measurements of creep crack growth rates in different environ-
ments show widely differing effects. Figure 10 shows a recent study
of crack growth in type 304 stainless steel in air and in vacuum
at 593°C. There was no significant effect of the air environment
on the crack growth rate in these tests.

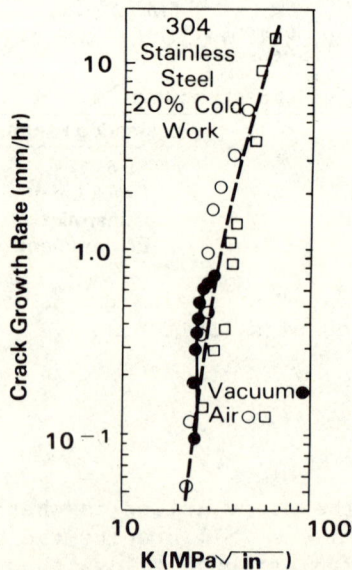

Fig. 10.   Crack growth rate of 304 stainless steel at 593°C in
           air and in vacuum (Ref. 44).

In contrast to these results, Figure 11 compares the crack growth rates of a superalloy in air and helium. In this instance the air environment produced an approximately 100-fold increase in crack growth rate. Interestingly enough allowing for some oxidation in air, the fracture surfaces appeared identical in both environments. Evidently the air environment did not introduce a new crack growth mechanism but accelerated the mechanism taking place in helium.

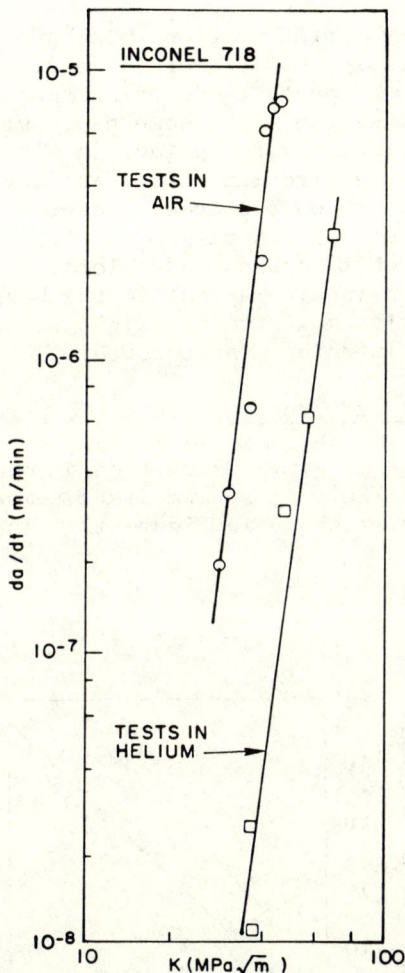

Fig. 11.   Crack growth rates in INCONEL 718 at 650°C in air and in high purity helium (Ref. 45).

To a first approximation, the distinctions in environmental effects exemplified by Figures 10 and 11 seem to depend upon strength level. That is, crack growth rates in low strength alloys

are less sensitive to environments than are high strength alloys.
That higher strength would make an alloy more susceptible to environmental degradation seems plausible, but it should be noted
that not all high strength alloys are equally susceptible.   In
addition heat treatment changes can alter the environmental sus-
ceptibility.   While the strength level is important, it is not
the sole reason and perhaps not always the major factor controlling the environmental effects.

A number of very tentative proposals have been put forward
to explain the aggressive effects of the environment.   At present
no one model appears adequate to describe the results, or to predict the possible consequences of some new environment.   From a
practical point of view, several points should be kept in mind.
The first is that the environment can have very large and often
very deleterious effects on the creep behavior.   Conventional
hot gas corrosion tests on unstressed specimens may be of no
value in predicting what happens under load.   Finally, alloys
optimized for service in air may not be the best materials for
service in other environments where oxygen is no longer the
dominant species influencing crack growth.

Composition and Microstructural Effects

As mentioned previously, segregation of minor elements to the
grain boundaries is likely at intermediate temperatures.   This
segregation tendency is shown in Figure 12.   As the solubility is

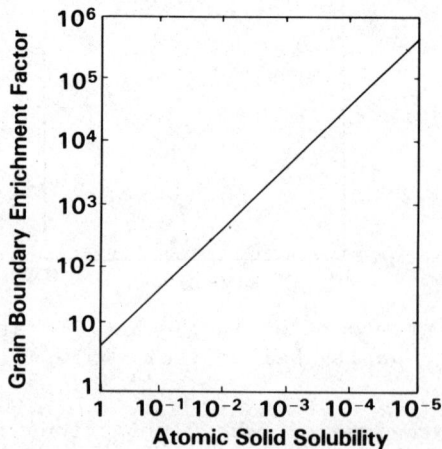

Fig. 12.   Grain boundary enrichment versus solubility (Ref. 46).

lowered very high concentrations of solute atoms may accumulate in the grain boundaries. At temperatures on the order of half the melting point many odd sized atoms have low solubilities, while the diffusion rates are high enough so that they can migrate to the boundaries in short times. Thus at these temperatures the grain boundaries often will have a significantly different chemistry than the matrix.

An extensive technology has been developed in the case of nickel-base superalloys to optimize the grain boundary chemistry. Two elements of particular interest are B and Zr. Early studies have shown that very small additions of these elements significantly retard cracking in the grain boundaries and considerably improve the ductility and rupture life. The reasons for the potent effects of these additions is still a matter of debate. Recently, crack growth rate measurements showed that B plus Zr additions raised the threshold stress intensity at which cracking initiated, but had no effect on the crack growth rates at higher stress intensities.[39]

Outside of nickel-base superalloys in air environments, there has been little systematic study and optimization of grain boundary chemistry in high temperature materials. Such work could be of considerable practical and theoretical interest. In particular, one might expect that developing materials to resist aggressive environments may depend very heavily upon control of the grain boundary chemistry.

Microstructural parameters can also influence creep crack growth. Work on Cr-Mo-V steels has shown, for example, that increasing ferrite contents in mixed bainite plus ferrite microstructures markedly lowered the crack growth rates.[47]

Most of the microstructural studies have dealt with nickel-base superalloys. In these materials increasing grain size has been found to retard both creep and fatigue creep propagation.[2] Grain shape is also important. In some alloys it is possible to control the processing history to produce serrated type grain boundaries, and as might be expected these types of structures show much better crack growth resistance than planar boundaries.[48] The grain aspect ratio is another significant parameter. In materials with pancake shaped grains, for example, cracks ran very rapidly parallel to the pancakes. Cracks oriented perpendicular to the pancakes however, would not propagate through the structure. Instead, these cracks turned 90° during the crack growth tests and then propagated down the elongated boundaries.[3]

Several studies have shown that the age hardening heat treatments can be changed to advantage. In particular, overaging has been found to improve the crack growth resistance.[2,49-51] Homogenization of slip usually has been cited as the reason for the improved properties, but corresponding changes in the grain boundary microstructure with overaging cannot be ruled out.

While microstructural changes can improve the resistance to creep cracking, it is important to note that these changes may affect the other mechanical properties in adverse ways. Table I is an attempt to summarize the general effects of microstructure on some of the relevant properties. Note that these are general trends, and exceptions to most of these effects probably can be found. The trade-offs in properties can be illustrated by the effects produced by changing the grain size (G.S.). Increasing grain size retards the creep crack growth rate à and the fatigue crack growth rate da/dN, but lowers the creep strength and enhances fatigue crack initiation. Thus the optimum grain size must represent the best compromise between competing factors such as these. Most composition and microstructural variables in high temperature materials probably have to be optimized with a view to the total service requirements.

Table 1. Effects of Microstructure on Elevated Temperature Properties of Ni-Base Superalloys (Ref. 3).

| | | PROPERTY | | |
| | | HIGH FREQUENCY FATIGUE | | CREEP |
MICROSTRUCTURE	CREEP à	INITIATION	da/dn	STRENGTH
Coarse ppt.	Lowers	Retards	No effect	Lowers
Coarse G.S.	Lowers	Shortens	Lowers	Lowers
Serrated G.B.	Lowers	No effect	No effect	No effect
Elongated G.B.	Lowers ⊥ Raises ‖	?	Lowers ⊥ Raises ‖	Increase ‖

CONCLUSION

Perhaps the only conclusion that can be made at this time is that the science of creep crack growth is in its infancy. Creep crack growth may occur surprisingly easily. Numerous measurements have been made, and we now have a rough idea of how various types of materials behave. There are many ideas about cracking mechanisms, some of which seem correct in detail, but no overall theory has emerged that covers the broad range of behavior observed. In some cases practical things can be done to improve the resistance to crack growth. The environment can play a very active role in crack growth, but we are only beginning to understand environmental effects. If significant progress can be made, the rewards could be very large, both in terms of achieving the best use of current materials and also in developing new materials for more severe high temperature applications.

REFERENCES

1. M. F. Ashby, C. Gandhi and D. M. R. Taplin, Acta Met. 27, 699 (1979).
2. S. Floreen, Met. Trans. 6A, 1741 (1975).
3. S. Floreen, AIME Symposium on Creep-Fatigue-Environment Inter-actions, in press.
4. L. S. Fu, Eng. Fract. Mech. 13, 307 (1980).
5. H. P. Van Leuwen, Eng. Fract. Mech. 9, 951 (1977).
6. R. Raj and M. F. Ashby, Acta Met. 23, 653 (1975).
7. R. Raj, Acta Met. 26, 995 (1978).
8. R. C. Koeller and R. Raj, Acta Met. 26, 1551 (1978).
9. P. W. Davies and R. Dutton, Acta Met. 14, 1138 (1966).
10. T. G. Nieh and W. D. Nix, Scripta Met. 14, 365 (1980).
11. D. Hull and D. E. Rimmer, Phil. Mag. 4, 673 (1959).
12. F. Dobes and J. Cadek, Met. Sci. Jour. 9, 355 (1972).
13. M. V. Speight and J. E. Harris, Met. Sci. Jour. 1, 83 (1967).
14. R. Raj, Acta Met. 26, 341 (1978).
15. M. V. Speight and W. Beere, Met. Sci. Jour. 9, 190 (1975).
16. M. V. Speight and W. Beere, Met. Sci. Jour. 12, 172 (1978).
17. G. M. Pharr and W. D. Nix, Acta Met. 27, 1615 (1979).
18. T. J. Chaung and J. R. Rice, Acta Met. 21, 162S (1973).
19. J. W. Hancock, Met. Sci. Jour. 10, 319 (1976).
20. W. Pavinich and R. Raj, Met. Trans. 8A, 1917 8(1977).
21. G. H. Edwards and M. F. Ashby, Acta Met. 27, 1505 (1979).
22. S. H. Goods and W. D. Nix, Acta Met. 24, 1041 (1976).
23. F. W. Crossman and M. F. Ashby, Acta Met. 23, 425 (1975).
24. J. A. Williams, Phil. Mag. 20, 635 (1969).
25. D. G. Morris and D. R. Harris, J. Mat. Sci. 12, 1587 (1977).
26. J. R. Haigh, Mat. Sci. & Eng. 20, 225 (1975).
27. B. F. Dyson, Met. Sci. 10, 349 (1976).
28. W. Beere and M. V. Speight, Met. Sci. 12, 172 (1978).
29. B. F. Dyson, Canad. Met. Quart. 18, 31 (1979).
30. W. Beere, Acta Met. 28, 143 (1980).
31. D. A. Miller and T. G. Langdon, Scripta Met. 14, 143 (1980).
32. D. A. Miller and T. G. Langdon, Scripta Met. 14, 179 (1980).
33. D. J. DiMelfi and W. D. Nix, Int. J. Fract. 13, 341 (1977).
34. K. Sadananda, Met. Trans. 9A, 635 (1977).
35. V. Vitek, Acta Met. 26, 1345 (1978).
36. W. Beere and M. V. Speight, Met. Sci. 12, 593 (1978).
37. R. N. Stevens, R. Dutton and M. L. Puls, Acta Met. 22, 629 (1974).

38. P. T. Heald, J. A. Williams and R. P. Harrison, Scripta Met.
     5, 543 (1971).
39. W. D. Nix, D. K. Matlock and R. J. DiMelfi, Acta Met. 25, 495
     (1977).
40. C. J. Neate, Eng. Fract. Mech. 9, 297 (1977).
41. S. Floreen, unpublished work.
42. D. B. Gooch, J. R. Haigh and B. C. King, Met. Sci. 11, 545
     (1977).
43. J. K. Tien and J. M. Davidson, "Advances in Corrosion Science
     and Technology," Plenum Press, NY (1980).
44. K. Sadananda and P. Shihinian, Met. Trans. 11A, 267 (1980).
45. S. Floreen and R. H. Kane, Fatigue of Eng. Mat. 2, 401 (1980).
46. E. D. Hondros, J. Phys. 36, Coll C4-117.
47. C. L. Jones and R. Pilkington, Met. Trans. 9A, 865 (1978).
48. J. M. Larson and S. Floreen, Met. Trans. 8A, 51 (1977).
49. K. Sadananda and P. Shihinian, Met. Trans. 8A, 439 (1977).
50. K. Sadananda and P. Shihinian, Met. Trans. 9A, 79 (1978).
51. R. B. Scarlin, Mat. Sci. Eng. 30, 55 (1977).

TEMPERATURE DEPENDENT DEFORMATION MECHANISMS OF

ALLOY 718 IN LOW CYCLE FATIGUE

T.H. Sanders, Jr., R.E. Frishmuth[*] and G.T. Embley[*]

Fatigue and Fracture Research Laboratory
Georgia Institute of Technology
Atlanta, Georgia   30332
*General Electric Company
Gas Turbine Division
Schenectady, New York   12345

INTRODUCTION

During the mid and late 1970's, the increased emphasis on efficient use of oil and gasified coal for power generation resulted in the emergence of new design technology. One aspect of this technology is the water cooling of rotating parts in power generating gas turbines. By means of water cooling, it is possible to efficiently burn low BTU content fuels at the required high temperatures (above 1370°C) while maintaining low metal temperatures in the parts.[1]

Typical design configurations for water cooled turbine buckets can result in metal temperatures of about 204°C (400°F) near the flowing water in the part to nearly 649°C (1200°F) at the metal surface adjacent to the hot gas path. It is desirable to know the low cycle fatigue (LCF) behavior of the turbine bucket material over this range to assist in parts life analysis. The goal of this paper

163

is to report on the findings of a series of LCF tests over the temperature range 204°C (400°F) to 649°C (1200°F) for Alloy 718. It will be shown that at high strain ranges, the lowest fatigue life occurred at the highest temperature as one would intuitively expect. However, at low strain ranges and long life, an inversion was noted and the longest cyclic lives occurred at higher temperatures. A detailed transmission electron microscope (TEM) study on strategic specimens revealed a systematic change in deformation mode accomanying the inversion in LCF behavior. This paper discusses the observed changes in deformation behavior and the relationship of deformation mode with strain, temperature and frequency. Other work in this alloy is reviewed and discussed in light of the change in deformation mode.

MATERIALS AND TEST PROCEDURE

The material used in this test program was wrought alloy 718 in the form of 1.905 cm (.75 inch) diameter barstock, which received the heat treatment shown in Table 1. The chemical composition and mechanical properties are provided in Table 2. An ASTM grain size of 10 was measured for this material. The specimen design used was a 0.635 cm (.25 inch) diameter circular specimen with a uniform gage and button-head ends as recommended in ASTM E606-77T.

Table 1.   Heat Treatment of Wrought Alloy 718
used in LCF Testing

- Solution treat at 1775°F for one
  hour and oil cool.

- Age at 1325°F, hold eight hours,
  furnace cool at rate of 100°F/hour
  to 1150°F.

All tests were conducted at Mar-Test, Inc., Cincinnati, Ohio. Testing was performed in air under total axial strain control using a closed loop, servocontrolled, hydraulically activated test machine. Sample heating was accomplished by means of an induction heater.

Throughout each test, axial force and plastic strain signals were continuously monitored. The plastic strain signal was obtained from the force and total strain through use of an analog strain computer. Initiation in this program was defined to be the cycles to obtain the first noticeable increase in plastic strain after the initial strain softening had occurred. Detailed descriptions of the test procedures used are contained in References 2 and 3. All testing was conducted at a frequency of 20 cycles per minute.

Table 2. Composition of Alloy 718

C	Mn	P	S	Si	Cr	Ni	Cu	Mo	Ti	Fe	Al	Co	B	(Cb + Ta)
0.05	0.01	0.005	0.004	0.12	18.28	51.93	0.03	2.88	1.03	20.06	0.47	0.12	0.005	5.01

Mechanical Properties:

Temp. (°F)	Yield Strength (0.2% Offset) (ksi)	Ultimate Tensile Strength (ksi)	Elongation (%)	Reduction in Area (%)
RT	168	194	21.3	43.6
1200	148.7	171.5	20.0	36.8

TEST RESULTS

    LCF test results obtained for temperatures up to 649°C (1200°F)
are listed in Table 3.   Examination of the recorded plastic strain
and load signals indicated that the material cyclically softened
and that all softening had occurred by the time $N_f/2$ was reached.

    The relationship between total axial strain, $\Delta\varepsilon$, cycles to
initiation, N_i, and temperature is plotted in Figure 1.   At cyclic
lives of less than about 10^4 cycles the lifetime decreases with
temperature.   Above 10^4 cycles this trend begins to reverse.   Above
10^5 cycles the strain required to produce failure decreases with
temperature from 538°C (1000°F) to 204°C (400°F).

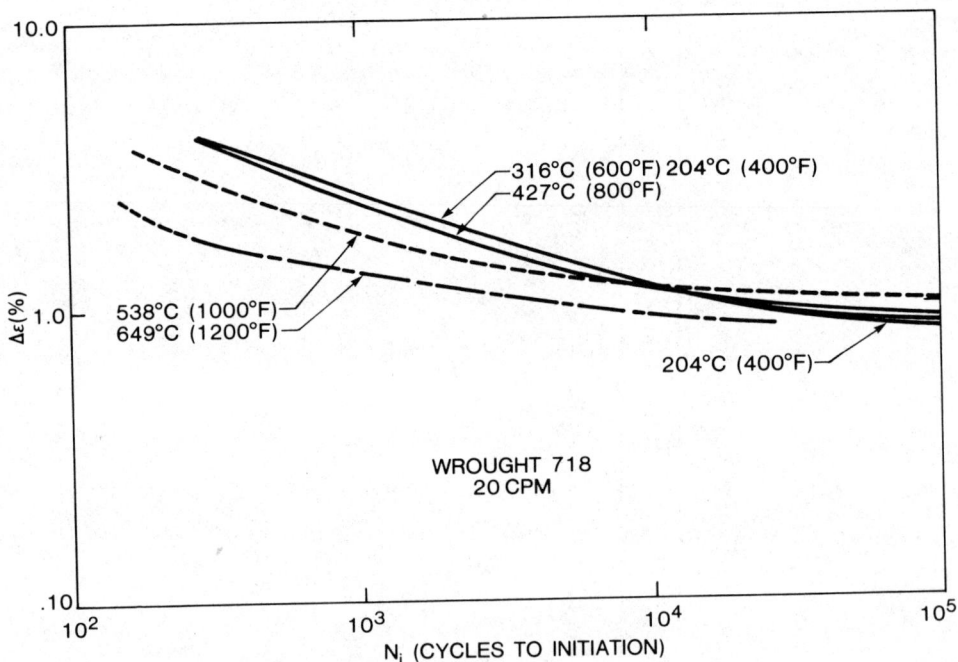

Figure 1.   Total strain range versus number of cycles to initiation
             for 718 at various temperatures.

    Figure 2 is a plot of plastic strain range, $\Delta\varepsilon_p$, versus cycles
to crack initiation, N_i.   This figure indicates that for temperatures
up to 427°C (800°F) dependence of N_i on plastic strain range $\Delta\varepsilon_p$ is
essentially unchanged.   Above 427°C, the material response changes
dramatically.   When N_i exceeds about 10^4 cycles, lifetime at a given
value of $\Delta\varepsilon_p$ is greater for the two high temperatures than for the
lower temperatures.

Table 3.   Wrought IN718 LCF Results - Continuous Cycling Tests

Spec. No.	Temp. [°F]	E [10^6 psi]	$\Delta\varepsilon$ [%]	$\Delta\sigma$ at start [ksi]	$\Delta\sigma$ [ksi]	σ_t [ksi]	σ_c [ksi]	$\Delta\varepsilon_p$ [%] meas.	N_i [cycles]	Test Time [minutes]
						at $N_f/2$				
40	R.T.	29.4	0.90	266.1	257.8	126.7	131.1	0.05	38,938	1,996.9
38	200	29.5	1.30	325.9	275.8	134.4	141.4	0.39	7,920	481.0
31 ✳	400	27.8	2.00	319.1	289.0	140.1	148.9	0.98	2,115	110.8
32	400	27.7	4.00	334.0	320.8	159.6	161.2	2.80	252	13.0
34 ✳	400	27.2	1.30	307.9	264.7	129.5	135.2	0.38	8,128	411.4
33 ✳	400	28.0	0.90	252.0	235.7	116.6	119.1	0.08	34,312	1,752.1
36	400	27.9	0.85	236.4	226.4	110.2	116.2	0.04	85,670	4,311.7
39	400	27.2	0.91	248.6	230.5	113.4	117.1	0.09	39,617	2,035.8
7	600	26.3	2.20	332.0	297.1	142.6	154.5	1.10	1,686	91.5
11	600	25.9	4.00	327.3	311.1	149.5	161.6	2.88	310	16.0
12 ✳	600	25.8	1.40	294.4	273.8[g]	132.7[g]	141.1[g]	0.38[g]	(a)	331.9
30 ✳	600	27.2	0.90	239.9	234.4	119.2	115.2	0.04	69,554	3,548.4
5	800	24.9	2.04	298.4	274.2	132.2	142.0	1.00	1,340	77.5
10	800	25.2	4.00	295.8	293.8	140.8	153.0	2.84	288	15.7
9 ✳	800	24.8	1.30	277.7	241.4	116.7	124.7	0.36	5,342	296.1
29 ✳	800	25.4	0.90	225.4	230.4[h]	116.1[h]	114.3[h]	0.01	-	(c)
4 ✳	1000	24.3	1.02	245.5	218.5[f]	105.4[f]	113.1[f]	0.13[f]	-	2,501.2
4 ✳	1000	-	2.00	286.5	257.5	124.7	132.8	0.96	556	37.8
6	1000	24.4	1.32	272.5	225.6	108.6	117.0	0.42	3,926	223.3
8	1000	24.4	3.20	300.6	260.5	125.0	135.5	2.20	224	13.1
37 ✳	1000	24.4	0.90	218.9	200.0	100.4	99.6	0.09	-	(e)
19	1000	25.1	2.00	298.2	238.3	114.8	123.5	1.08	799	47.9
1	1200	22.8	0.90	204.8	171.4	84.1	87.3	0.18	(a)	776.6
2 ✳	1200	23.0	1.72	243.5	203.6	99.0	104.6	0.86	292	22.6
3 ✳	1200	23.4	1.25	249.5	188.4	91.8	96.6	0.48	1,696	104.8
35 ✳	1200	23.6	0.90	209.6	183.4	93.9	89.9	0.16	23,244	1,180.7

Notation explained in Table

Test Conditions

$\alpha = 8.9 \times 10^{-6}/°F$
$A = \infty$
Freq. = 20 cpm

Remarks:

(a) Not available - inking problem
(b) Failed out of gage in uniform section
(c) Stopped test - runout at 105,734 cycles (5,286.7 minutes)
(d) Failed at radius - uniform section interface
(e) Stopped test - runout at 167,284 cycles (8,3364.2 minutes)
(f) With $\Delta\varepsilon = 1.02\%$; run continued with $\Delta\varepsilon = 2.00\%$
(g) at $N_f/4$
(h) At termination

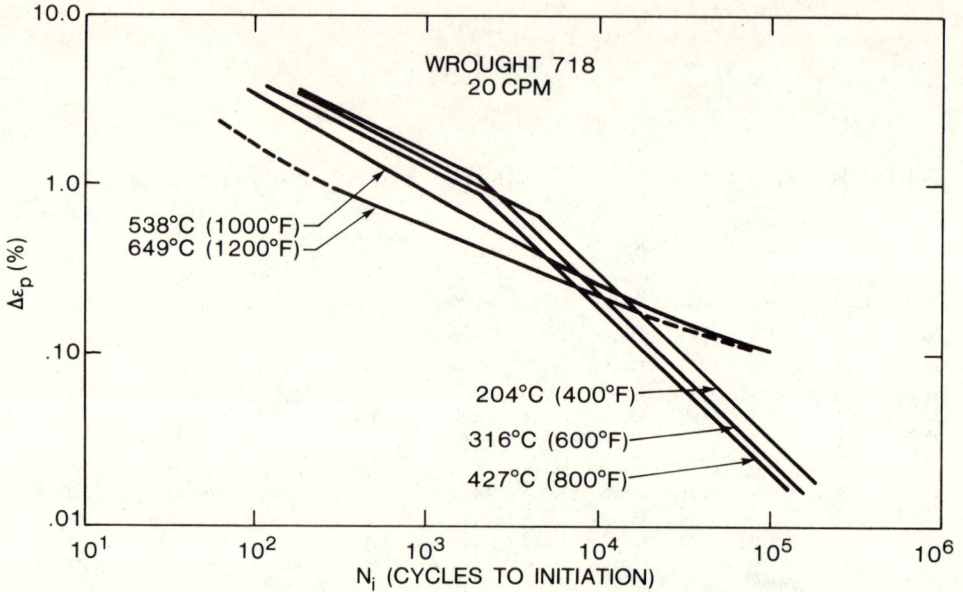

Figure 2.   Plastic strain range versus number of cycles to initiation
            for 718 at various temperatures.

        Figure 2 also shows that for temperatures of 427°C (800°F) and
below, the familiar Coffin-Manson relationship:[4,5]

$$\Delta\varepsilon_p = \varepsilon_f'(N_f)^{-c}$$ [1]

does not describe the relationship between plastic strain and life
throughout the range examined. However, Equation [1] can be used to
describe behavior in two distinct regions of life. That is, one set
of constants for equation [1] at high strain and another set for low
strain.   At temperatures of 538°C and 649°C, Equation [1] describes
behavior throughout the range examined, although some nonlinearity
(Ln $\Delta\varepsilon_p$ vs. Ln N_i) was observed at low values of $\Delta\varepsilon_p$.

        Table 4 provides values of ε_f' and c (Equation 1) for each temp-
erature.   The exponent c was established by using not only data
generated in this study but also data obtained for similarly heat
treated Alloy 718 which were obtained from References 6 and 7.
Figure 3 is a plot of data generated in Reference 6 together with
data from the present study which illustrates the strong bilinear
characteristic of the plastic strain life plot at 427°C.

Table 4. Parameters plastic strain-life coefficients
for wrought alloy 718

Temperature (°C)	Portion of Curve	Parameter $\varepsilon_f'(\%)$	C
204	Upper	44.8	.50
204	Lower	3190	1.0
306	Upper	47.8	.50
316	Lower	2649	1.0
427	Upper	42.0	.50
427	Lower	1923	1.0
538	---	62.5	.60
649	---	20.3	.50

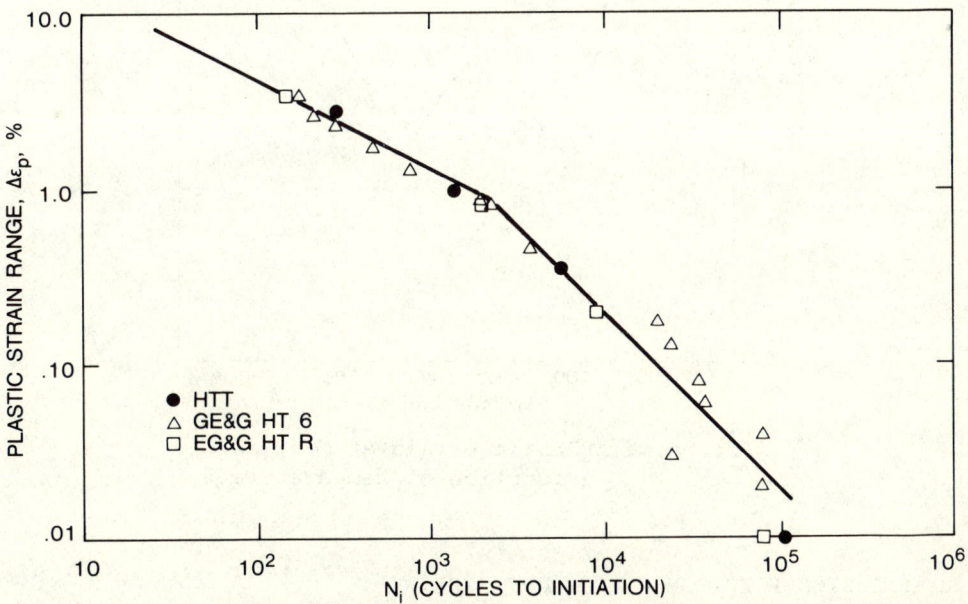

Figure 3. Plastic strain vs. life at 427°C. Wrought alloy 718

Figure 4 is a plot of plastic strain, $\Delta\varepsilon_p$ versus temperatures for various total strain values. This plot shows the dramatic change in plastic strain with temperature at low total strain. This dip in plastic strain appears at 427°C (800°F) and its severity decreases markedly as total strain increases from 0.9% to 1.3% and above. This occurs because yield strength is fairly constant up to 427°C but elastic modulus decrease with temperature. Thus, for a constant total strain range the plastic component decreases with temperature.

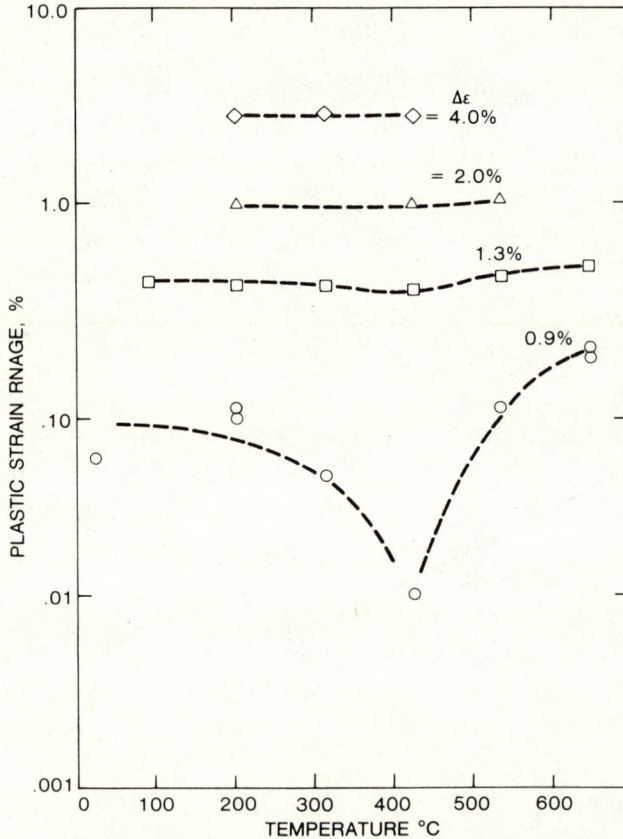

Figure 4.   Plastic strain at $N_f/2$ as a function of temperature.

TRANSMISSION ELECTRON MICROSCOPY STUDIES

TEM foils were prepared from LCF specimens marked with an asterisk in Table 3. Wafers approximately 0.5 mm thick were sectioned 2 mm below the fracture surfaces and discs 3 mm inches diameter were

punched from the wafer. The discs were electropolished by a twin
jet polishing technique in a 2:1 methanol: nitric acid solution
cooled to -36°C. The foils were examined in a JEOL-JEM 100C electron microscope.

As reported by other authors[8] the TEM investigations confirmed
the presence of microtwins which formed during the deformation process. Principally, twinning was confined to the lower temperatures
204°C (400°F), 316°C (600°F), and 427°C (800°F). The density of
microtwins appeared to increase with increasing plastic strain
range as reported by Fournier and Pineau.[8] The characterization of
the planar deformation was based on two relevant observations.
When the volume fraction of the planar defects was high, as in the
case shown in Figure 5, ($\Delta\varepsilon_{tot}$ = 2.0%, T = 204°C) diffuse intensity
along matrix <111> directions was observed in selected area diffraction (SAD) patterns. The thickness of these planar defects, t,
is significantly smaller than the length and width; consequently,
the two dimensional characteristic of these defects caused strong
two dimensional diffraction effects. Since the plane of the defects
was parallel to (111) of the matrix, the streaks would be perpendicular to these planes, and thus lie along <111> directions.

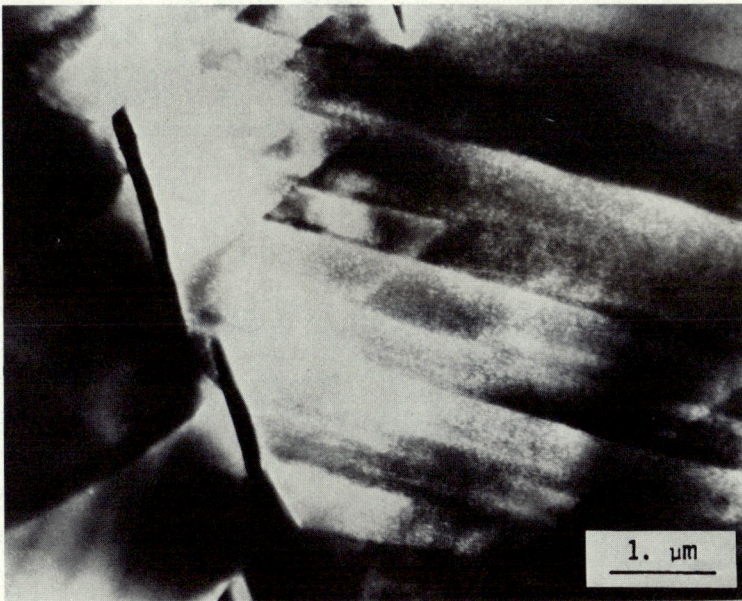

Figure 5. Electron micrograph showing the planar deformation
features present inspecimen 31.

The second observation was the presence of extra reflections in the SAD patterns which would eliminate the possibility of stacking faults. Consequently, the defects were characterized as being fine twins which form during the deformation process.

Decreasing the strain-range appeared to increase the average spacing between the twin bands. For example, this can be seen by comparing Figure 5 with Figure 6, ($\Delta\varepsilon_{tot}$ = 1.3%, T = 204°C).

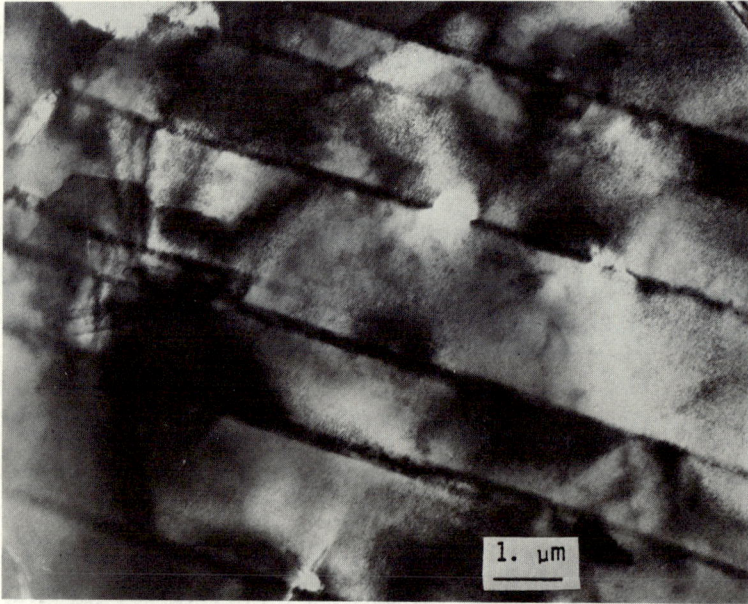

Figure 6.   Deformation structure in specimen 34.

Although micrographs are not shown in this paper, evidence of twinning was also observed in foils prepared from LCF specimens tested at 316 and 427°C.

At the two higher temperatures (538°C and 649°C), the primary mode of deformation was thought to be by slip, Figure 7 ($\Delta\varepsilon_{tot}$ = 2.0%, T = 538°C) and Figure 8 ($\Delta\varepsilon_{tot}$ = 9.0%, T = 649°C), since streaking and extra reflections were not observed in the SAD patterns. Furthermore, when slip bands were observed, often incoherent particles were associated with these bands, suggesting that dislocations were generated at the interface between these particles and the matrix. Occasionally, however, a favorably oriented grain may twin at the higher temperature. If after the twinning event occurs, successive deformation leads to slip, which is the predominate mode of deformation, the twin will become damaged, Figure 8.

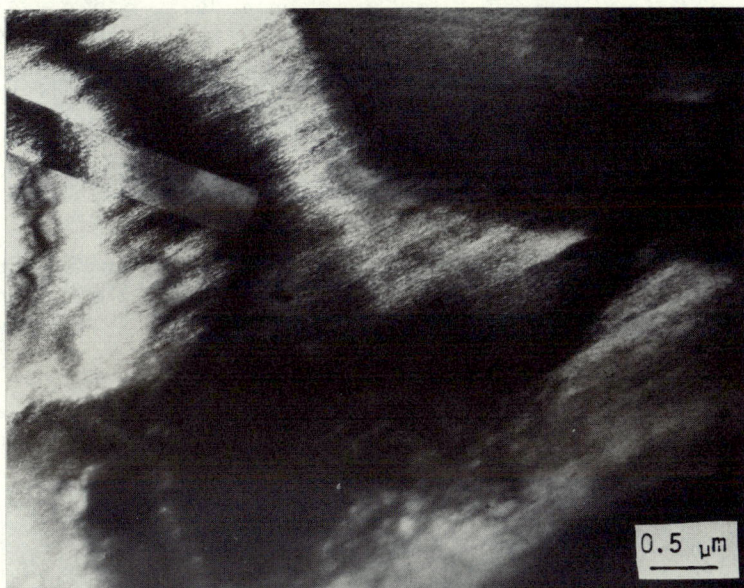

Figure 7. Deformation in specimen 4. Note the difference in the appearance of the deformation as compared to Figure 5.

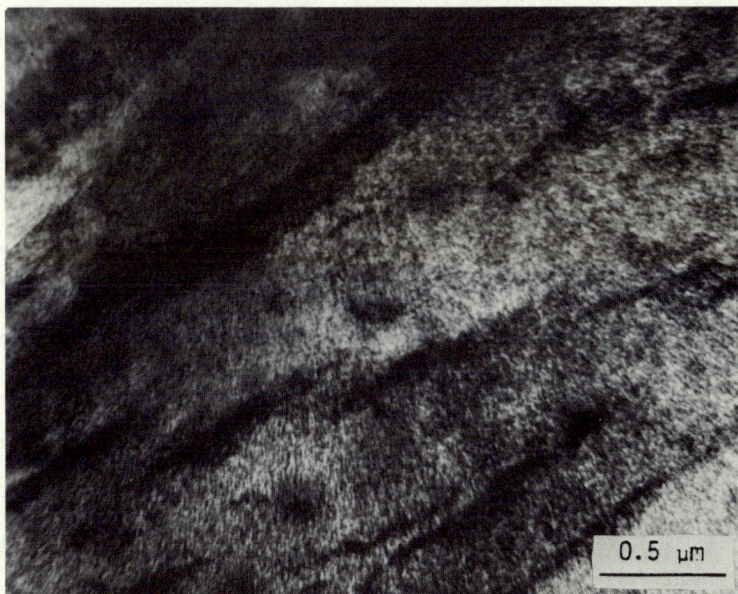

Figure 8. Deformation structure in specimen 35.

The observation of damaged twins has been made in other systems
which undergo deformation by twinning and slip.[9] The damage of the
twin does not occur uniformly since the growth of a deformation
twin generates lattice defects which are not uniformly distributed
in the vicinity of the twin interface.

SCANNING ELECTRON MICROSCOPY STUDIES

A limited SEM study showed that on specimens which were tested
at temperatures which favored twinning, evidence of twin bands could
be seen on the fracture surfaces, Figures 9 and 10. At high plastic
strain ranges, very fine, intersecting twin bands could be seen,
Figure 9, and on a specimen cycled at a similar temperature but at a
comparatively lower strain range, the spacing of the twin bands increased, Figure 10. These SEM observations complemented the TEM observations. Also, associated with the coarse twin bands were small
secondary cracks as shown at position "A" in Figure 10.

DISCUSSION

The results reported here are consistent with other investigations, which have shown that the nature of the deformation process
can be strongly affected by plastic strain amplitude, temperature
and frequency. Changes in deformation process can affect cyclic
lifetime with environment often playing an additional role. Aspects
of these investigations will be considered here in light of the results of this current study.

Investigators have used the analysis of Coffin and Manson[4,5] to
describe the SCF behavior of materials. Out of this research one
fact continually emerges: numerous alloy systems do not strictly
follow the Coffin/Manson relationship described by Equation [1].
There are three possible explanations which can be presented to
account for such a deviation (observed here for temperatures below
427°C). First, the nature of the deformation process may be affected by the plastic strain amplitude. For example, Saxena and
Antolovich[10] studied the effect of stacking fault energy (SFE) on
the fatigue deformation process in the Cu-Al alloy system. These
authors observed that for high SFE alloys, the fatigue life plot
was linear in logarithmic coordinates. However, for the low SFE
alloys the linear relationship between $Ln \Delta\epsilon_p$ and $Ln N_i$ showed a
distinct break in slope. They attributed the break to the inability of the dislocations to cross-slip at low plastic strain amplitudes when the SFE was low.

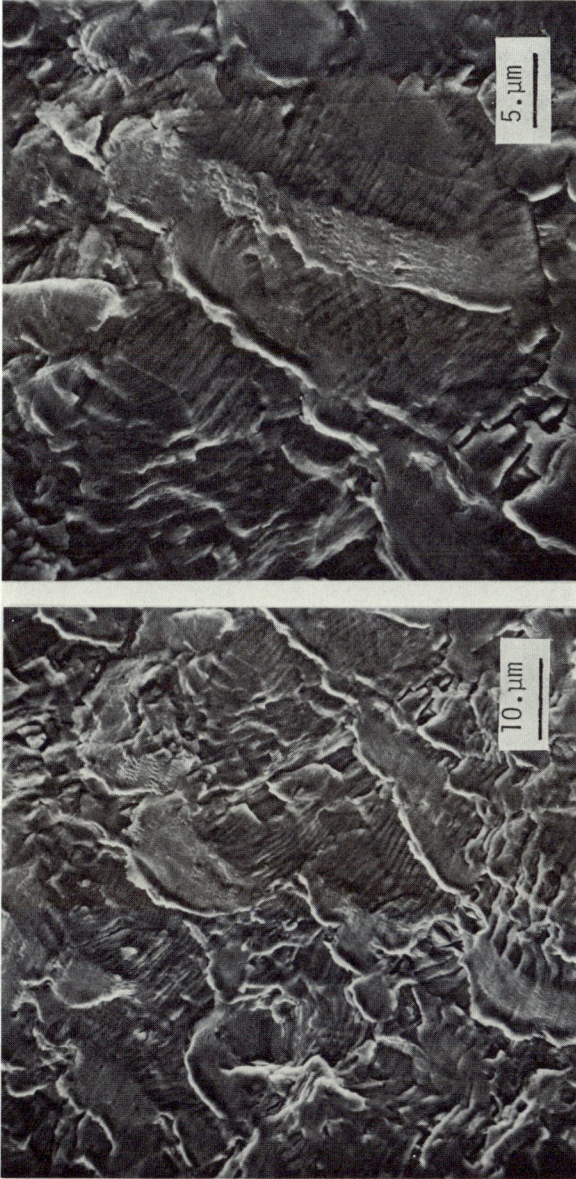

Figure 9. Fracture surface of specimen 31. Evidence of planar features on the fracture surface

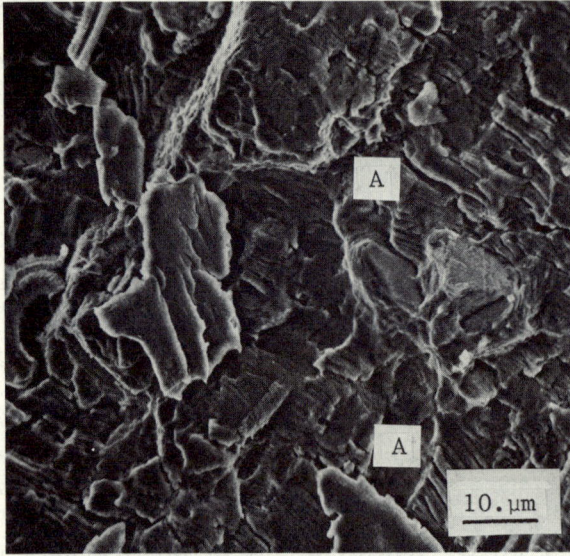

Figure 10.   Fracture surface of specimen 30.   Note the
presence of small cracks at location A.

In accordance with Saxena and Antolovich, changes in the deformation process alter the strain-life response.   Since we have
observed a difference in the deformation structure between the
three lowest temperatures and the two highest temperatures we may
postulate that the cyclic plastic behavior at 204, 316, and 427°C
is dominated by deformation twinning and consequently the strain-
life response for all those specimens are similar.   However, at
538 and 649°C the predominate mode of deformation is slip and thus,
the specimens tested at these temperatures will have a different
strain-life response than those tested at the three lower temperatures.

Alternatively, Coffin has attributed the presence of a break
in the curve to an environmental effect.[11]   Coffin subscribes to
the view that low cycle fatigue is principally a crack propagation
process.   Numerous investigators have shown the presence of microcracks as early as 10% of the life of the specimen; therefore, this
view is realistic.   At a specific frequency and elevated temperature
large plastic strains produce a ductile mode of fracture which is a
higher energy fracture process than intergranular fracture which
occurs at low plastic strains.   Thus at high temperatures, a break
would be anticipated.   One might carry this view one step further.

At low plastic strain amplitudes ($\Delta\varepsilon_p/2$), the life of the specimen is much greater than at a high $\Delta\varepsilon_p/2$, thus the time at elevated temperature (or in a corrosive environment) is longer at a low $\Delta\varepsilon_p/2$. That is, at high strains, short time, the effect of either temperature or environment would be less than if the specimen were cycled at a comparatively low $\Delta\varepsilon_p/2$. Consequently, the failure mechanism changes with total exposure time and $\Delta\varepsilon_p/2$. At high $\Delta\varepsilon_p/2$ the failure is predominately mechanical where as to low $\Delta\varepsilon_p/2$ the process is accelerated by the contribution of either temperature or environment. According to Coffin, we would expect to see the break in the strain-life plot occur with an increase in temperature, Figure 11. However, analysis of the SCF data presented in Figure 2 shows that as temperature is increased to 538°C the break in the curve disappears. This result would tend to contradict that the presence of a break in the curve is purely an environmental effect since this effect would be expected to become dominant at high rather than low temperature. The degradation in fatigue life at 538 and 649°C does indicate however that environmental effects are present at these temperatures.

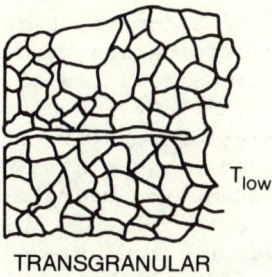

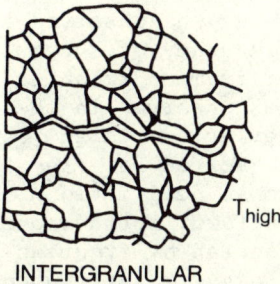

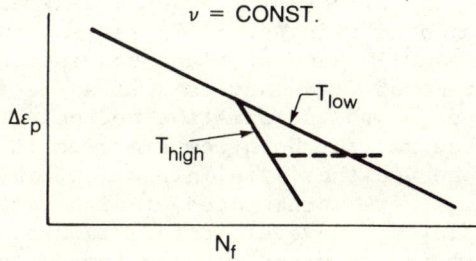

Figure 11.   Schematic representation of fracture mode change with temperature and strain range[8].

A third explanation that can account for a break in the LCF
curve is associated with the distribution of deformation.[12,13] The
presence or absence of localized deformation is significant for two
reasons: (1) these localized regions can act as initiation sites
for fracture and (2) the magnitude of microplastic strain cannot be
accurately measured by a conventional extensometer. Because the
extensometer measures the average rather than the localized strain
in the deformation bands, the Coffin-Manson plots for microstructures which deform homogeneously at high plastic strains and het-
erogeneously at low plastic strain must exhibit nonlinear behavior
when the data are plotted on logarithmic coordinates. The concept
of strain localization may account for the break in the curve when
specimens were tested at the three lowest temperatures. The very
localized nature of deformation twinning may account for this nonlinear behavior. As observed in this research and in the work of
Fournier and Pineau[8] decreasing the strain amplitude increases the
spacing between twin bands. The widely spaced, coarse twin bands
may act as stress concentrators. Also, cracking along the twin-
matrix interface in SEM was observed. Thus, as strain-amplitude
is reduced, deformation becomes localized and the macroview of the
deformation structure gradually departs from the microview. Therefore, if smaller gradations of $\Delta\varepsilon_p$ were used, a gradual rather than
sharp departure from linearity would probably be observed (See
Figure 3).

Thus far, discussion has centered upon the relationship between plastic strain amplitude, temperature and deformation mechanism. The effect of frequency generally has been associated with
environmental effects.[11] Investigations of Alloy 718 by Fournier
and Pineau[8,14] have provided considerable information relative to
the effect of frequency as well as temperature and plastic strain
amplitude on the deformation mechanism. In Reference 8, microtwins
were observed for both temperatures (RT and 550°C) examined. They
also observed that more intense deformation bands (formed by
twinning) were accumulated at high temperatures (550°C) than at low
temperatures. However, deformation by twinning was promoted by a
decrease in frequency at high temperature. The present results
support this conclusion since very little twinning was observed at
a frequency of 20 cpm (compared to a maximum frequency of 3 cpm in
the Reference 8 tests). This range of behavior from no twinning
to intense deformation bands is exactly what was observed in the
Reference 14 study of fatigue crack growth behavior. At high
stress intensity range (ΔK), no twinning was observed for frequencies of 30 cpm and higher, whereas the intense deformation
bands were noted for frequencies of 3 cpm and below. Furthermore,
crack growth rate at high ΔK levels was not affected by frequency
for rates above 30 cpm, whereas they increased significantly when
the frequency was decreased to 3 cpm (with an associated change in
deformation mechanism). These observations are important when

considering the effect of hold time and wave shape at high temperatures.   Data obtained from Reference 6 for 1% plastic strain
and 649° are used in Figure 12 to show that the effect of decreasing frequency using a square wave (with a tension or compression
hold time) is considerably less severe than if a triangular wave
form is used.   The strain rate was not changed when the square
wave form was used.   Thus, it appears that the combination of a
change in deformation substructure and environmental effects may
have a more severe effect on life than environmental effects alone.

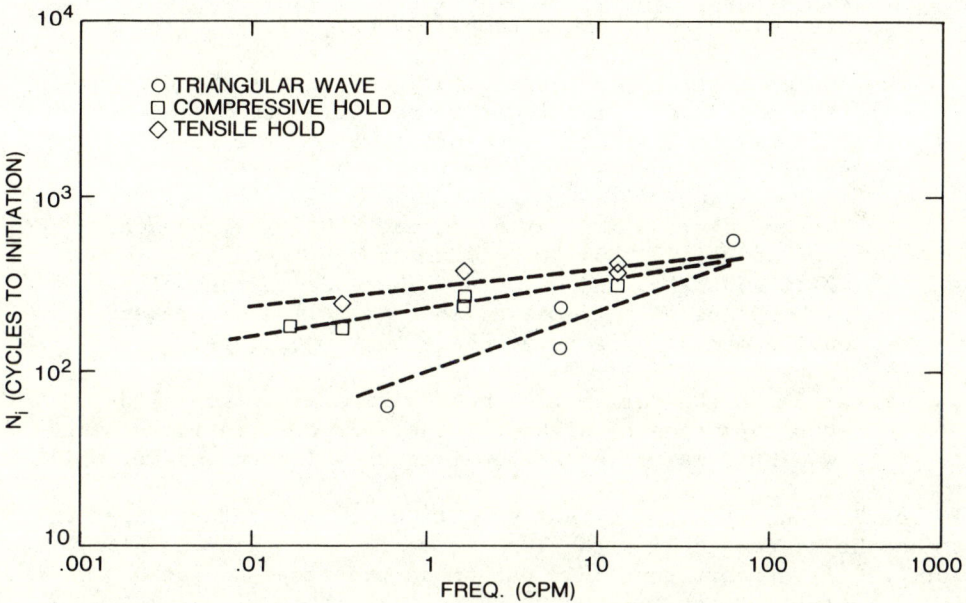

Figure 12.   Effect of wave shape and frequency on life
of wrought Alloy 718 at 649°C (EG&G heat 2)

SUMMARY AND CONCLUSIONS

    To summarize this investigation of the deformation behavior of
Alloy 718 in low cycle fatigue, the following points can be made:

- The relationship between total strain and fatigue life
  for continuous cycling at 20 cpm was characterized as
  a function of temperature.

- At 427°C and below the cross-over of total strain-life
  curves (Figure 1) was associated with a change in plastic
  strain as a function of total strain (see Figure 4).

- The plastic strain-life curves (Figure 2) for 427°C and below were similar in character. The nonlinear (or bi-linear) nature of these curves is thought to be associated with a change in deformation character from homogeneous to hetergeneous. The dominant deformation mode was microtwinning.

- For temperatures of 538°C and 649°C the plastic strain-life curves were also considered to be nonlinear (although supporting data was limited). However, in this case, the charge in slope with decreasing $\Delta\varepsilon_p$, on a logarithmic plot of strain versus cycles, was the opposite of that for lower temperatures.

- As total strain decreased, fatigue life at 538°C changed from poorer than the lower temperature results to better. This was consistent with a crossover of the plastic strain-life curves.

- The deformation mode at 538°C and 649°C was characterized as slip. Thus the differences between behavior at 427°C and below and 649°C and above can be associated with change in deformation mode as well as increased environmental effects at elevated temperature.

- At 649°C the deformation mechanism shifts from slip band formation at high frequency (20 cpm) to increasing twin band formation at low frequency (3 cpm and below).

In conclusion, it appears that some type of temperature dependent phenomena is controlling the LCF deformation behavior in Alloy 718. Further, it appears that this phenomena undergoes a distinct change in character between 427°C and 538°C (See Figure 4). This point could lead to further development of this alloy or more accurate constitutive models for engineering use.

ACKNOWLEDGEMENTS

    This work has been conducted under United States Department of Energy Contract Number EX-76-C-01-1806. The financial support of DOE is gratefully acknowledged.

cess disclosed in this paper or represents that its use by such third party would not infringe privately owned rights.

The authors would like to thank W.J. Ostergren and W.C. Chambers of the General Electric Gas Turbine Division for their encouragement and support. The early work in pointing out the unusual behavior in this alloy and in generating interest in further study was done by J. Conway of Mar-Test, Inc., Cincinnati, Ohio. His contributions and assistance are sincerely appreciated.

REFERENCES

1. Caruvana, A., Manning, G. B., Day, W.H. and Sheldon, R. C., "Evaluation of a Water Cooled Gas Turbine Combined Cycle Plant," ASME Paper #78-GT-77.

2. Conway, J. B., Stentz, R. H., Berling, J. T., Fatigue, Tensile, and Relaxation Behavior of Stainless Steels, United States Atomic Energy Commission, TID 26135, 1975.

3. Slot, T., Stanty, R. H., Berling, J. T., "Controlled Strain Test Procedures" in Manual on Low-Cycle Fatigue Testing, STM STP 465, Page 100, 1969.

4. Coffin, L. F., Jr., "A Study of the Effects of Cyclic Thermal Stresses on a Ductile Metal," Trans. ASME Volume 76, pp. 923-949 (1954).

5. Manson, S. S., Behavior of Metals under Conditions of Thermal Stress, NACA Technical Note 2933 (1954).

6. Korth, G. E., Smolik, G. R., Status Report of Physical and Mechanical Test Data of Alloy 718, United States Department of Energy Report TREE-1254, March, 1978.

7. Brinkman, C. R. and Korth, G. E., "Strain Fatigue and Tensile Behavior of Inconel 718 from Room Temperature to 650°C," ASTM, Journal of Testing and Evaluation, Volume 2, No. 4, July 1974, pp. 249-259.

8. Fournier, D. and Pineau, A., "Low Cycle Fatigue Behavior of Inconel 718 at 298°K and 823°K," Mat. Trans. A, 8A, July 1977, pp. 1095-1105.

9. Chakrabortty, S. B., Mukhopadhyay, T. K. and Starke, E. A., Jr., "The Cyclic Stress - Strain Response of Titanium - Vanadium Alloys," Acta Met Volume 26, pp. 909-920, 1978.

10. Saxena, A. and Antolovich, S. D., "Low Cycle Fatigue, Fatigue Crack Propagation and Substructure in a Series of Polycrystallic Cu-Al Alloys," Met. Trans., Volume 6A, p. 1809, 1975.

11. Coffin, L. F., Jr., "A Note of Low-Cycle Fatigue Laws," Journal of Materials, June 1971, pp. 388-402.

12. Sanders, T. H., Jr. and Starke, E. A., Jr., "The Relationship of Microstructure to Monotinic and Cyclic Straining in Two Al-Zn-Mg Precipitation Hardening Alloys," Met. Trans., 7A, 1976, p. 1407.

13. Sanders, T. H., Jr., Mauney, D. A. and Staley, J. T., "Strain Controlled Fatigue to Interpret Fatigue Initiation of Aluminum Alloys," _Fundamental Aspects of Structural Alloy Design_, Edited by Jaffee, R. I. and Wilcox, B. A., Plenum Publishing (1977).

14. Clavel, M. and Pineau, A., "Frequency and Wave-Form Effects on the Fatigued Crack Growth Behavior of Alloy 718 at 298°K and 823°K," Met. Trans., Volume 9A, April 1978, pp. 471-479.

DEFORMATION INDUCED MICROSTRUCTURAL CHANGES IN

AUSTENITIC STAINLESS STEELS

John Moteff

Materials Science and Metallurgical Engineering Dept.
University of Cincinnati
Cincinnati, OH 45221

INTRODUCTION

Microstructural changes that occur in metals and alloys as a
result of plastic deformation, and at times leading to fracture, are
presented. Since the number of alloys that can be considered are
quite numerous, one system has been selected and discussed in some
detail. The dislocation microstructure of the austenitic stainless
steels, AISI 304 and 316, are presented for specimens tested in the
tensile, creep, fatigue and time-dependent fatigue modes. Microstructural changes characteristic of the plastic zone in the region
of the tip of a fatigue crack is also presented. In addition, the
density changes due to the deformation induced micro-cracking, and
subsequent strain rate dependence, at grain boundaries are discussed.

The basic understanding of the mechanisms of deformation and
fracture, especially at elevated temperatures, is vitally important
in life prediction, accelerated mechanical property testing, and the
development of new high performance engineering load bearing structural components. This is because high temperature plastic flow and
fracture combines a number of highly complex phenomena which can be
rate controlling depending on the microstructure, impurity content
and segregation, environment, and the stress-strain-temperature-time
history. Empirical characterization of deformation and fracture behavior leads, in too many instances, to many adjustable parameters
which renders this approach of little practical value.

Deformation and fracture of engineering materials at elevated temperatures is quite sensitive to microstructure. Basic research, however, should seek to identify the appropriate mechanisms which are common to most materials, recognizing that the regime of temperature and stress in which a particular mechanism dominates, differs from one material to another. It is of some comfort to know that deformation behavior of widely different materials (i.e., ice, lead, aluminum, copper, molybdenum, tungsten, etc.) show remarkable similarities when the data is normalized by the temperature dependent shear modulus (G) for the case of the applied true shear stress (τ) and by the absolute melting temperature (T_m) for the test temperature (T). This τ/G versus T/T_m plot, referred to as a <u>deformation map</u>, has been shown by Ashby (Ashby, 1972) to be a descriptive presentation of the dominant deformation mechanisms as different fields in this two dimensional plot. Much research is needed to study the basic mechanisms and to establish how the microstructure shifts the transitions from one deformation mode to another. The corresponding fracture behavior is also well characterized by means of a <u>fracture-mechanism map</u> (Ashly et al., 1979) using τ/G versus T/T_m and showing the various fields representing cavities, wedge cracking, dynamic fracture, ductile transgranular fracture, transgranular creep fracture, intergranular creep fracture and rupture.

For the metallic alloy systems, one may classify (Wolf, 1979) them according to their distinguishing microstructures. These may be: (1) Solid solution and precipitate strengthened alloys with a small volume fraction of second phase particles. Special consideration should be given to the particles in the grain boundary since the low ductility at elevated temperature results from separation at the grain boundaries that is initiated from interfaces of particles; (2) Solid solution and precipitate strengthened alloys which contain a very large volume fraction of a second phase such as the nickel base superalloys. The matrix second phase is often coherent and is usually different from the grain boundary second phase which consists mostly of metallic carbides. Whereas the creep properties of the material are controlled primarily by the matrix phase, the fracture behavior is strongly influenced by the second phase present in the grain boundary; (3) Dispersion strengthened alloys with or without additional cold work such as TD nickel. These alloys may be different from those in (1) and (2) in the sense that the particles may not be very strongly adhered to the matrix since they are introduced into the material often by mechanical mixing rather than by precipitation; and (4) Directionally solidified multiphase composite alloys, rapidly quenched and hot isostatically pressed (HIP) alloys.

In addition to the consideration of materials and microstructures, the subjects of (1) fracture under quasi-stationary loading to include (a) cavity nucleation, (b) growth and linking of cavities in grain boundaries in quasi-homogeneous stress fields, and (c) time dependent growth of macro-cracks, (2) mixtures of steady loading, cyclic loading and transient loading to include (a) initiation and

early growth of cracks, and crack propagation, (3) phenomena peculiar to irradiation environment, (4) service life predictions and
(5) deformation instabilities and localization must also be further
understood.

In view of the wide distribution of material types with their
respective characteristic deformation behavior, it is the purpose of
this paper to demonstrate the important deformation-microstructure
(dislocations) relationships of only one alloy which is typical of a
solid solution with a small volume fraction of a second phase. This
alloy system will be the austenitic stainless steels (AISI 304 and
316).

DISCUSSION OF EXPERIMENTAL DATA

General

Most of the data that is presented will be on the influence of
stress and temperature on the dislocation microstructure of plastically deformed test specimens. The plastic strain (ε) may be given
as

$$\varepsilon \simeq \alpha b \rho d \tag{1}$$

where ρ is the dislocation density, d is the average distance the
dislocations move under an applied shear stress, b is the Burger's
vector and α a constant. The strain rate ($\dot{\varepsilon}$) is given by

$$\dot{\varepsilon} \simeq \alpha b \rho v \tag{2}$$

where v is the average dislocation velocity. It is clear, then, that
both strain and strain rate are dependent on two variables, the dislocation density and average distance and the dislocation density and
average velocity, respectively. However, in a macroscopic tensile
experiment only the average product of the two variables is determined.

In pure metals and simple solid solution alloys, the dislocation density (ρ) will increase with an applied shear stress (τ) by a
multiplication process according to the following relationship:

$$\rho \simeq [(\tau-\tau_o) / \alpha G b]^2 \tag{3}$$

G is the shear modulus and τ_o a constant.

Cells and Subgrains

For the case of the austenitic stainless steels and many other alloys belonging to this class, the dislocation density may increase up to some critical level and then take on a three dimensional modulated dislocation microstructure. These structures appear as small cells and have a size (λ) that is normally inversely dependent on the square of the applied shear stress,

$$\lambda = 175 \ (\tau/Gb)^{-2} \tag{4}$$

At the higher temperatures and for slower strain rates (i.e., lower stress levels) the edge dislocations have the ability to climb over small obstacles and to form a well-ordered network wall of dislocations. These dislocation configurations may be referred to (Michel, 1973) as subgrains and have the stress dependency as follows:

$$\lambda = 12(\tau/Gb)^{-1} \tag{5}$$

A plot of the cell or subgrain size as a function of the applied tensile or creep stress (σ) is shown in Figure 1. Based on the experimental data (Michel, 1973), the size of the modulated dislocation structure is primarily a function of the applied stress and essentially independent of the test temperature, other than the differentiation between the tendency for cell and subgrain formation.

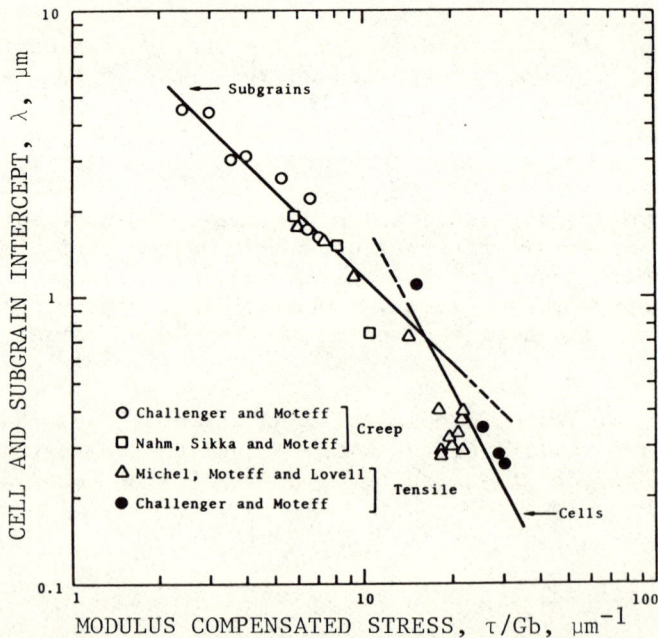

Figure 1.   Relation between average subgrain or dislocation cell intercept and maximum true tensile stress or initial applied creep stress.

The cell/subgrain size data given in Figure 1 are based on end of life tests, i.e. tensile and stress-rupture tests corrected for true stress and with the TEM specimens taken away from the fracture region. In order to study the development of the cell/subgrain dimensional changes with the true stress as a function of the life of the test, a series of tensile tests were performed (Foulds et. al., 1980) at the same temperature and strain rate but to different strain levels. The true uniform strain levels of the annealed AISI 304 stainless steel specimens were 3.7, 8.5, 17.4 and 30 percent. The free dislocation density increased from about 5.0 E+08 up to 1.0 E+10 dislocations/cm^2 from the pre-tested microstructure to the 3.7 percent strain level. Subsequent strain produced dislocation cells which progressively reached a saturation size of about 0.34 μm at a true strain level of 17.4 percent.

Fatigue-No Hold Time

The development of the dislocation microstructure (Nahm, 1977) as a function of the fraction of the fatigue life of AISI 304 stainless steel at 650°C and a total strain range of 2% is shown in Figure 2. The strain rate was 4 E-03 sec^{-1}. It is shown (Nahm, 1977) that the saturation stress level (~42 ksi) is reached at about 10 percent of the fatigue life, a circumstance at which the cell size (~0.66 μm) also reaches a saturation value. The additional energy of the deformation process (the remaining 90 percent of life) is attributed to both the propagation of the fatigue crack and supplying the driving force for the relative rotation of the cells (i.e., misorientation angle increasing). By counting the fatigue striations on the surface of the fracture, it is estimated that the fatigue crack, for these test conditions, was nucleated at about the same time (80 cycles) that the dislocation cells reached an equilibrium size.

Fatigue-Hold Time

The classical hold time experiments by Berling (Berling et. al., 1969) on AISI 304 stainless steel specimens have been the basis (i.e., appropriate bench mark data) for the correlation of the frequency modified (Coffin, 1972) model of plastic-elastic fatigue, the strain-range partitioning (Manson, 1973) model, Ostergren's damage (Ostergren, 1976) relationship, the damage rate equation (Majumdar, 1976) and many others.

TEM results (Kenfield, 1974) of those specimens used for the bench mark data for correlation with the above models show that significant changes indeed occur in the microstructure as a result of the various tensile hold times. The cell sizes show increases (factor of three) as a result of stress relaxation with the longer (10 hours) hold times. At certain conditions a pronounced segregation

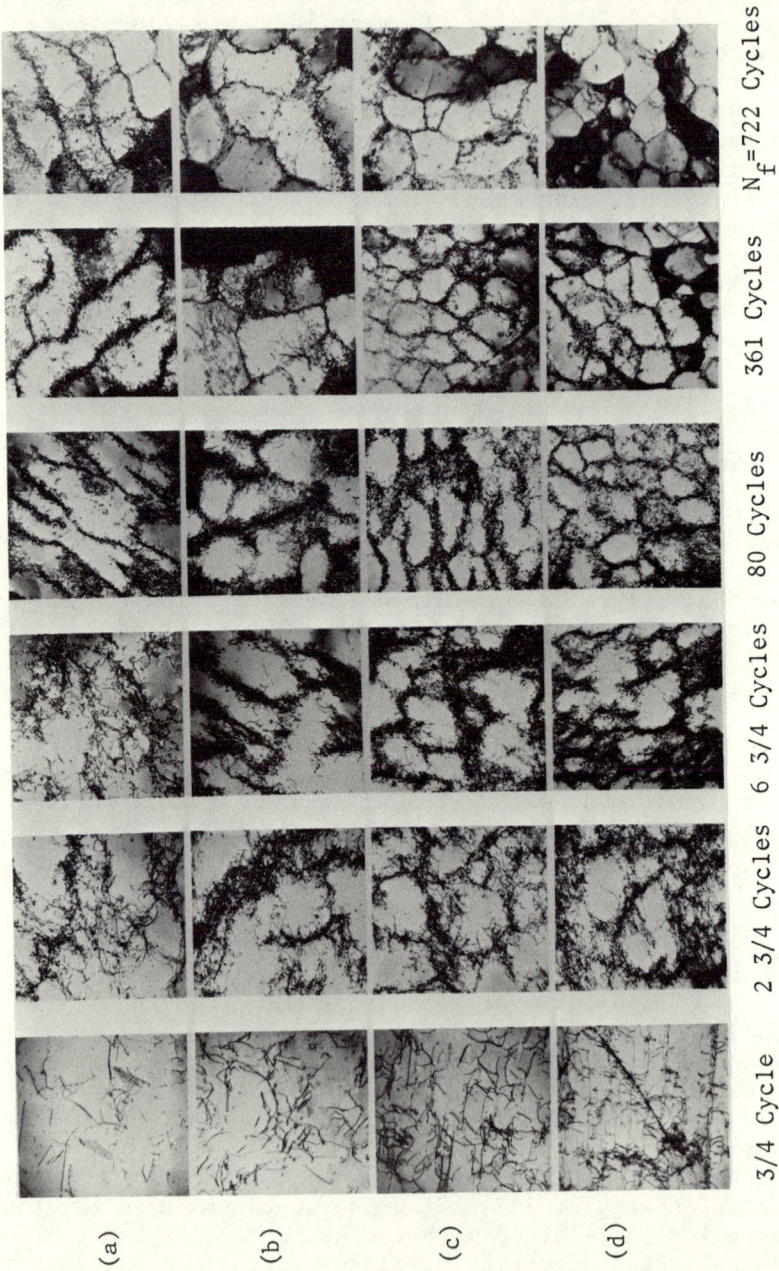

3/4 Cycle   2 3/4 Cycles 6 3/4 Cycles 80 Cycles   361 Cycles   N$_f$=722 Cycles

Figure 2. Low cycle fatigue (total strain range ~ 2%) in AISI 304 stainless steel at 649C, showing substructural development: during life. Each column shows the variation in sub-structure (a-d) that can occur within a given specimen after the indicated number of cycles.

of carbon in the form of carbides which decorate the dislocations, which are convenient sinks for these point defects, and thereby tending to stabilize some of the microstructure. There should be no question that the microstructural changes would influence the response of the material to external loads and that this circumstance should be included as an explicit function in the mathematical relationships of constitutive equations.

A study of the development of the microstructure, as a function of the fraction of the fatigue life, was recently completed (Ermi, 1979). Specimens, each tested to different fractions of the fatigue life at 593°C and at 0.5, 1.0 and 2.0% total strain ranges for zero and one minute hold times, have been used for transmission electron microscopy evaluations. The rapid hardening occurs in the first 10 cycles of this test. Figure 3 shows the results of the TEM studies. Here, the cell sizes decrease during the rapid hardening range until a saturation stress level is achieved. A relationship coupling the shear stress, strain and corresponding cell size is given as

$$\tau_m/G = 0.036(b\gamma p/\lambda')^{0.25} \tag{6}$$

where τ_m is the maximum shear stress and γ_p is the plastic shear strain.

Microstructure at Crack Tip

A fatigue crack growth specimen (James, 1972) of AISI 304 stainless steel tested at 538°C and at a stress ratio of R = 0.05 was used to evaluate the microstructure in the region of the crack tip (Ermi, 1981). A profile of microstructural changes, resulting in an increase in the hardness, was obtained by a systematic mapping of the hardness (see Figure 4) in the region of the crack tip. Detailed TEM investigations show the microstructural changes that occur at different positions away from the crack. The dislocation configurations and corresponding microhardness values are in reasonable agreement with data obtained from tensile results having similar dislocation features. Well defined cells form very close to the crack tip where the stress level is sufficiently high to form cells. The corresponding stress levels could be estimated from the cell size versus stress level relationship shown in Figure 1. The value of the cell size close to the crack tip is in reasonable agreement with fatigue process zone size ℓ in the relationship (Saxena et. al., 1975)

$$da/dN = 4[(0.7A)/E\sigma_{ys}{}^{1+S}\varepsilon_f]^{1/\beta}[1/(\ell^{1/\beta-1}]\Delta K(2+S)/\beta \tag{7}$$

where β is the exponent in the Coffin-Manson equation and S and A are defined by the equation:

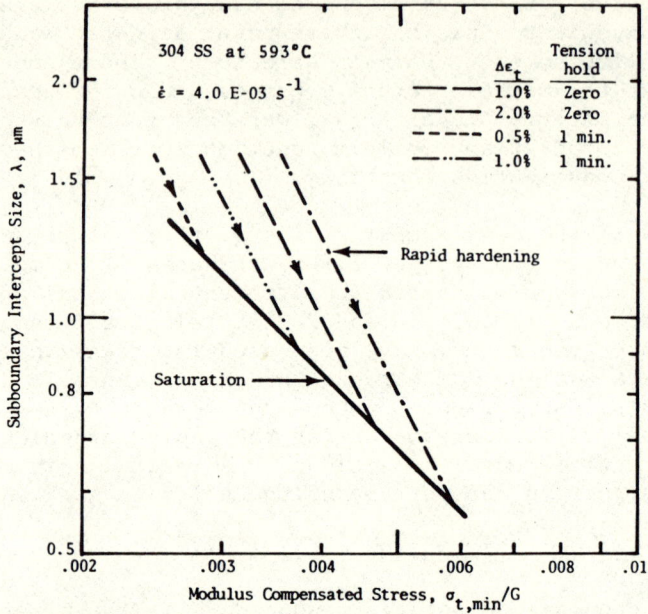

Figure 3.   Subboundary intercept size versus modulus compensated
            minimum tensile stress for 304 SS during rapid hard-
            ening and saturation.

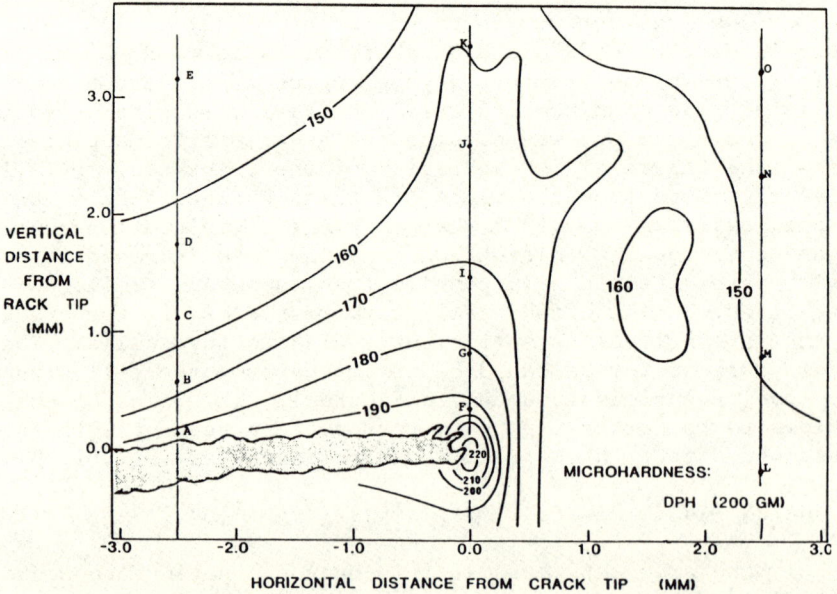

Figure 4.   Isomicrohardness Profiles in Region Around Fatigue
            Crack Tip.

$$R_p^f = A(\Delta K/\sigma_{ys})^{2+S} \tag{8}$$

where R_p^f is the fatigue plastic zone size and S is a small constant indicating (Saxena et al., 1975) a variation in real materials from the theoretically predicted second power dependence of R_p^f on ΔK.

Deformation Induced Void Formation.

The influence of deformation rate on the stable intergranular crack propagation behavior of AISI 304 stainless steel, as reflected in crack width, length, and angular orientation parameters was examined (Nahm, 1973). Specimens deformed to failure in the slow tension and creep-rupture modes at 650°C were studied. The results indicate that a rapid, step-wise crack propagation between grain-boundary triple junctions does not occur for these specimens, but that the triple junctions do provide a significant barrier to crack propagation. The crack angular orientation and width, as a function of deformation rate (4 E-08 to 8 E-4 sec^{-4} sec^{-1}), were concluded to be the parameters which reflect the crack growth rate for the test conditions employed in this work. The characteristic tendency for the cracks to form at 90 degrees to the tensile stress axis is quite apparent for the slower deformation rates.

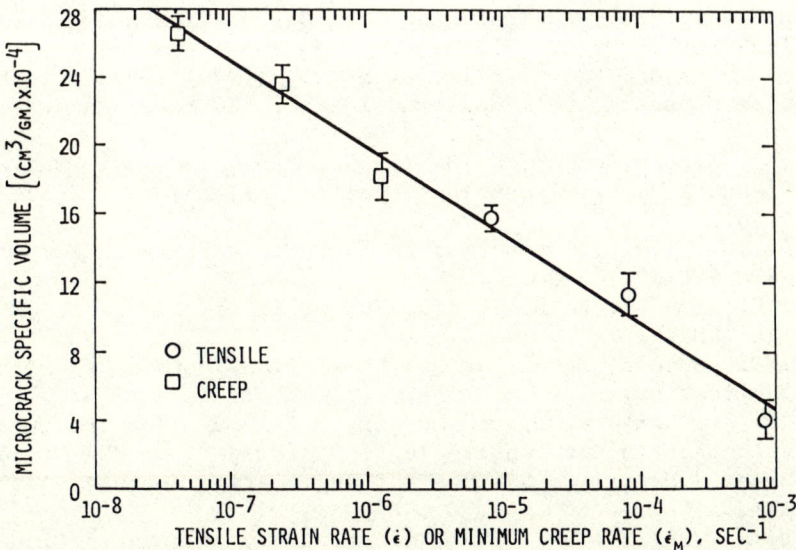

Figure 5.   Microcrack specific volume versus deformation rate for 304 SS at 650°C.

Bulk density measurements were made on the above specimens by the immersion density technique. Here the sample was weighed in air and then in a fluid. The results presented in Figure 5 as a micro-crack specific volume (cm^3 gm^{-1}) change as a function of the deformation rate. The specimens tested at the lower strain rates show the larger values of the microcrack specific volume. This clearly shows that the crack density is a strong function of the strain rate, in general agreement with the equation (Williams, 1967)

$$dc/dt \simeq (\mu \sigma D^2 \dot{\varepsilon}^2 t)/[4\pi(1-\nu)\gamma] \tag{9}$$

which describes the stable growth of wedge cracks in terms of the change in crack length, c, with respect to time, t. In this expression D is the grain size, γ the effective surface energy, ν is Poisson's ratio and μ the shear modulus.

ACKNOWLEDGEMENTS

This work was supported, in a large part, by the Division of Nuclear Power of the U.S. Department of Energy. The author appreciates the contributions of and the association with all the graduate students and Post-Doctoral Fellows who were assigned to these programs during the past ten years.

REFERENCES

Ashby, M. F. (1972), "A First Report on Deformation Mechanism Maps," Acta Met., 20, 887-97.
Ashby, M.F., Gandhi, C., Taplin, D. M. R., (1979), "Fracture-Mechanism Maps and Their Construction for FCC Metals and Alloys," Acta. Met., 27, 699-729.
Berling, J. T. and Conway, J. B. (1969), Proc. 1st Int. Conf. on Pressure Vessel Technology, Delft, Holland, Part 2, 1233-1246.
Coffin, L. C., (1972), Symposium on Fatigue at Elevated Temperature, ASTM STP 520, 5-34.
Ermi, A. E., (1979), Ph.D. Thesis, University of Cincinnati, Cincinnati, Ohio.
Ermi, A. E., Moteff, J. and James, L. A. (1980), to be submitted for publication.
Foulds, J. R., Ermi, A. M. and Moteff, J. (1980) "Substructural Development in Hot Tensile Testing of Type AISI 304 Stainless Steel," Materials Science and Engineering Journal, 45, 137-141.
James, L. A. (1972), "The Effect of Frequency Upon the Fatigue-Crack Growth of Type 304 Stainless Steel at 1000F," ASTM STP-513, 218-229.

Kenfield, T. A., (1974) "The Effect of Creep-Fatigue Interaction on
        the Substructure and Fracture Characteristics in 304 Stain-
        less Steel," M.S. Thesis, University of Cincinnati

Majumdar, S. and Malya, P. S. (1976), Proc. International Confer-
        ence on Materials -II, Boston, 924.

Manson, S. S., (1973), Fatigue at Elevated Temperatures, ASTM STP-
        520, 744-782.

Michel, D. J., Moteff, J. and Lovell, A. J., (1973), "Substructure
        of Type 316 Stainless Steel Deformed in Slow Tension at
        Temperatures Between 21 and 816 C," Acta Met., 21, 1269-
        1277.

Nahm, H., Moteff, J., and Diercks, D. R. (1977), "Substructural
        Development During Low Cycle Fatigue of AISI 304 Stainless
        Steel at 649°C," Acta Met., 25, 107-116.

Ostergren, W., (1976), Journal of Testing and Evaluation, ASTM, 4,
        327.

Saxena, A., and Antolovich, S. D., (1975), "Low Cycle Fatigue,
        Fatigue Crack Propagation and Substructures in a Series
        of Polycrystalline Cu-Al Alloys," Met. Trans., 6A, 1809-
        1828.

Vocé, E., (1948), J. Inst. Metals, 79, 1.

Williams, J. A., (1967), Phil. Mag. 15, 1289.

Wolf, S. M., Editor, (1979), Time Dependent Fracture of Materials
        at Elevated Temperature, Proceedings of Workshop, Division
        of Materials Science, U.S. DOE, CONF 790236, June, 1979,
        177-184.

FATIGUE AND FRACTURE RESISTANCE OF STAINLESS STEEL WELD

DEPOSITS AFTER ELEVATED TEMPERATURE IRRADIATION

J. R. Hawthorne

Naval Research Laboratory
Washington, D.C.

ABSTRACT

The fatigue crack growth and fracture resistance of Type 308-16 austenitic stainless steel weld deposits were investigated for 427 and 649°C neutron irradiation conditions. The welds (63.5 mm thick) were made by the shielded metal arc process and depicted variations in delta ferrite content from ferrite number 5.2 to 19.0. Specimen irradiations were conducted in the EBR-11 reactor in flowing sodium and static sodium environments.

Fatigue crack growth resistances at 427 and 649°C were determined using single-edge-notch (SEN) cantilever fatigue specimens tested in air using a zero-tension-zero loading cycle. Crack growth rates (da/dN) were related to the stress intensity factor range (ΔK). Effects of a tension-hold time of 0.5 minute were explored relative to weld behavior under continuous load cycling conditions. Fracture resistance at elevated temperature was investigated through notch ductility and dynamic fracture toughness determinations by Charpy-V and fatigue precracked Charpy-V test methods respectively.

Neutron fluences in the range 1 to 1.5 x 10^{22} n/cm^2, E>0.1 MeV, were found to have a large detrimental effect on fatigue crack growth resistance for the 649°C irradiation condition but a beneficial effect for the 427°C irradiated condition. A large detrimental effect of 427°C irradiation on elevated temperature fracture resistance was observed. The study also revealed that delta ferrite content and fatigue loading patterns can have a major influence on postirradiation fatigue crack growth trends.

INTRODUCTION

Proposed designs of advanced nuclear power systems will make extensive use of welded austenitic stainless steels. In support of this application, in-depth studies of elevated temperature mechanical properties are being made. The temperature range of interest extends to 650°C.

The present study focuses on the influence of delta ferrite content on the fatigue crack growth resistance and fracture resistance of E308-16 stainless steel weld deposits before and after nuclear irradiation. The investigations were prompted not only by the projected need of welding in breeder and fusion reactor systems but also by the early observations of large variations in fatigue and fracture properties among welds in exploratory tests.[1,2] A factor of ten difference in fatigue crack growth (FCG) rate, for example, was found between two supposedly identical weld deposits. Likewise, large differences in Charpy-V (C_V) notch ductility between welds and between parent metal (high) and weld metal (low) were noted. The isolation of contributing variables was therefore undertaken to improve weld consistency thereby assisting the planned material applications. The studies of suspect metallurgical factors have, as a long term objective, the development of guidelines for optimizing welds for the advanced system requirements.

Delta ferrite content is one of several welding variables having potential for influencing weld metal behavior. This report builds on earlier NRL studies of the as welded condition in which the influences of delta ferrite content on FCG resistance under continuous fatigue cycling and weld metal notch ductility and strength were assessed.[3,4] The investigations did not reveal a major effect on these properties in the nonirradiated material state; however, positive indications of a delta ferrite contribution on fatigue resistance were recorded in initial tests of the irradiated condition. The present investigation shows more clearly the combined effect of neutron irradiation and delta ferrite content on weld properties. In addition, the study explores the significance of superimposed load hold times to FCG resistance and the significance of the notch ductility degradation by irradiation in terms of reduced dynamic fracture toughness.

MATERIALS

The range of delta ferrite content of most interest to reactor applications is approximately from 5 to 15 percent. Materials employed in this investigation and the predecessor investigations[3,4] were from a series of four 63.5 mm thick shielded metal arc welds (Type 304 base plate, Type 308-16 filler) which encompassed this range. The welds were obtained from the Arcos Corporation by contract; the electrode composition and coatings used were those developed by Arcos for a prior Metal Properties Council (MPC) project.

Chemical compositions of the weld deposits are listed in Table 1. Welding parameters and conditions are given in reference 3. Each weld was a full-thickness weld (19 mm minimum weld width); the root regions were air-arc back gouged and ground to all weld metal after layer seven. Welding was accomplished under full mechanical restraint; however, opposite faces were welded alternately in a sequence designed to minimize unbalanced stresses. Delta ferrite contents of the individual welds in ferrite number, as determined by Magne-Gage, were 5.2, 10.4, 15.7, and 19.0, respectively. The welds were not given a post weld thermal treatment.

SPECIMEN DESIGN AND TESTING

Fatigue Tests

A single-edge-notch (SEN) cantilever specimen of the design shown in Figure 1 was used for the FCG determinations. The plane of the fatigue crack was oriented parallel to the welding direction and perpendicular to the weldment surface. All specimens were composite specimens made by joining (electron beam or metal inert-gas welding) end tabs to a center test section 55 x 64 x 13 mm in size. Comparisons of welded vs non-welded specimens of similar materials (AISI Type 316 plate and welds) have indicated that test results from each are comparable.[5]

All tests were conducted in air using a zero-tension-zero loading cycle with and without a 0.5 minute tension hold period. Specimen temperatures were provided by induction heating and were monitored continuously by thermocouples. When a high rate of fatigue crack growth became evident, tests normally were interrupted during non-working hours and were resumed only after the specimens had again reached temperature. No noticeable effect of this procedure was seen in the data.

Crack length measurements were accomplished by means of a traveling microscope at a magnification of X35 or by means of a high-resolution, closed-circuit television system. The television system was used for those tests conducted remotely in the NRL hot-cell facility. Rates of crack growth were established from plots

Table 1. Chemical Compositions of the Type 308-16 Shielded Metal Arc
Weld Series with Variable Delta Ferrite Content

NRL Weld Code	Delta Ferrite Content[a]	Chemical Composition (wt-%)[b]								
		C	Mn	Si	P	S	Cr	Ni	Mo	N
V41[c]	5.2	0.056	1.88	0.32	0.024	0.011	19.71	10.35	0.05	0.068
V42	10.4	0.060	1.54	0.31	0.029	0.009	19.90	9.25	0.05	0.074
V43	15.7	0.060	1.65	0.32	0.029	0.011	20.89	9.11	0.06	0.079
V44	19.0	0.060	1.38	0.43	0.028	0.010	21.08	8.93	0.08	0.084

[a]Weld deposit ferrite number (avg); Magne-Gage determination.
[b]Composition based on standard WRC weld test pad (courtesy Arcos Corporation); core wire for all electrodes from same steel melt.
[c]0.07% Cu in weld deposit

of crack length vs number of cycles using the ASTM-recommended
incremental polynomial method. The method basically involves
computer fitting, by least squares criteria, seven consecutive
data points (N_{i-3} to N_{i+3}) to a second order polynomial. The
polynomial in turn is differentiated to yield da/dN to the N_i^{th}
point.

The SEN specimen crack growth rates (da/dN) were related to
stress-intensity factor range (ΔK) using the expression for K for
pure bending developed by Gross and Srawley:[6]

$$K = \frac{6PL}{(BB_n)^{1/2} \cdot W^{3/2}} \cdot Y,\tag{1}$$

where $Y = 1.99\,(a/W)^{1/2} - 2.47(a/W)^{3/2} + 12.97(a/W)^{5/2} - 23.17$
$(a/W)^{7/2} + 24.80\,(a/W)^{9/2}$, and where P is the cyclic load, L is
the distance from the crack plane to the point of load application,
a is the total length of notch and crack, W is the specimen width,
B is the specimen thickness, and B_n is the net thickness between
the specimen side grooves. A correction for plasticity at the
crack tip was not made. Tests normally were terminated when the
total flaw length, a, reached about 38 mm.

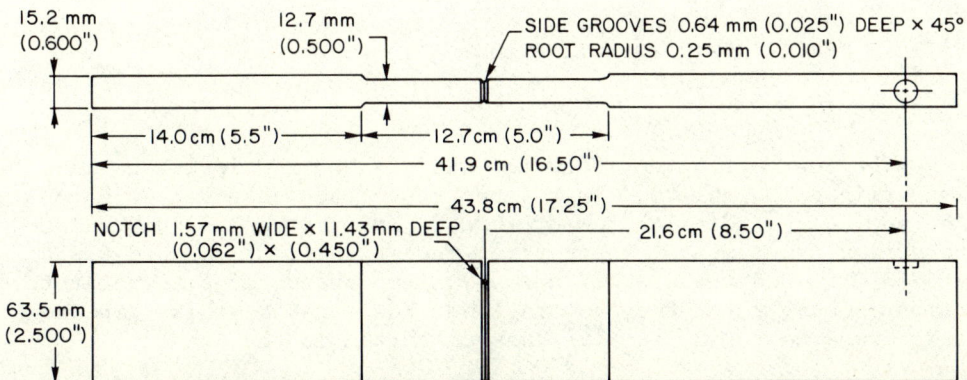

Fig. 1. Design of the single-edge-notch (SEN) cantilever fatigue
test specimen.

Fracture Resistance Tests

    Standard Charpy V-notch (C_V) specimens, ASTM Type A, were used
for the notch ductility determinations. Tests were conducted in
accordance with ASTM Recommended Practice E-23. Fatigue precracked
Charpy-V (PCC_V) specimens were used for the dynamic fracture toughness (K_J) determinations. Specifications for fatigue precracking
called for a specimen crack length-to-width ratio (a/W) of 0.5:
the maximum allowable stress intensity (K_1) during the last increment (0.76 mm) of fatigue crack growth was 22 MPa$\sqrt{m}$ or 20
ksi$\sqrt{in}$.. PCC_V testing was in conformance with standard procedures
developed by the Electric Power Research Institute (EPRI) for K_J
determinations.[7] All K_J determinations given in this report are
based on energy absorbed to maximum load corrected for specimen
and test machine compliance. In this regard, K_J values as computed would tend to overestimate the K_J at crack initiation if
some stable, i.e., rising load, crack extension takes place before
the attainment of maximum load.

MATERIAL IRRADIATION

    Material irradiations were conducted in the EBR-II reactor in
two experiments. One experiment, number H-8, utilized a controlled
temperature heat pipe irradiation assembly in which the specimens
were immersed during irradiation in static sodium at $\sim$649°C. This
temperature condition, as noted above, lies at the upper end of
the temperature range of interest. The second experiment, H-11,
was not of a controlled temperature design but placed the specimens in direct contact with the flowing reactor sodium coolant at
$\sim$427°C. Radiation effects processes at this nominal temperature
are significantly different than at 649°C.[8] Target neutron
fluences (n/cm^2, E > 0.1 MeV) were 1 x 10^{22}n/cm^2 and 1.5 x 10^{22}
n/cm^2, respectively. Other experiment details are provided in
Table 2.

FATIGUE CRACK GROWTH INVESTIGATIONS

Test Matrix

    The test matrix employed for the current FCG investigations
is outlined in Table 3. Test temperatures are observed to match
the nominal irradiation temperatures. In the case of unirradiated
condition assessments, the earlier investigations included tests
at 260°C in addition to 427 and 649°C for broad range temperature
comparisons. A tendency toward lower FCG rates at this temperature compared to 427 or 649°C was discerned in the data.

    Reference tests are those conducted with continuous cycling
at 10 cpm in a tension-zero-tension sawtooth mode. Hold time
tests used the same loading and unloading rates as the reference

Table 2. Material Irradiation Experiments

Exp. No.	EBR-II Subassembly	Subassembly Design	Specimen Types	Irrad. Temp($^\circ$C)[a]	Fluence Target (10^{22}n/cm^2)[b]	Period of Irradn(h)	Period in Reactor (h)
H-8	X-266 (8F5)[c]	Heat Pipe	SEN,PCC$_v$	649	1.0	3110	~6000
H-11	X-322 (3E2+2D1)[c]	Open	SEN,PCC$_v$ C$_v$, Tensile	427	1.5	2100	~2500

a – nominal irradiation temperature
b – E > 0.1 MeV
c – fuel core position

Table 3. Experimental Test Matrix (SEN Specimens)

Weld Code	Ferrite Number	Test Temperature			
		427°C		649°C	
		(0.0M Hold)[a]	(0.5M Hold)	(0.0M Hold)[a]	(0.5M Hold)
Unirradiated Condition					
V41	5.2	6	5	7	–
V42	10.4	6	8	3	7
V43	15.7	5	–	6	–
V44	19.0	6	7	3	12
Irradiated Condition [b,c]					
V41	5.2	–	4	–	–
V42	10.4	5	4	2	1
V44	19.0	4	5	1	2

[a] Prior test series, continuous cycling mode
[b] 427°C tests for experiment H-11
[c] 649°C tests for experiment H-8

tests but included a 0.5 minute hold in tension for a 2 cpm cycling rate.

Results

Data developed by tests at 649 and 427°C are presented in Figures 2-4 and Figures 5-7, respectively. For reference, delta ferrite content variations within the range investigated did not result in a major difference in FCG rates in the unirradiated condition at either 260, 427, or 649°C when cycling was continuous at 10 cpm, i.e., without a tension hold period[4] (see Figs. 2 and 5).

Comparisons of the new data against the reference data provide these additional observations:

Tension hold time - A 0.5 M tension hold time produces a detrimental effect on FCG resistance at 649°C but not at 427°C in the unirradiated condition (see Figs. 2 and 5, respectively). For the irradiated condition, a detrimental effect was observed for the tension hold time in 649 and 427°C tests (see Figs. 3 and 6), the magnitude of which at the higher temperature is clearly dependent on delta ferrite content.

Neutron exposure - Neutron irradiation at 649°C produces an increase in FCG rate at 649°C for both the continuous cycling mode and for cycling with the 0.5 M tension hold time (Figs. 3 and 4). Compared to reference condition tests, an elevation in FCG rates on the order of a factor of 1⁻ was induced by the 649°C radiation exposure in combination with the tension hold time. In direct contrast to this observation, neutron irradiation at 427°C produced a reduction in FCG rate (improvement at 427°C) for the continuous cycling mode. The data for weld V44, however, suggest that this benefit may be reduced or fully lost if a tension hold time is applied (Fig. 7). These observations, of course, are made for the specific fluence levels investigated.

Delta ferrite content - An increase in delta ferrite content from FN 10.4 to FN 19.0 is shown to be detrimental to FCG resistance at 649°C in the unirradiated condition when cycled with a tension hold time but not with continuous cycling at 10 cpm. In the 649°C irradiated condition, delta ferrite content is seen to be a significant factor in 649°C FCG resistance in either cycling mode. On the other hand, the 427°C FCG trends, with one possible exception, do not indicate an effect of ferrite level on FCG properties for either unirradiated, irradiated, continuous cycling or tension hold time conditions. The possible exception noted is

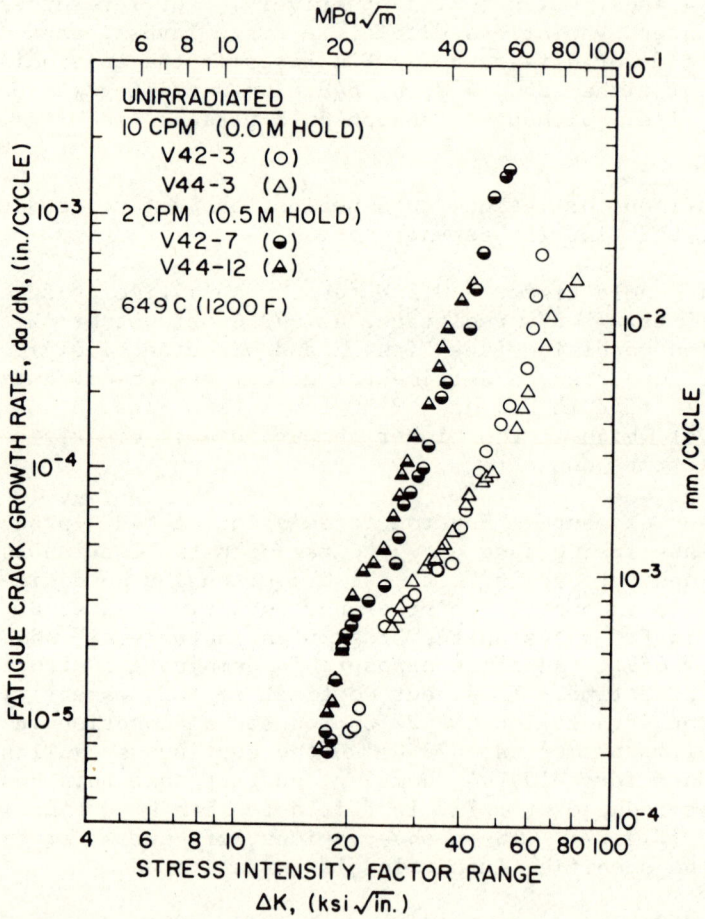

Fig. 2.   Fatigue Crack Growht Rates of the Welds V42 and V44 at
          649°C in the Unirradiated (As-Welded) Condition. Data for
          Continuous Cycling Conditions (10 cpm) and Cycling with a
          0.5 Minute Tension Hold Time (2 cpm) are Shown.

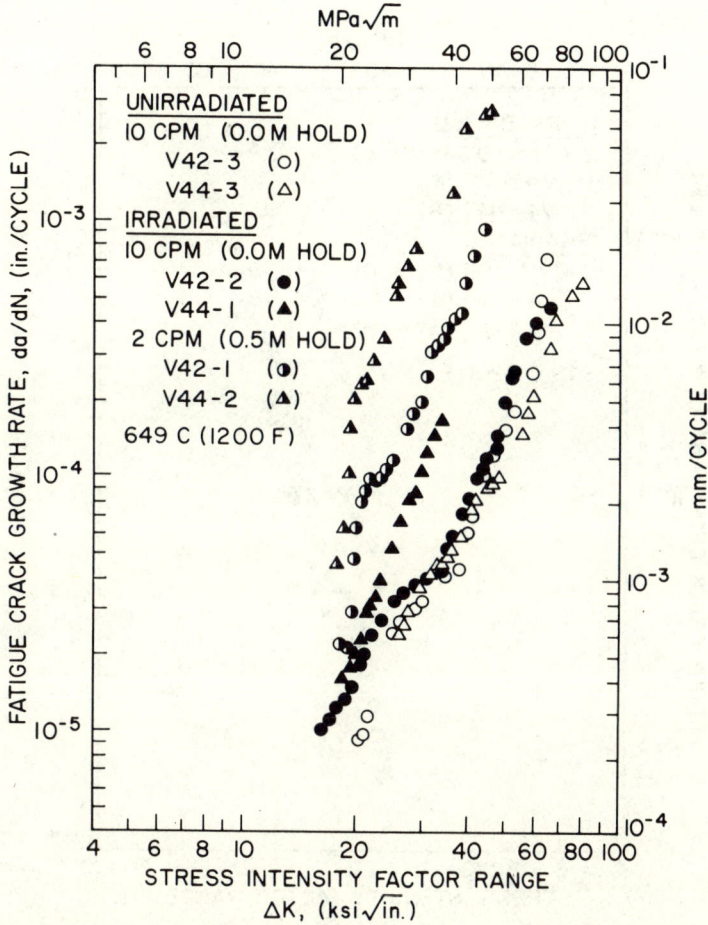

Fig. 3.   Fatigue Crack Growth Rates of the Welds V42 and V44 at
           649°C in the Unirradiated and 649°C Irradiated Conditions.
           Data for the Irradiated Condition Refer to COntinuous
           Cycling Conditions and for Cycling with a 0.5 Minute
           Tension Hold Time; the Data for the Unirradiated Con-
           dition are for the Continuous Cycling Mode.

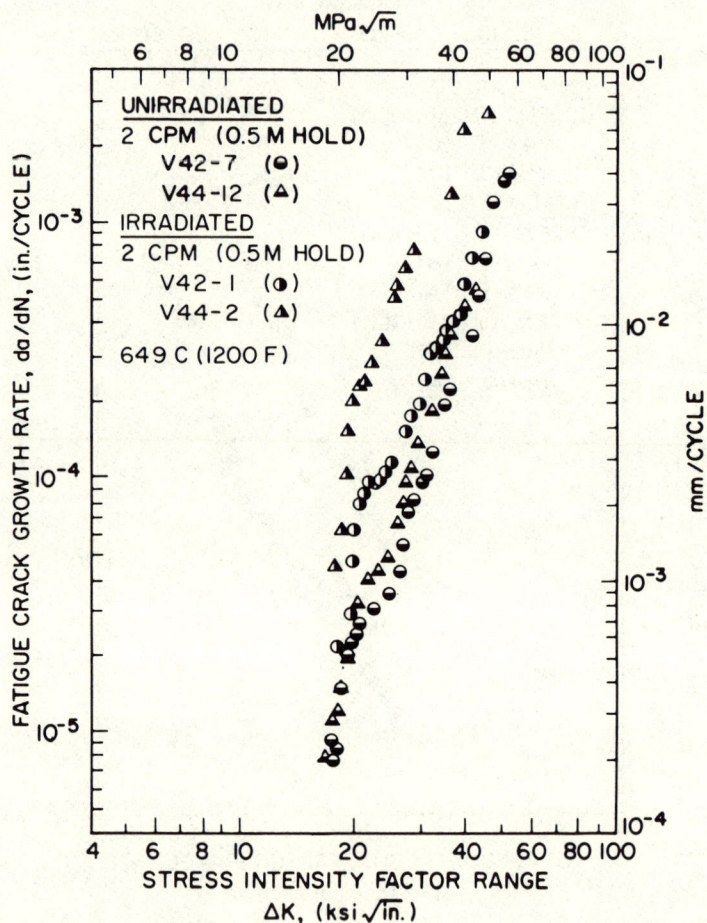

Fig. 4. Fatigue Crack Growth Rates of the Welds V42 and V44 at
649°C in the Unirradiated and 649°C Irradiated Conditions
for the Case of Fatigue Cycling with a 0.5 Minute Tension
Hold Time Only.

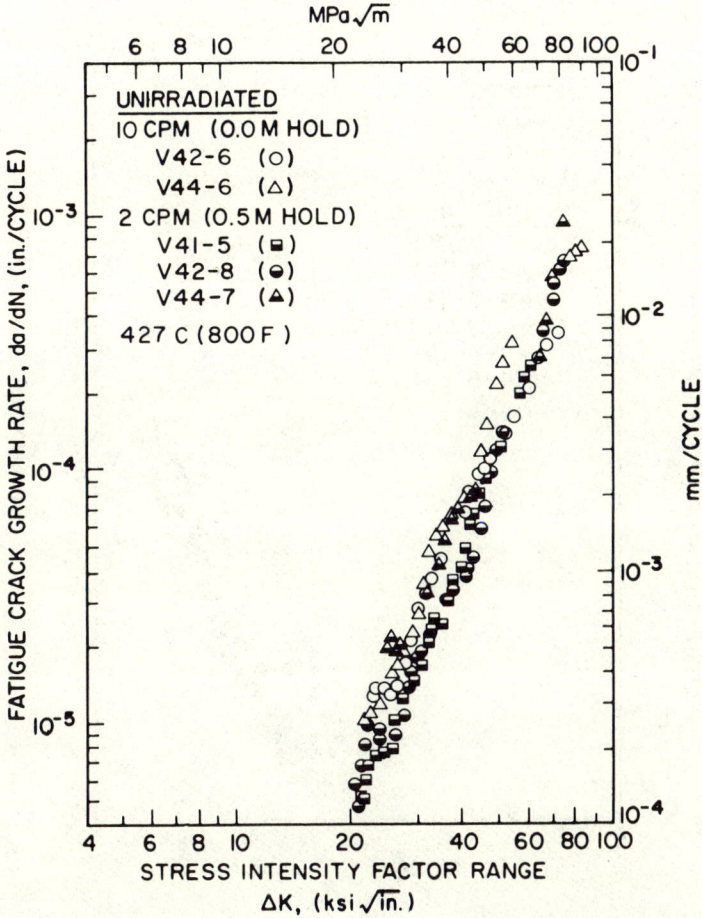

Fig. 5. Fatigue Crack Growth Rates of the Welds V41, V42 and V44
         at 427°C in the Unirradiated Condition. Data for Continuous
         Cycling Conditions and Cycling with a 0.5 Minute Tension
         Hold Time are Shown. (Reference Condition Data for Weld
         V41 Fell Within the Data Scatter Band and Have Been Omitted
         for Clarity).

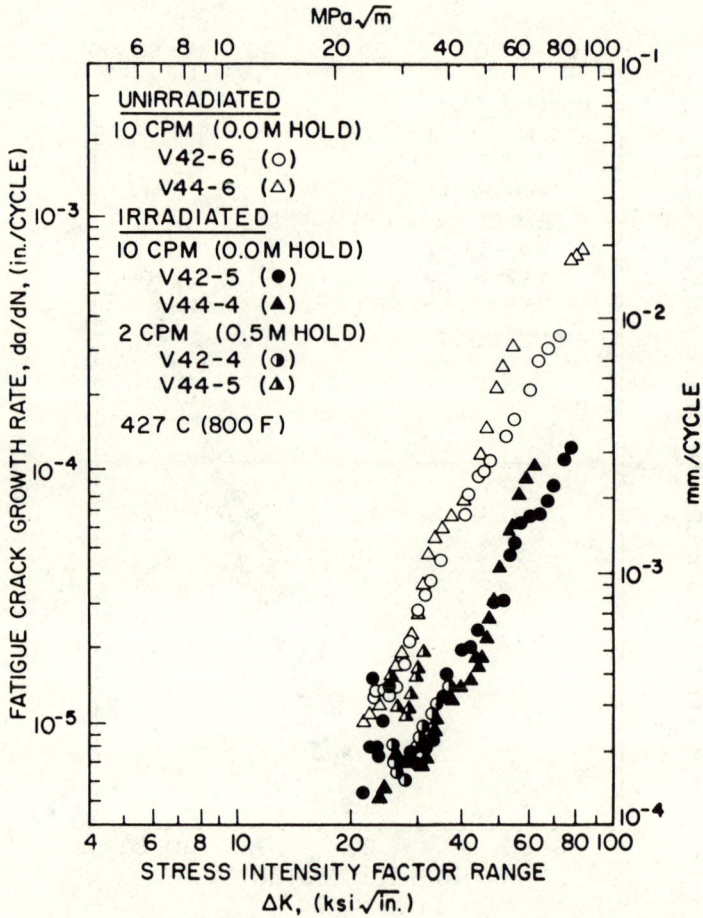

Fig. 6. Fatigue Crack Growth Rates of the Welds V42 and V44 at
427°C in the Unirradiated and 427°C Irradiation Conditions.
Data for the Irradiated Condition Refer to Continuous
Cycling Conditions and for Cycling wiht a 0.5 Minute Tension
Hold Time, the Data for the Unirradiated Condition are for
the Continuous Cycling Mode.

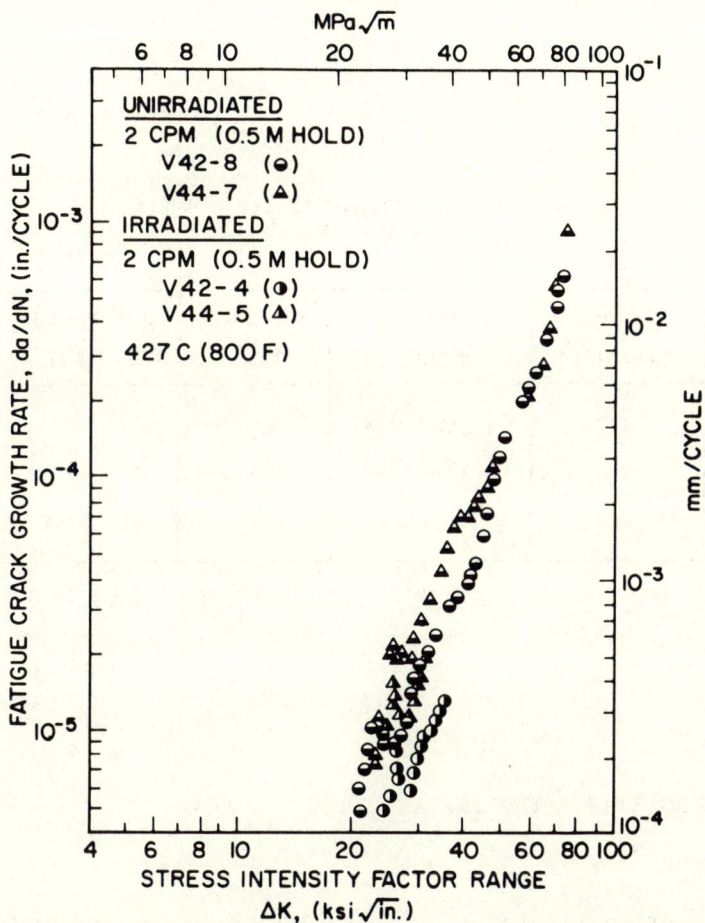

Fig. 7. Fatigue Crack Growth Rates of the Welds V42 and V44 at 427°C in the Unirradiated and 427°C Irradiated Conditions for the Case of Fatigue Cycling wiht a 0.5 Minute Tension Hold Time Only.

postirradiation testing with a tension hold time. Specifically, a detrimental effect of the higher delta ferrite
level is suggested by the data comparisons for weld V44;
however, the data obtained with specimen V44-5 are very
limited. This particular test was terminated early because
the advancing crack developed a small curvature away from
the normal (straight) crack path.

Table 4.   Experimental Test Matrix
(C_v and PCC_v Specimens)

Exp. No.	Irrad. Temp. (°C)	Weld Code	Specimen Type	Test Temperature (°C) 371	427	482	566
H-11	427	V41,V42,V44	C_v	X		X	
		V41,V42,V44	PCC_v		X		
H-8	649	V42,V44	PCC_v				X

FRACTURE RESISTANCE INVESTIGATIONS

Test Matrix

The test matrix for the C_v and PCC_v tests is shown in Table 4.
Post irradiation C_v test temperatures for experiment H-11 were
selected to bracket the prior irradiation temperature because reference condition tests indicated a general independence of C_v
energy absorption on temperature in the range of 260 to 593°C. (A
lower C_v energy absorption was found at 24°C). Reference condition PCC_v tests showed a similar independence of behavior in this
temperature range. Post-irradiation PCC_v tests of experiment H-11
were conducted at the prior irradiation temperature of 427°C; however, tests for experiment H-8 were conducted at temperatures somewhat below the exposure temperature of 649°C because of equipment
limitations.

Results

Experiment H-11 – The C_V data for the preirradiation and post-irradiation conditions are presented in Figure 8 and are summarized in Table 5. The results clearly show a very large detrimental effect on notch ductility produced by the 427°C irradiation exposure. Indicated reductions in C_V energy absorption range from 73 to 82 percent. The PCC_V test data for the same conditions are given in Table 6 and show large reductions in K_J ranging from 70 to 79 percent. Equally important, several of the PCC_V specimens fractured before attaining a condition of general yielding, i.e., elastic fracture behavior exhibited. Table 7 reports postirradiation strength values for the welds V42 and V44. Here, a doubling of the yield strength and a 50 percent increase in tensile strength is indicated. These increases would be reasonable projections of the strength elevation for weld V41 also. (Limited irradiation space precluded inclusion of a tensile specimen of this weld). In comparison with the irradiation effect on strength thermal conditioning at a somewhat higher temperature of 482°C for 2500 hours in the absence of irradiation produced small reductions in yield strength and a somewhat larger reduction in tensile strength (Table 7). From the ratio of postirradiation fracture toughness and yield strength, it would appear that all of the welds after irradiation at 427°C would exhibit elastic fracture behavior in relatively thin section sizes, i.e., less than 25.4 mm thickness.

On balance, the data indicate a slightly greater irradiation effect (percentage property change) with increasing ferrite content. However, the spread among all postirradiation C_V (and K_J) values is relatively low. In turn, the percentage variations become of less significance for the fluence condition evaluated.

Experiment H-8 – Postirradiation PCC_V test results have been included in Table 6. Within the data scatter for individual welds, an effect of delta ferrite content is not discerned. Although a reduction in fracture toughness with irradiation is evident, the K_J values in general are much higher than those observed for the welds with 427°C irradiation. Changes in strength with irradiation were very small for the materials based on postirradiation hardness determinations (V42: RB 94.5; V44:RB 96.0). The postirradiation K_J values of 150 MPa $\sqrt{m}$ thus signify high fracture resistance retention.

DISCUSSION

The data secured with this investigation clearly indicate a need for more detailed studies of fatigue and fracture resistance behavior of Type E308-16 welds for the full range of projected applications in nuclear power systems. For one, the investigations revealed a shift from "improvement" in FCG resistance to

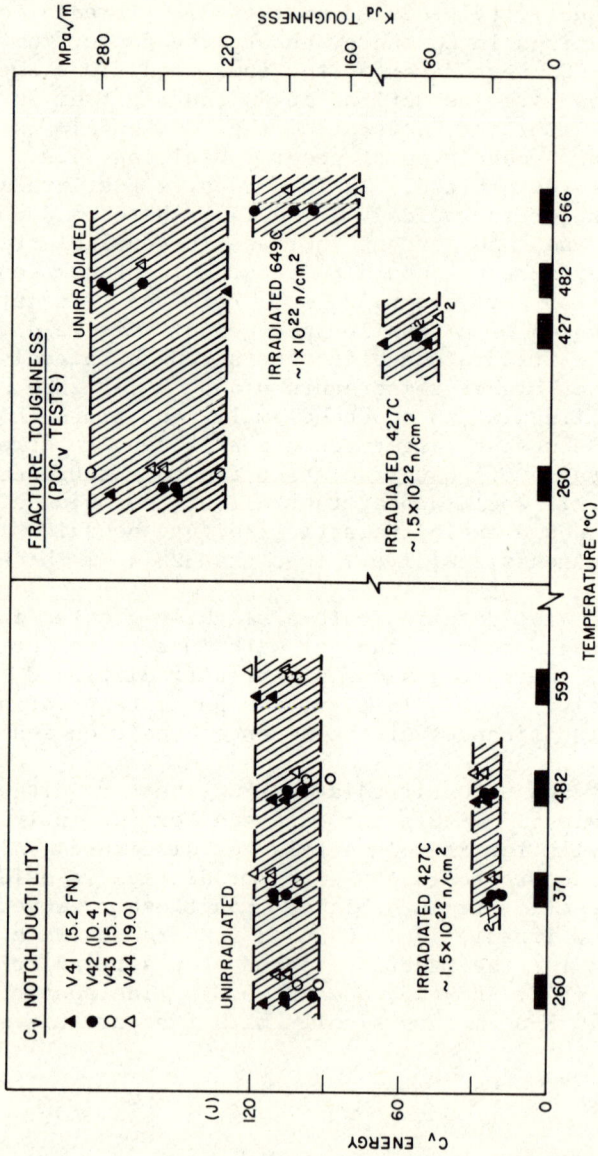

Fig. 8.   Notch Ductility and Fracture Toughness Properties of the Weld
          Series Before and After Irradiation.

Table 5. Charpy-V Notch Ductility of the Weld Series in Unirradiated and Irradiated Conditions.

Weld Code	Ferrite Number	C_v Energy Absorption (J)[a]					
		Unirradiated Condition				427°C Irradiated Condition[b]	
		260°C	371°C	482°C	593°C	371°C	482°C
V41	5.2	106	100	106	111	24	24
		114	110	111	118	24	30
V42	10.4	95	104	99	---c	19	23
		106	110	104	---	23	26
V43	15.7	92	100	87	102	---	---
		100	111	98	104	---	---
V44	19.0	104	111	102	106	20	26
		108	118	---	122	23	30

a 1 J = 0.738 ft-lb

b ~ 1.5 x 10^{22} n/cm^2, E > 0.1 MeV

c not determined

Table 6. Dynamic Toughness of the Welds V41, V42, and V44 in Unirradiated and Irradiated Conditions (PCC$_V$ Test, J-Integral Analysis).

Weld Code	Ferrite Number	Dynamic Fracture Toughness K_J (MPa $\sqrt{m}$) [a]			
		Unirradiated Condition		427°C Irradiated Condition [b]	649°C Irradiated Condition [c]
		260°C	482°C	427°C	566°C
V41	5.2	242	211	62[d]	---
		274	278	85	---
V42	10.4	243	259	65	185
		249	279	68	207
					177[e]
V44	19.0	248	257	54[d]	152
		255	---	56[d]	189

[a] 1 MPa $\sqrt{m}$ = 0.91 ksi $\sqrt{in}$.
[b] ~1.5 x 10^{22} n/cm^2, E > 0.1 MeV
[c] ~1.0 x 10^{22} n/cm^2, E > 0.1 MeV
[d] Specimen fractured before general yielding
[e] 482°C test

Table 7. Tensile Properties of the Weld Series in Unirradiated, Irradiated and Thermally Aged Conditions.

Weld Code	Ferrite Number	Unirradiated Condition[a]						427°C Irradiated Condition[b]			
		Yield Strength (MPa)			Tensile Strength (MPa)			Yield Strength	Tensile Strength	Elongation (%)[c]	Hardness (Rc)[d]
		260°C	371°C	482°C	260°C	371°C	482°C	427°C	427°C	427°C	24°C
V41	5.2	382 (365)[e]	350	325 (310)[f]	473 (485)[e]	476	430 (412)[f]	---	---	---	30.4
V42	10.4	420 (403)[e]	394	358 (329)[f]	521 (519)[e]	514	478 (425)[f]	679	703	7.1	32.9
V43	15.7	415 (415)[e]	378	362 (342)[f]	520 (522)[e]	494	482 (447)[f]	---	---	---	---
V44	19.0	447 (416)[e]	378	376 (362)[f]	563 (564)[e]	535	517 (475)[f]	811	811	5.8	33.9

[a] 1 MPa = 0.145 ksi
[b] 1.5 x 10^{22} n/cm^2, E > 0.1 MeV (5.74mm gage diameter specimens)
[c] Elongation in 25.4 mm
[d] Average of duplicate specimens
[e] Thermally conditioned at 260°C for 2500 hours
[f] Thermally conditioned at 482°C for 2500 hours

"impairment" with irradiation depending on irradiation and test
temperature conditions. Equally important, a clearer understanding of the detrimental effect of the 0.5 M tension hold time
on FCG resistance is needed. From the standpoint of fracture resistance, the low C_V energy absorption and the low K_J values found
in PCC_V tests after $427^\circ C$ irradiation can be a cause for concern.
Tests with larger size specimens for determination of J-R curve
characteristics are needed to confirm and assess the engineering
significance of the elastic fracture trend indications.

    Studies of the mechanisms underlying the observed property
changes have been initiated,[9] using TEM and SEM procedures. For
the unirradiated state and continuous cycling conditions, clear
differences among the welds were not found for either 427 or
$649^\circ C$ tests. Also, sigma phase was not observed in the microstructure of the one weld (FN 15.7) examined thus far in the
reference condition, (FCG tests of the unirradiated condition at
10 cpm, however, were completed in a cycling time of 115 hours or
less). Selected area diffraction analyses did reveal that small
precipitates were formed during fatigue cycling and were $M_{23}C_6$
carbide particles. TEM and SEM investigations of irradiated
samples are just getting underway. For the irradiation conditions
used here, only limited void formation is normally expected and
should have a negligible effect on fatigue and fracture resistance
properties.

    Long term thermal exposure in the range of 500 to $900^\circ C$ can
cause the transformation of ferrite (ductile) to sigma phase
(brittle) in addition to the carbide precipitation reported by
reference 9. Accordingly, the total time at temperature, i.e.,
the residence time in reactor plus the total time of cycling, can
be quite important in property trend assessments. The effect of
$482^\circ C$ thermal conditioning on tensile properties was shown in
Table 7. Comparison tests of fracture resistance using dynamic
tear test samples aged at $482^\circ C$ for 2400 hours, gave indications
of reduced energy absorption for welds with FN $\geq$ 10.4 but the
changes were small. In-depth studies of the effects of long term
(up to 10,000 hours) $593^\circ C$ thermal conditioning on delta ferrite
content and weld metallographic features have been made by the MPC
study.[10] The study used material from multilayer weld test pads
prepared with the same welding electrode/electrode coating compositions used here. However, the MPC test pads had lower ferrite
levels than the thick section NRL welds, an unexpected occurrence
at that time. Weld pad ferrite contents were on the order of FN 2,
4, 9 and 16. The effect of aging on microstructure was found to
differ depending on the ferrite level. Specifically, in the two
higher ferrite content materials, the measured decrease in ferrite
content with aging resulted primarily from a transformation to
sigma phase. In the two lower ferrite content materials, the decrease in ferrite content resulted primarily from a formation of

carbides and possibly, additional austenite; little of the ferrite transformed to sigma in these materials. In all cases, some ferrite was observed after 10,000 hours at 593°C.

The following interpretations of the FCG data trends, in terms of probable mechanisms and radiation effects on the mechanisms, can be made with the aid of the MPC observations:

(1)   The equally detrimental effect of a tension hold time on 649°C FCG resistance of welds V42 and V44 in the unirradiated condition may be due to sigma phase formation if the percent sigma formed (or its contribution) is independent of ferrite content in the range of FN 10.4 to 19.0.

(2)   The unequal detrimental effect of irradiation on 649°C FCG resistance of welds V42 and V44 under continuous cycling may indicate radiation-enhanced sigma phase formation in weld V44 or possibly a change in mechanisms in weld V42.

(3)   For the 649°C irradiation condition, the equal increase in FCG rates for tension hold time tests over continuous cycling tests for welds V42 and V44 suggests a direct effect of the tension hold component rather than a time-at-temperature effect. That is, the increase in time at temperature by hold time testing (< 100 hours) was only a small fracture of the (prior) time at temperature during irradiation (3110 hours).

(4)   The negligible effect of the tension hold time on 427°C FCG resistance of the welds V41, V42 and V44 in the unirradiated condition may indicate the absence of sigma phase formation for all ferrite levels at this temperature. The maximum duration of these tests was ∿525 hours.

(5)   The beneficial and equal effect of 427°C irradiation on the FCG resistance of the welds V42 and V44 under continuous cycling may indicate radiation enhanced carbide precipitation.

(6)   For the 427°C irradiation condition, the unequal increase in FCG rates for tension hold time tests over continuous cycling tests for welds V42 and V44 may indicate two competing mechanisms dependent on ferrite level.

It is expected that the Magne-Gage and fatigue tests of thermally conditioned SEN specimens now in progress will help clarify the respective roles of thermal conditioning and irradiation exposure in FCG trends.

Referring next to postirradiation fracture resistance, the elevation in yield strength is believed to be largely responsible for the reduction in C_v energy absorption observed with $427^{\circ}C$ irradiation. Carbide formation may have contributed to this elevation. The relatively small effect on PCC_v fracture toughness by $649^{\circ}C$ irradiation, on the other hand, would be considered inconsistent with a significant transformation of delta ferrite to sigma phase. However, K_J values relate to crack initiation rather than crack propagation. Crack propagation resistance assessments would require J-R curve test determinations.

CONCLUSIONS

In summary, the following general conclusions and primary observations were drawn from the experimental data and analyses presented here:

(1) Delta ferrite content can have an appreciable influence on elevated temperature FCG resistance after 427 or $649^{\circ}C$ irradiation, depending on fatigue cycling conditions. This welding variable, however, does not appear to affect FCG resistance in the unirradiated condition under continuous cycling conditions or cycling with a tension hold time.

(2) Delta ferrite content has only a small or negligible effect on elevated temperature fracture resistance after 427 or $649^{\circ}C$ irradiation to fluences on the order of 1 to $1.5 \times 10^{22} n/cm^2$, $E > 0.1MeV$.

(3) Irradiation at $649^{\circ}C$ to $\sim 1 \times 10^{22}$ n/cm^2, in general, is detrimental to $649^{\circ}C$ FCG_2 resistance. Irradiation at $427^{\circ}C$ to $\sim 1.5 \times 10^{22} n/cm^2$, in contrast, proved beneficial to $427^{\circ}C$ FCG resistance, but had a large detrimental effect on fracture resistance.

(4) Postirradiation FCG resistance trends at $427^{\circ}C$ indicate two competing mechanisms which individually may stem from carbide precipitation versus sigma phase formation.

(5) A 0.5 M tension hold time, with one exception noted, appeared to be detrimental to FCG resistance after 427 or $649^{\circ}C$ irradiation.

(6)   Postirradiation FCG resistance, on balance, is greater
      for a delta ferrite content of FN 10.4 than delta fer-
      rite content of FN 19.0.

ACKNOWLEDGMENTS

This investigation was conducted for the DOE under Interagency
Agreement EX-76-A-27-2110. The continuing support of this agency
is sincerely appreciated.

The author expresses his appreciation to H. E. Watson, W. E.
Hagel, and E. Woodall for their assistance and participation in
the experimental phases of the investigations.

REFERENCES:

1.   Hawthorne, J. R. and Watson, H. E., "Significance of Welding
       Variables, Long Term Aging and Tension Hold Times on
       Fatigue Crack Growth in Type 316 Stainless Steel Welds,
       "1976 ASME-MPC Symposium on Creep-Fatigue Interaction,
       ASME-MPC-3, pp. 417-432, 1976.
2.   Hawthorne, J. R. and Watson, H. E., "Notch Toughness of
       Austenitic Stainless Steel Weldments with Nuclear Ir-
       radiation," Welding Journal, Vol. 52, Research Supplement
       255-s-260-s, June 1973.
3.   Hawthorne, J. R. and Menke, B. H., "Influence of Delta Ferrite
       Content and Welding Variables on Notch Toughness of Austen-
       itic Stainless Steel Weldments," Structural Materials for
       Service at Elevated Temperatures in Nuclear Power Operation,
       ASME-MPC-1, pp. 351-364, 1975.
4.   Hawthorne, J. R., "Significance of Delta Ferrite Content to
       Fatigue Crack Growth Resistance of Austenitic Stainless
       Steel Weld Deposits, NRL Report 8201, March 1, 1978.
5.   Shahinian, P., Smith, H. H., and Hawthorne, J. R., "Fatigue
       Crack Propagation in Stainless Steel Weldments at High
       Temperature," Welding Journal, Vol. 51, Res. Supl. 527-
       s-532-s, November 1972.
6.   Gross, B., and Srawley, J. E., "Stress-Intensity Factors for
       Single-Edge-Notch Specimens in Bending or Combined Bending
       and Tension by Boundary Collocation of a Stress Function,"
       NASA-TN-D-2603, National Aeronautics and Space Agency,
       January 1965.
7.   Ireland, D. R., Server, W. L., Wullaert, R. A., "Procedures
       for Testing and Data Analysis," Task A-Topical Report,
       Effects Technology, Inc., ETI Technical Report 75-43,
       October 1975.
8.   Appleby, W. K., Bloom, E. E., Flinn, J. E. and Garner, F. A.

in: Radiation Effects in Breeder Reactor Structural
Materials, M. L. Bleiberg and J. W. Bennett, Eds. 1977,
p. 509.

9. Provenzano, V., Hawthorne, J. R. and Sprague, J. A., "Fracto-
graphic Analysis of Elevated Temperature Fatigue Crack
Propagation in AISI Type 308 Weld Deposits," 1978 ASME-
MPC Symposium on Properties of Steel Weldments for
Elevated Temperature Pressure Containment Applications,
ASME-MPC-9, pp. 63-75, 1978.

10. Edmonds, D. P., Vandergriff, D. M. and Gray, R. J., "Effect
of Delta Ferrite Content on the Mechanical Properties of
E308-16 Stainless Steel Weld Metal-111. Supplemental
Studies, 1978 ASME-MPC Symposium on Properties of Steel
Weldments for Elevated Temperature Pressure Containment
Applications, ASME-MPC-9, pp. 47-61, 1978.

HIGH-TEMPERATURE STATIC FATIGUE IN CERAMICS

R.N. Katz, G.D. Quinn and E.M. Lenoe

Army Materials and Mechanics Research Center
Watertown, Massachusetts 02172

I INTRODUCTION

During the past decade, the growing utilization of high-performance ceramic materials in energy conversion and conservation applications, optical communications systems, and metal working applications has generated much work on fatigue of ceramics. Kossowsky (1) has shown cyclic fatigue effects in hot-pressed silicon nitride at temperatures between 1000°C and 1200°C, while no fatigue behavior was observed at low temperatures, i.e., 200°C. Seaton and Katz (2) have shown that acoustic fatigue caused strength degradation in TiB_2, whereas acoustic loading of the same magnitude did not affect hot-pressed silicon nitride. Quinn (3) has demonstrated that thermal fatigue of hot-pressed SiC becomes significant between 500 and 10,000 thermal excursions to 1,370°C.

Ceramics can also fail from static fatigue. Static fatigue can occur via several mechanisms and has been observed even under normal ambient conditions. Consider the classic ceramic static fatigue behavior for Al_2O_3 and B_4C shown in Figure 1 (4). Such behavior may be due to stress corrosion cracking caused by the action of water vapor in the environment on a glass in the ceramics grain boundary. Such stress corrosion type of static fatigue behavior in glasses has come into recent prominence for predicting the life of fiber optic wave guides and vitreous bonded grinding wheels (5). Our laboratory has focussed upon static fatigue of high-temperature ceramics which are candidate materials for heat engine applications.

221

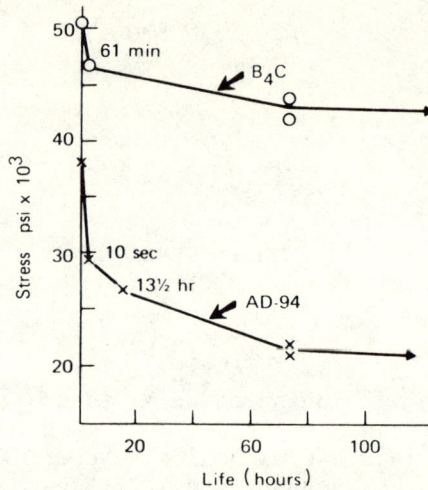

Fig. 1.   Static Fatigue of Ceramic Material (after Acquiviva
          and Chait, 1972).

High-performance ceramics based on the nitrides and carbides
of silicon are being actively evaluated in a large number of heat
engine and industrial heat exchanger demonstration programs.   In
most of these applications the ceramic components will be exposed
to high temperatures, in oxidizing environments, for times ranging
from a few thousand to several tens of thousands of hours.   For
successful application, it is essential that the designer have a
full understanding of the time, temperature, and stress dependence of the strength (and/or retained strength after environmental
exposure) for these materials.   Unfortunately, the available data
on silicon nitride and carbide ceramics are very sparse.   Recognizing the need for a more extensive data base in this
important area, the U.S. Department of Energy funded our laboratory and others to address the problem.   This paper presents the
methodology used and some of the principal results obtained by
our laboratory during the past four years.   More detailed results
are presented in references 3, 6, 7, and 8.

One of the major reasons for the scarcity of such data was
(and remains) the extremely rapid rate of materials development
and improvement for these nitrides and carbides during the past
decade.   Since major improvements in materials and materials
processing were occurring approximately every 6 to 12 months, there
were problems with the economic costs of carrying out extensive
testing on each materials variant, and also problems in obtaining
sufficient samples for tests due to the developmental nature of
many of the materials.   Therefore, a screening test was required to
survey the time, temperature, and stress behavior of a material
which would use relatively few specimens.   This lead to the
development of the stepped-temperature stress-rupture (STSR) test

described below. For materials which showed promise upon screening and for which sufficient material was available, conventional
stress-rupture (static fatigue) testing was carried out. Such
tests can be used to determine the time of survival of a material
at a given temperature, stress, and stressed volume. Alternatively,
assuming slow crack growth, key parameters for life prediction
can be extracted by the appropriate mathematical treatment of the
stress rupture data (9). A third area of concern has been to determine if, or to what degree, combined high-temperature oxidative
exposure and thermal cycling will reduce the retained strength of
the candidate materials. A discussion of these three areas of
study follows.

II THE STEPPED-TEMPERATURE RUPTURE TEST

The STSR test is an extension of the common flexural
stress-rupture test to include a range of temperatures. One
typical temperature cycle chosen for our requirements (and which
conveniently fits into the usual laboratory work week) is illustrated in Figure 2. A specimen is loaded into a furnace equipped
with a four-point bend fixture (spans 3.8 x 1.9 cm; see Ref. 3
for details) and the furnace is heated to $1000^{o}C$ in air with no
load applied to the sample. At $1000^{o}C$ a deadweight load is
applied. Should the sample survive 24 hours at that temperature,
the furnace is then heated (in $\cong$ 1/2 h) to $1100^{o}C$ and again allowed
to soak for 24 hours. This cycle is repeated for $1200^{o}C$, $1300^{o}C$,
and $1400^{o}C$, but in the last case, the soak is maintained for 60
hours. Throughout the test, the sample is subjected to a constant
deadweight load. If a sample breaks, the furnace is cooled and
unloaded and the time of failure is denoted by an arrow on the
STSR plot. The arrow is labeled with the stress that was applied
to the sample. A series of tests was executed with differing loads
corresponding to stress levels calculated from the elastic beam
formula.

For example, as illustrated in Figure 2, a sample loaded to
486 MPa survived the $1000^{o}C$ soak, but failed at 10 hours into the
$1100^{o}C$ soak. Another sample, loaded to 324 MPa, survived $1000^{o}C$
and $1200^{o}C$ soaks, but failed during heatup to $1300^{o}C$. As a last
example, one sample survived the entire heat cycle, while sus-
taining a stress of 35.1 MPa.

Although the temperature history of an STSR specimen is more
complex than that in a conventional stress-rupture test, the key
point is that any unusual temperature sensitivity will be identified quickly with a minimum number of specimens. Figure 2 illustrates a "normal", well-behaved material, i.e., time-dependent
failures occur at lower stresses as one increases temperature
and the transitions occur rather gradually as one goes from one

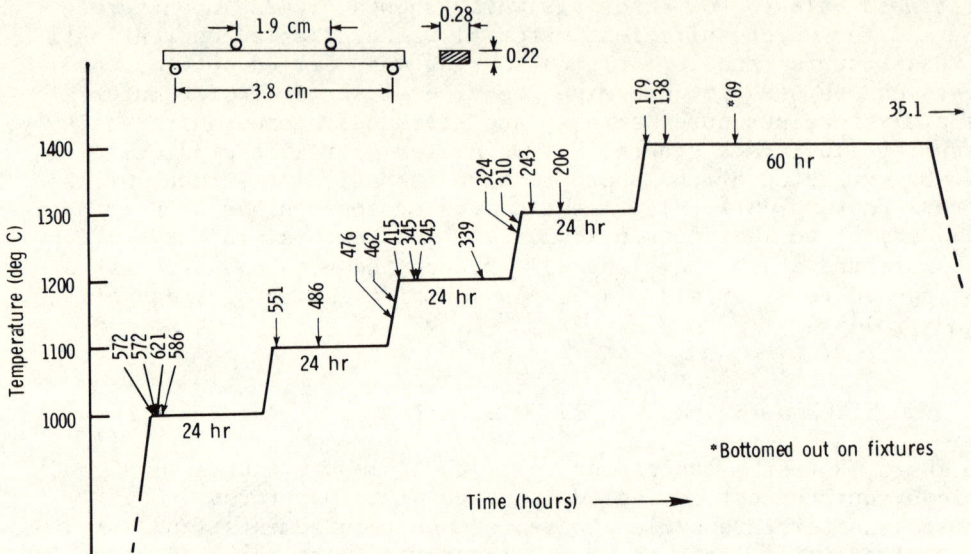

Fig. 2.   Stepped Temperature Stress Rupture Data for NC 132 HPSN.
           Arrows Indicate Failure Times, Arrow Labels Specify Stress
           on Sample in MN/m^2 (KSI).

temperature to the next. Such behavior is not always observed.
One striking example is hot-pressed Si_3N_4 with 13% Y_2O_3. The Y_2O_3
additive is effective in promoting sintering yet minimizing the
high-temperature creep that is characteristic of the MgO additive
grades of Si_3N_4 (10). Initial test results were promising; however, a catastrophic instability at 1000°C was identified for some
compositions (11-12). The 13% Y_2O_3 material was susceptible to
this instability (11) and, as a result, the manufacturer withdrew
it from the market. Nevertheless, it was chosen for the present
study to verify the ability of the STSR test to isolate such areas
of instability.

Conventional stress-rupture tests of 13% Y_2O_3 material at 1200°C
revealed very little time-dependent behavior: samples loaded to
500 MPa tended to break during loading or survive hundreds of hours
without failure (at which point the specimen was unloaded intact).
Thus one would normally expect that this material would be out-
standing at loads below 500 MPa and temperatures below 1200°C. The
STSR results shown in Figure 3 illustrate otherwise. Of eight
samples tested, seven failed in the 1000°C range at stresses below
350 MPa. The 69 MPa sample failed at 19.7 hours and had gross
secondary cracks through the entire sample. All these samples would
likely have survived the 1200°C stress-rupture trials for hundreds
of hours. The STSR trials succeeded in identifying the unstable
temperature regime in this case.

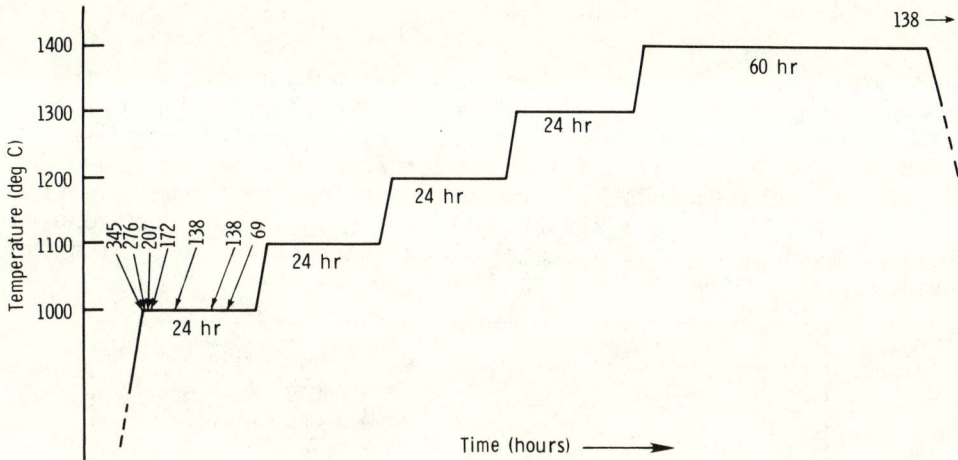

Fig. 3.   Stepped Temperature Stress Rupture Data for Hot-Pressed
           Silicon Nitride with Yttria Additive.

Evaluation of three grades of RBSN (ρ 2.53 to 2.77 g/cc) indicated
that the most critical temperatures from a time-dependent strength
standpoint are the 1000°C to 1100°C and 1300°C to 1400°C ranges (8).
Time dependence in the 1000°C to 1100°C range has been attributed
to oxidation (13) and in the 1100°C to 1300°C range to creep (14).
Again the STSR test quickly isolated the critical temperature and
stress ranges where time-dependent behavior occurred even though
different mechanisms were involved.

III   STRESS-RUPTURE TESTS

As previously stated, S-R testing enables one, in many cases, to
evaluate materials at temperatures, stresses, and stressed volumes
approximating those to be encountered in service.   Our laboratory
is presently engaged in such a materials evaluation of four candidate
silicon-based ceramics for application as erosion-resistant trailing
edges in a small radial gas turbine nozzle.   S-R testing was carried
out to times in excess of 1000 hours.   As a result of these tests,
three materials were judged acceptable from the standpoint of time-
dependent strength retention.   Final materials selection will be
made on the basis of erosion and cost data.   Such systems-specific
applications of S-R testing are relatively unusual for Si_3N_4 or SiC
ceramics at this point in time.

A more general use of S-R testing is to obtain data for life
prediction calculations (15).   The reciprocal slope of the S-R curve
(in log-log form) is the exponent n in the slow crack growth power
law:

$$V = AK^n \tag{1}$$

where V is the crack growth rate, A is a constant, and K is the stress intensity factor. S-R data at 1200°C obtained on NC 132 hot-pressed Si_3N_4 by various investigators (3, 16-18) with different lots of material and differing test geometries is shown in Figure 4. The data and the value of n obtained by these investigators (n ≅ 9 to 13) are signficantly more consistent than results obtained in double torsion tests or variable strain rate strength testing (n ≅ 5 to 100).

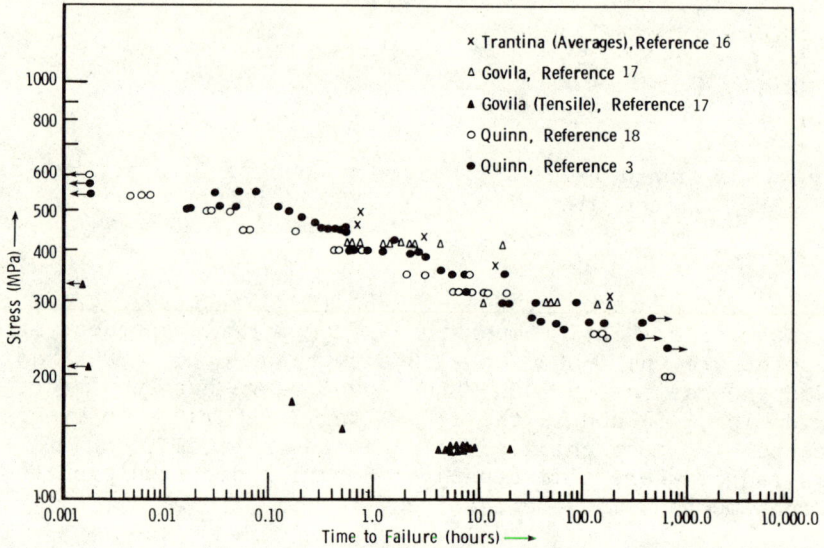

Fig. 4.   Stress Rupture Data for NC 132 at 1200°C.

Because of the above considerations, we have concentrated on the use of S-R testing to generate our life prediction data base. Figures 5 through 7 summarize some of the S-R data obtained on Si_3N_4 and SiC in our laboratory during the past four years.

IV   COMBINED-EXPOSURE/THERMAL CYCLING EFFECTS

Most silicon nitrides and carbides exhibit a decrease in RT strength after exposure to oxidation in the 1000°C to 1400°C range for several hundred hours (19). If static oxidation presents a problem, what will occur under cyclic oxidation? Work is currently underway at AMMRC (8), AiResearch (20), Volkswagen (21), and elsewhere to study effects on retained strength after cyclic exposure of engine ceramics at high temperatures, in oxidizing environments, and with large members of thermal cycles. Benn and Carruthers at AiResearch (20) utilize a combustor rig with an

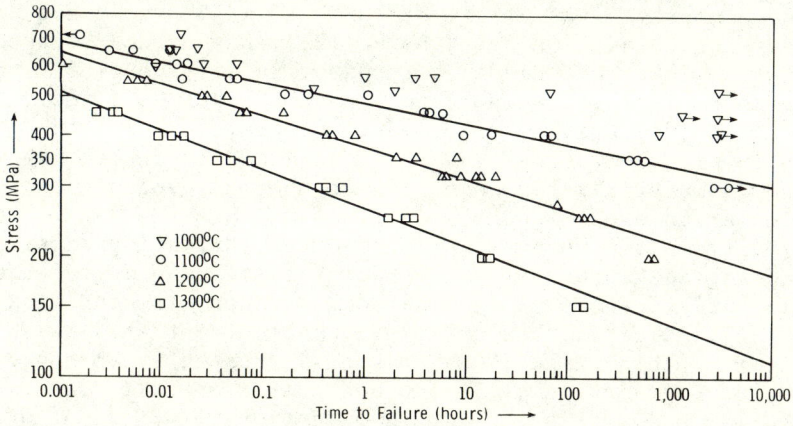

Fig. 5.   Stress-Rupture of NC 132 (HPSN) as a Function of Temperature.

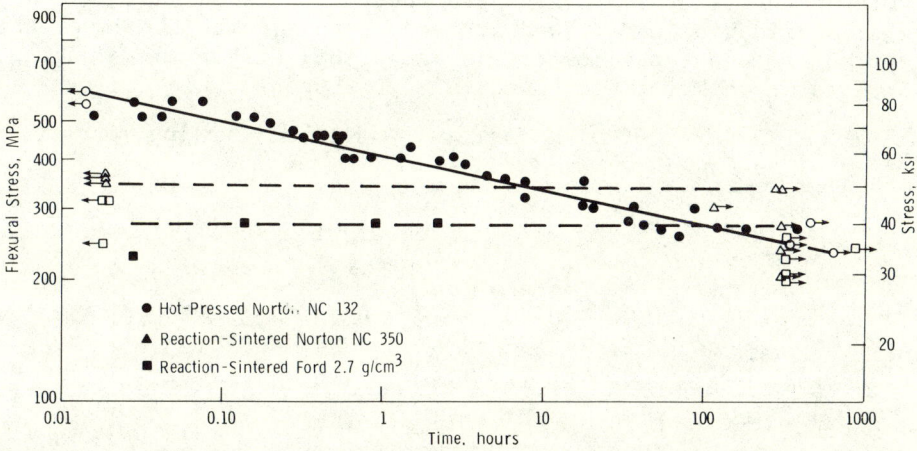

Fig. 6.   Flexural Stress Rupture at 1200°C for Three Silicon Nitrides.

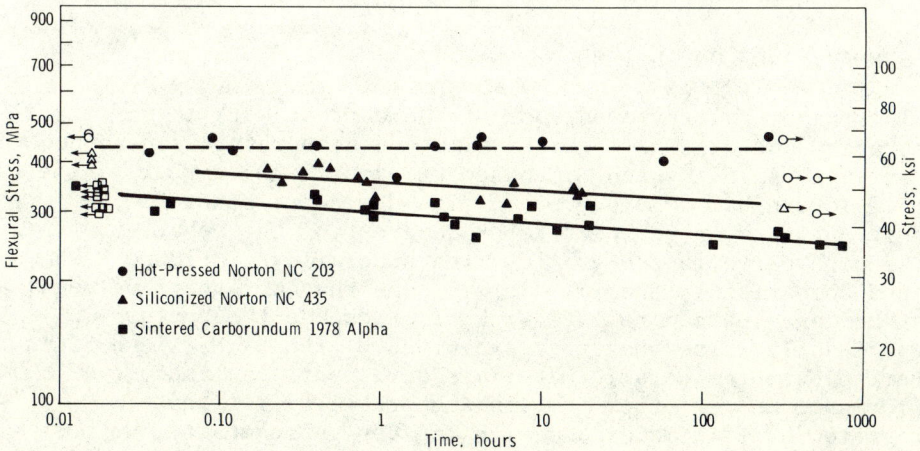

Fig. 7.   Flexural Stress Rupture at 1200°C for Three Silicon Carbides.

oxidizing combustion gas at temperatures to 1375°C combined with
cyclic air quenching. Quinn at AMMRC (3,6) has used a stepped
temperature oxidation cycle in a furnace, in air, coupled with
multiple thermal shocks accumulated in a MAPP gas/O_2 flame.
Siebels at VW (21) utilizes stepped temperature cycling in a
furnace, in air, between RT and 900°C to 1260°C. Table 1 shows
the range of materials response to such thermal cycling studies.
Of particular interest is the comparison of the RBSN data. In
spite of different manufacturing procedures, differing densities,
and differing test procedures, the percent strength degradation
for similar times falls in the 15% to 20% range. Recent data of
Siebels (21) indicates that newer grades of RBSN tested at VW now
also fall into this range (RBSN previously tested at VW showed
much greater degradation). It appears that a mutually consistent
data base may be emerging on cyclic oxidation of RBSN. At present,
multiple data bases do no exist for the response of other engine
ceramics to cyclic oxidation. Also specimen data on the effect
of cyclic oxidation under load are nonexistent, with the important
exception of recent cyclic tests of an actual stator in an engine
test rig at Ford Motor Co., under NASA/DOE sponsorship (22).

Table 1. Combined Thermal Exposure - Thermal Cycling Tests of
Engine Ceramics to 2500°F.

Material	Virgin MOR (ksi)	Exposed MOR (ksi)	% Change in MOR	Laboratory and Test Condition
NC-132 HPSN	104	50.5	-51	
NC-203 HP SiC	99	102	+3	
NC-350 RBSN	43	35	-19	AMMRC - 360 hour/500 cycles in air and flame [3]
KBI-RBSN	30	24	-20	
Silcomp - Si/SiC	47	32	-32	
Ford - RBSN	42	36	-15	
ACC RBSN-101	37.4	29.4	-21	AiResearch - 350 hour/1700 cycles in combustor gas [20]
Sintered α-SiC	45.8	45.9	0	

V   SUMMARY AND CONCLUSIONS

This paper has reviewed some of the current work in our
laboratory and elsewhere on the time and temperature dependence
of strength in Si_3N_4 and SiC-based ceramics. While most of these
materials exhibit some degree of time-dependent behavior, we
should put this in perspective by contrasting the behavior of
these high-performance ceramics with superalloys. Figure 8 shows
such a comparison. One immediately sees that designers of metal
systems have learned to live with time-dependent behavior
significantly worse than that exhibited by the Si_3N_4's and SiC's.
Nevertheless, the ceramic materials developers can improve
performance and in order to gain acceptability for these
alternate, brittle materials with the design community, we must
do so.

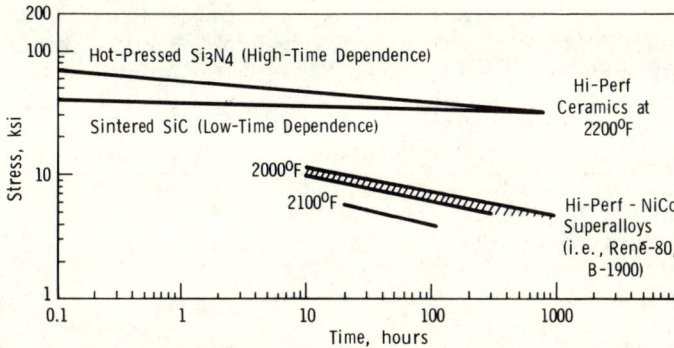

Fig. 8.   Stress-Rupture Behavior of Hi-Performance Ceramics
Compared to Hi-Performance Superalloys.

The major conclusions are as follows:

1)   The STSR test has proven its value as a method of screening
for time-dependent behavior in a short time with few specimens.

2)   The S-R test is highly reproducible from laboratory to
laboratory and test configuration to test configuration. This
reproducibility manifests itself in the consistency of the slow
crack growth exponent n.

3)   The limited data obtained so far shows the importance of
combined oxidative exposure/thermal cycling tests. It is important
that such testing be extended to include tests with an applied
load, in contrast to the tests described above, which were carried
out without a load.

Many of the concepts introduced in this paper regarding the use
of static fatigue (S-R) testing as opposed to alternative techniques
are amplified in a paper by Quinn and Quinn (18) subsequent to
this paper, and in a comprehensive review of static fatigue in
SiC and Si_3N_4 ceramics (23).

ACKNOWLEDGMENT

The authors would like to acknowledge the U.S. Department of
Energy, which provided partial support for this work under DOE/AMMRC
Interagency Agreement E (49-28)-1017.

REFERENCES

1.   R. Kossowsky, "Creep and Fatigue of Si_3N_4 as Related to Micro-
structures" in Ceramics for High Performance Applications, ed.
J.J. Burke, A.E. Gorum, and R.N. Katz, Brook Hill Publ., Chestnut
Hill, MA. pp. 347-371, (1974).

2. C.C. Seaton, and R.N. Katz, J. Amer. Ceram. Soc.,56 (1973) 283

3. G.D. Quinn,"Characterization of Turbine Ceramics After Long-Term Environmental Exposure", AMMRC TR 80-15, April 1980.

4. S.J. Acquaviva, and R. Chait, "Static and Cyclic Fatigue of Ceramic Materials",AMMRC TR72-9 (1972).

5. Reliability of Ceramics for Heat Engine Applications, National Materials Advisory Board, Report #NMAB-357 (1980).

6. G.D. Quinn, R.N. Katz, and E.M. Lenoe, Proceedings of the DARPA/NAVSEA Ceramic Gas Turbine Demonstration Engine Program Review MCIC 78-36, August 1977, 715-737.

7. G.D. Quinn and R.N. Katz, Am. Cer. Soc. Bull., 57 (1978) 1057-8.

8. G.D. Quinn and R.N. Katz, J. Am. Cer. Soc., 63 (1980), 117-119.

9. K. Jakus, and J.E. Ritter, J. Am. Cer. Soc., 63 (1980), 117-119.

10. G.E. Gazza, J. Am. Cer. Soc., 56 (1973) 662.

11. A.F. McLean, E.A. Fisher and R.J. Bratton, "Brittle Materials Design, High Temperature Gas Turbine", AMMRC CTR 75-28 (1975)138-39.

12. F.F. Lange, S.C. Singhal, and R.C. Kuznicki, J. Am. Cer. Soc., 60 (1977) 249-52.

13. A.F. McLean, E.A. Fischer and R.J. Bratton, Ford Motor Co., Detroit, MI, "Brittle Materials Design, High Temperature Gas Turbine", AMMRC CTR 75-8, April 1975, Pg. 81-84.

14. U. Din and P.S. Nickolson, J. Am. Cer. Soc., 58 (1975) 500-502

15. R. Davidge, J. McLaren, and G. Tappin, J. Mat. Sci, 8 (1973) 1699-1705.

16. G.G. Trantina, J. Am. Cer. Soc. 62 (1979) 377-380.

17. R. Govilla, "Ceramic Life Prediction Parameters", AMMRC TR 80-18, May 1980.

18. G.D. Quinn and J.B. Quinn, "Slow Crack Growth in Hot-Pressed Silicon Nitride", to be published in Fracture Mechanics of Ceramics 6, Plenum Press, New York, 1982.

19. D.G. Miller, C.A. Anderson, S.C. Singhal, F.F. Lange, E.S. Diaz and R. Kossowsky, Westinghouse Electric Corp., Pittsburgh, PA.,"Brittle Materials Design, High Temperature Gas Turbine - Materials Technology", AMMRC CTR 76-32, Vol. 4, December 1976.

20. K.W. Benn and W.D. Carruthers, "3500 Hour Durability Testing of Commercial Ceramic Materials", 5th Interim Report on Contract DEN 3-27, June 15, 1979.

21. J.E. Siebels "Oxidation and Strength of Silicon Nitride and Silicon Carbide", Presented at the 6th Army Technology Conference: Ceramics for High Performance Applications-III, Orcas Island, WA July 1979.

22. W. Trela, "Status of the Ford Program to Evaluate Ceramics for One-Piece Stator Applications in Gas Turbine Engines" in proceedings of the 17th DOE, Highway Vehicle Systems Contractors Coordination Meeting, DOE Conf. 791082, October 1979.

23. G.D. Quinn, "Review of Static Fatigue in Silicon Nitride and Silicon Carbide", to be published in Ceramic Proceedings, 1982.

ENVIRONMENT, FREQUENCY AND TEMPERATURE EFFECTS ON FATIGUE IN

ENGINEERING PLASTICS

R.W. Hertzberg and J.A. Manson

Materials Research Center
Lehigh University
Bethlehem, Pennsylvania

INTRODUCTION

In recognition of the trend toward lighter and higher strength structures, there has been growing interest in the commercial application of plastics in numerous engineering systems. These include ground and air transport vehicles and gas and water pipelines. In addition, engineering plastics have found a place in the manufacture of such exotic components as prosthetic devices. To be sure, the usage of these polymeric materials cannot precede a proper evaluation of such mechanical properties as tensile and yield strength, ductility, toughness, creep and fatigue response. At the same time, it is necessary that these mechanical properties be examined as a function of major test variables such as environment, temperature and test frequency; these variables take on major importance in polymeric solids owing to their viscoelastic character.[1] In this brief overview, we shall attempt to highlight some theoretical and experimental results that bear on the impact of test environment, temperature and cyclic frequency on polymer fatigue behavior. In doing so, we intend to emphasize that changes in these variables affect the viscoelastic character of the polymer material itself as well as any possible surface-environment rate-activated processes.

EARLIER STUDIES

When unnotched polymer samples are load-cycled in a reversed bending mode, typical S-N curves[2] can be generated which depict the relationship between cyclic stress amplitude and cycles to failure (Fig. 1). At first glance it would appear that a material such as an epoxy resin possessed superior overall fatigue resistance to that

231

of, say, nylon 66 and polycarbonate. In fact, the results shown
in Fig. 1 are weighted heavily by the larger number of cycles necessary to initiate a crack in the epoxy resin under the test conditions associated with these experiments (unnotched specimens with
the test frequency set at 30 Hz). When crack propagation rates are
compared, however,[1, ch.3] the relative ranking of these materials
with regard to fatigue resistance is reversed (Fig. 2); fatigue
cracks are seen to propagate most rapidly in the epoxy resin.
Furthermore, the relative ranking for total fatigue life in Fig. 1
is strongly influenced by the aforementioned test frequency. At
30 Hz, the amount of hysteretically induced heating is so great in
a number of polymers that they literally melt during the course of
the test. Since the epoxy resin exhibits little damping, this test
specimen does not heat up. Therefore, the results in Fig. 1 reflect
a mixture of mechanically and thermally-induced damage mechanisms.
Were the test frequency to be lowered to preclude thermal-type failures, the relative fatigue resistance ranking of the materials shown
in Fig. 1 would surely be different.

Ferry analyzed the circumstances surrounding thermal fatigue
failures and was able to describe the energy dissipation rate for a
specified stress range by

$$\dot{E} = \pi f j'' (f,T) \sigma^2/2 \tag{1}$$

where $\dot{E}$ is the energy dissipation rate per unit time, f, the frequency, J'' is the loss compliance, and σ the peak shear stress.
(In tension, the loss tensile compliance, D'' replaces J''.) Others
have shown that fatigue life does, indeed, increase when the test
frequency in decreased.[3] It is interesting to note that the temperature rise in the fatigue sample often increases in an autoac-
celerating fashion such that the damaging high temperatures are
experienced only near the end of the test.[3,4] As such, the fatigue
life of a material susceptible to hysteretic heating could be
improved by conducting the test in an interrupted fashion with rest
periods allowing for cooling of the sample. For this reason, estimates of fatigue life of samples subjected to block loading are not
amenable to analysis using the Miner-Palmgren analysis procedure.

FATIGUE CRACK PROPAGATION STUDIES

The effect of test frequency on fatigue crack propagation (FCP)
in polymeric solids provides a dramatic comparison with the results
described above (e.g., recall Eq. 1). Some materials such as polycarbonate reveal no FCP sensitivity to frequency[5] while poly(vinyl
chloride) (PVC) reveals a clear-cut decrease in FCP rates[6] with
increasing test frequency (Fig. 3a,b). Several attempts have been
made to rationalize these data as described elsewhere.[1, ch.3] One
such correlation that appears to hold considerable promise involves
a relationship between the frequency sensitivity factor (FSF) for a

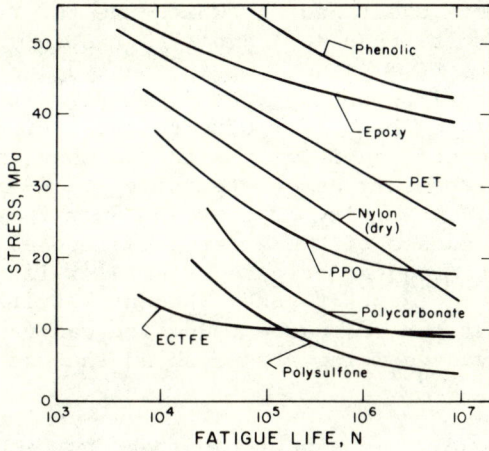

Fig. 1. Representative stress–log cycle life plots for numerous
         engineering plastics [Riddell (2)].

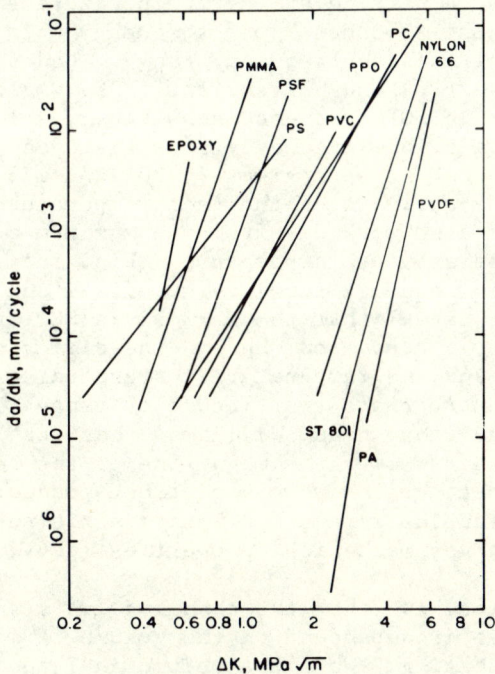

Fig. 2. Fatigue data for several polymeric solids. Note superior
         fatigue resistance in crystalline polymers (Nylon 66,
         ST801, PA, PVDF). Epoxy (A/E = 1.0), PMMA (M_V = 1.25 x 10^6),
         PSF (M_V = 5 x 10^4), PS (M_V = 2.7 x 10^5), PVC (M_w = 2.25 x 10^5),
         PPO, PC (M_V = 4.9 x 10^4), Nylon 66 (M_w = 17,000, 2.2% H_2O),
         ST801 (0% H_2O), PA (M_n = 7 x 10^4), PVDF (M_w = 2.2 x 10^5)
         [Hertzberg and Manson (1)].

given polymer (FSF = the factor by which the FCP rate changes per decade change in test frequency) and the frequency of movement of main chain segments (i.e., the jump frequency) responsible for generating the principal secondary transition peak (β peak) at a common test temperature[5,7,8] (Fig. 4). Of particular significance, the greatest FCP frequency sensitivity occurred when the test frequency was close to the frequency associated with the β-process. That is, maximum frequency sensitivity of these polymers was associated with a condition of resonance of the externally imposed test machine frequency with the frequency of the material's internal segmental motion corresponding to the β-peak. Recent results with nylon 66 have suggested that the resonance condition can be generalized to refer to any relevant damping peak such as the one associated with the glass transition.[9]

From Fig. 4 one would expect the frequency sensitivity of poly-carbonate and polysulfone to increase were it possible to conduct fatigue tests at frequencies in the range of 10^6 Hz. This point cannot be verified due to test machine limitations. Instead, it is possible to choose a particular test temperature for each material that would bring the β-process jump frequency into resonance with the allowable test machine frequency range. Under such conditions, both polycarbonate (PC) and polysulfone (PSF) would be expected to exhibit a maximum in FCP frequency sensitivity. Indeed, PC and PSF which revealed negligible frequency sensitivity at 23°K, each exhibited a maximum in FSF at approximately 200 and 175°K, respectively[7] (Fig. 5a,b). Conversely, when the test temperature in PMMA was reduced, FSF decreased markedly from its maximum corresponding to tests performed at ambient temperature[7] (Fig. 5c). Of particular significance, the frequency sensitivity of all engineering plastics examined thus far reveal that the maximum in FCP-frequency sensitivity corresponds to a condition wherein the difference between the test temperature and the temperature corresponding to the β damping peak within the appropriate test frequency range is minimized[8] (Fig. 5d). On the other hand, Williams[10] has concluded that the magnitude of FSF depends on the magnitude of the relevant damping peak. In both instances, however, a strong connection has been established between the fine scale viscoelastic response of the polymer and the gross mechanical properties (i.e., FCP rates).

The authors believe that the changing importance of frequency in polymer fatigue is dependent on the volume of material associated with hysteretic heating. When the volume is large as in the case of an unnotched sample, increased frequency will lead to heating of the entire gage section of the specimen and associated loss of stiffness. Fatigue life would be expected to suffer (recall Eq. 1). If, on the other hand, the heated zone is localized near the crack tip, then the crack tip would be expected to blunt. A lower effective crack tip stress intensity factor would result along with lower fatigue crack propagation rates. If little or no hysteretic heating were

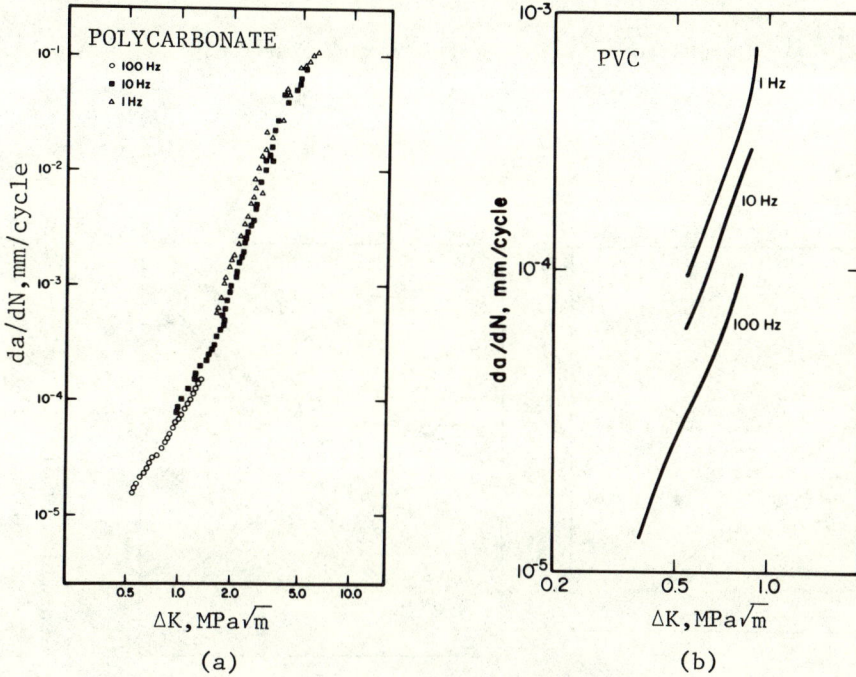

Fig. 3. Effect of cyclic frequency on FCP rates in a) polycar-
         bonate and b) poly(vinyl chloride) [Skibo et al. (6,7)].

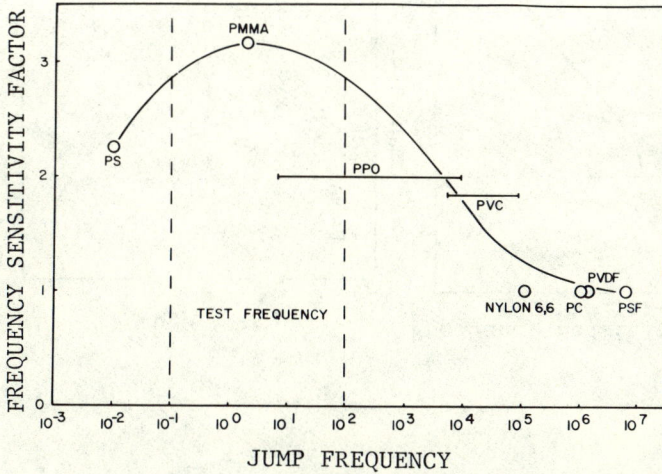

Fig. 4. Relationship between FCP frequency and the room tempera-
         ture jump frequency for several polymers [Hertzberg,
         Manson, Skibo (7)].

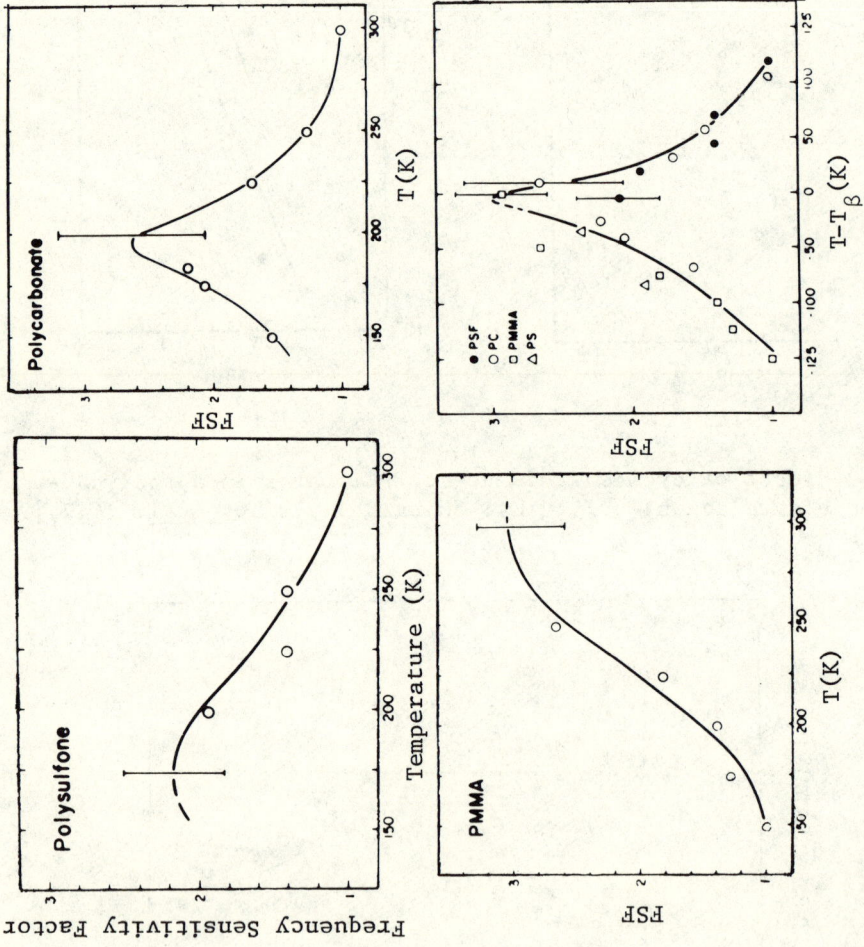

Fig. 5. Effect of temperature on frequency sensitivity in (a) polysulfone, (b) polycarbonate, and (c) poly(methylmethacrylate) [Skibo, Hertzberg and Manson (7)]; (d) frequency sensitivity factor relative to normalized β-transition temperature, $T - T_\beta$ [Hertzberg, Manson and Skibo (8)].

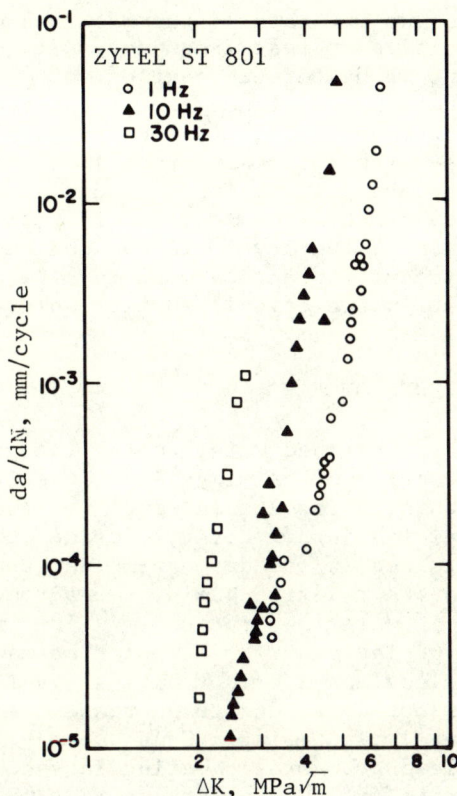

Fig. 6.   Effect of cyclic frequency on FCP rates in impact-
          modified nylon 66 [Skibo et al. (11)].

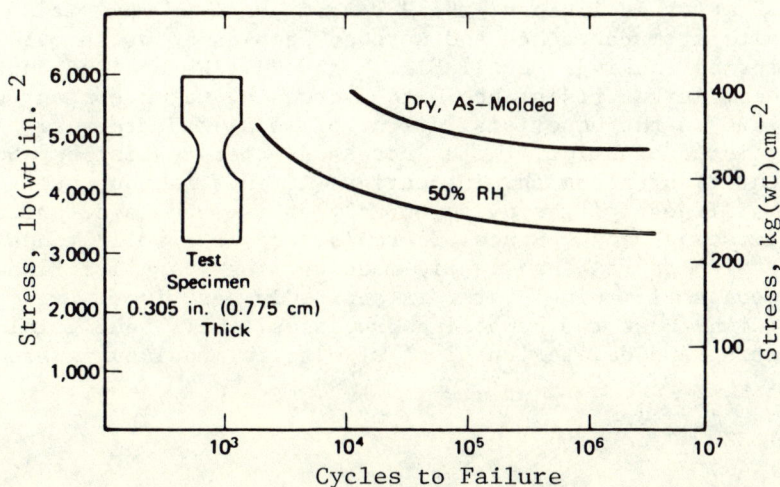

Fig. 7.   Flexural fatigue life of nylon 66 as-molded and equili-
          brated to 50% relative humidity [Kohan (12)].

to take place, then the FCP rate of the material would not be expected to be sensitive to test frequency. It should be pointed out that if the zone of hysteretic heating in a precracked sample extended to an appreciable distance from the crack tip, then negative FSF would be expected. That is, one would predict that FCP rates would increase with increasing test frequency. Indeed, this was found to be the case[11] in an impact-modified nylon 66 blend (Fig. 6). Large scale heating in excess of $100^{\circ}C$ was observed in this material. Clearly, the key to an evaluation of the frequency sensitivity in engineering plastics must involve a comparison between the volume of hysteretically-heated material and the volume of the test sample.

ENVIRONMENTAL INFLUENCE ON FCP

The influence of absorbed water on the fatigue response of nylon 66 (N66) is considered briefly here since it provides another useful comparison between results based on unnotched S-N test results and FCP experimental findings. First, consider the S-N curve for N66 (Fig. 7) when tested in the dry, as-molded condition versus N66 equilibrated with a 50% relative humidity environment (corresponding to approximately 2.5% water absorption).[12] The presence of 2.5% water in the nylon 66 (arising from a polar attraction between the nylon 66 and H_2O molecules) does not have a significant effect on the elastic modulus but does tend to decrease the yield strength and increase the level of damping. Consequently, the higher level of damping should lead to overall heating in the unnotched sample and result in inferior fatigue resistance (Fig. 7). When the FCP resistance of these materials is examined,[13,14] a different set of results is found. Figure 8a shows that the crack growth rate resistance of nylon 66 increases initially with absorbed water. Note that when 2.5% water is imbibed, the FCP resistance is maximized. Again, the results from unnotched and notched samples do not provide the same material ranking (recall Fig. 1 and 2). We believe that the improvement in FCP resistance with increasing water content up to 2.5% is due to the beneficial effect of localized crack tip heating and associated blunting. With increased water content beyond 2.5% the extent of blunting should continue to increase but now the overall stiffness of the nylon 66 polymer is seen to deteriorate at an accelerating pace. Since FCP resistance in a solid tends to decrease with decreasing elastic modulus, the overall influence of large amounts of imbibed water is decidedly negative. The competition between these two competing processes (i.e., beneficial crack tip blunting and deleterious loss of elastic modulus) is seen in Fig. 8b.

CONCLUSIONS

The fatigue crack propagation response of engineering polymers is seen to be sensitive to test temperature, frequency and environ-

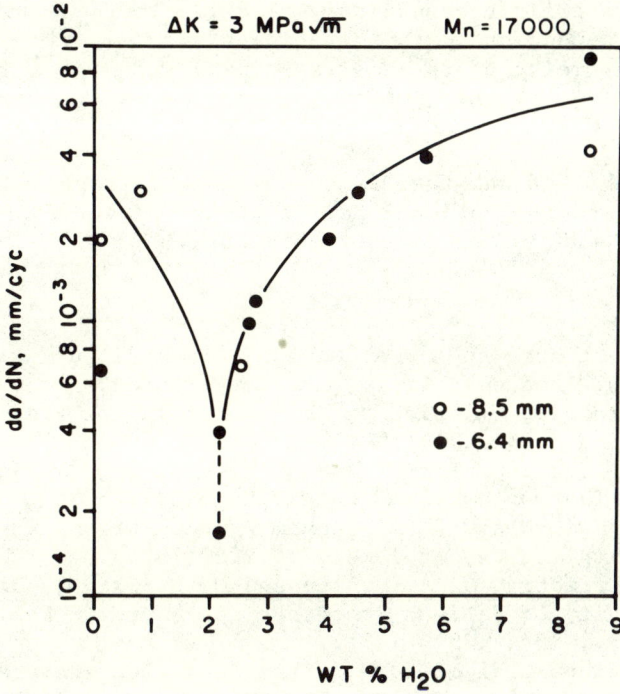

Fig. 8a.   FCP response of nylon 66 as a function of moisture
content in two material supplies (0, 8.5 mm; ●, 6.4
mm) at $\Delta K = 3$ MPa $\cdot$ m$^{\frac{1}{2}}$ [Bretz (14)].

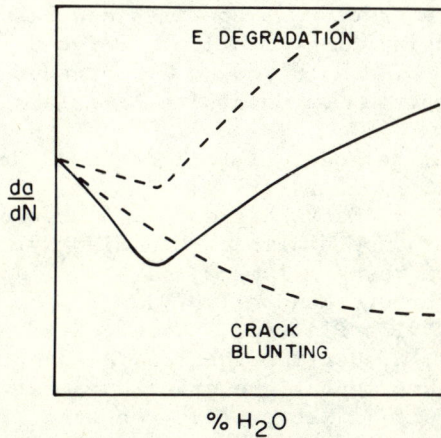

Fig. 8b.   Schematic diagram of the effect of crack blunting and
modulus degradation on crack-growth rates in nylon 66
as a function of moisture content (13).

ment.   The specific response of a material to these variables is
found to be dependent on the material's viscoelastic properties
and on the geometry of the test specimen.

ACKNOWLEDGEMENTS

    The authors thank the Office of Naval Research and the Poly-
mers Program of the National Science Foundation (Grant No. DMR-
8106489) for partial support of this work.

REFERENCES

1. R. W. Hertzberg and J. A. Manson, "Fatigue in Engineering
    Plastics," Academic Press, New York (1980).
2. M. N. Riddell, "A Guide to Better Testing of Plastics,"
    Plast. Eng. 30(4):71 (1974).
3. M. N. Riddell, G. P. Koo and J. L. O'Toole, "Fatigue Mechan-
    isms of Thermoplastics," Polym. Eng. Sci. 6:363 (1966).
4. R. J. Crawford and P. P. Benham, "Some Fatigue Characteristics
    of Thermoplastics," Polymer, 16:908 (1975).
5. R. W. Hertzberg, J. A. Manson and M. D. Skibo, "Frequency
    Sensitivity of Fatigue Processes in Polymeric Solids," Polym.
    Eng. Sci., 15(4):252 (1975).
6. M. D. Skibo, "The Effect of Frequency, Temperature and Mate-
    rials Structure on Fatigue Crack Propagation in Polymers,
    Ph.D. Dissertation, Lehigh Univ. (1977).
7. M. D. Skibo, R. W. Hertzberg and J. A. Manson, "The Effect of
    Temperature on the Frequency Sensitivity of Fatigue Crack
    Propagation in Polymers," Proceedings 4th Int'l. Conf. Fracture
    1977, Vol. 3 ICF4, p. 1127, Waterloo, Canada (1977).
8. R. W. Hertzberg, J. A. Manson and M. D. Skibo, A Correlation
    Between Fatigue Fracture Properties and Viscoelastic Damping
    Response in Engineering Plastics," Polymer, 19(3):358 (1978).
9. R. W. Lang, J. A. Manson and R. W. Hertzberg, Proceedings--
    U.S.-Italy Symposium on Composite Materials, Capri, June 1981,
    in press.
10. J. G. Williams, "A Model of Fatigue Crack Growth in Polymers,"
    J. Mater. Sci., 12:2525 (1977).
11. M. D. Skibo, R. W. Hertzberg and J. A. Manson, "Fatigue Crack
    Propagation Response of Impact-Modified Nylon 66," Deformation,
    Yield and Fracture of Polymers, p. 4.1, Plastics & Rubber
    Institute, Cambridge-London (1979).
12. M. I. Kohan, "Nylon Plastics," Wiley, New York (1973).
13. P. E. Bretz, R. W. Hertzberg and J. A. Manson, "Influence of
    absorbed moisture on fatigue crack propagation behaviour in
    polyamides. Part 1: Macroscopic response," J. Mat. Sci.,
    16:2061 (1981).
14. P. E. Bretz, R. W. Hertzberg, J. A. Manson and A. Ramirez,
    "Effect of Moisture on Fatigue Crack Propagation in Nylon 66,
    in "Water in Polymers," S. P. Rowland, ed., ACS Symposium
    Series 127, Paper 32, p. 531 (1980).

CREEP-FATIGUE EFFECTS IN STRUCTURAL MATERIALS USED IN

ADVANCED NUCLEAR POWER GENERATING SYSTEMS

C. R. Brinkman

Metals and Ceramics Division
Oak Ridge National Laboratory*
Oak Ridge, Tennessee   37830

ABSTRACT

We review various aspects of time-dependent fatigue behavior
of a number of structural alloys in use or planned for use in
advanced nuclear power generating systems. Materials included are
types 304 and 316 stainless steel, Fe-2 1/4 Cr-1 Mo steel, and
alloy 800H. Examples of environmental effects, including both
chemical and physical interaction, are presented for a number of
environments. The environments discussed are high-purity liquid
sodium, high vacuum, air, impure helium, and irradiation damage,
including internal helium bubble generation.

INTRODUCTION

Time-dependent fatigue behavior of structural steels proposed
for use in advanced nuclear power generating systems has been under
investigation for approximately 15 years in the United States.
However, while a cursory understanding of the complex mechanisms
involved has been achieved, the ability to extrapolate relatively
short-term uniaxial laboratory-generated data with confidence to
design times of 30 to 40 years is not yet possible. Therefore,
our objective is to review state-of-the-art understanding of time-
dependent fatigue behavior of several structural alloys by using
recently generated data and published results. These alloys are
being characterized for use in a number of nuclear fission and

*Operated by Union Carbide Corporation under contract W-7405-eng-26
with the U.S. Department of Energy.

fusion devices. Ongoing efforts in data generation for model development purposes will also be briefly discussed.

The materials to be considered will be those associated with the American Society of Mechanical Engineers (ASME) Boiler and Pressure Vessel Code, Case N-47 and are as follows: types 316 and 304 stainless steel, Fe-2 1/4 Cr-1 Mo steel, and alloy 800H.

Specific examples used here to show environmental interaction and the complexities of time-dependent fatigue involve liquid sodium, high vacuum, air, impure helium, and irradiation damage.

ENVIRONMENTAL INTERACTION AND TIME-DEPENDENT FATIGUE IN FAST BREEDER REACTOR SYSTEMS

High-purity liquid sodium is the heat transfer medium or core coolant for many planned as well as operating fast breeder reactor systems. In part this results from its excellent heat transfer characteristics and lack of deleterious effects on the mechanical properties of structural steels, provided that the oxygen[1] and carbon contents,[2] as well as other liquid-metal embrittling elements such as lead, tin, and antimony, are carefully controlled.[3] However, factors such as composition of the individual structural alloy, exposure temperature and time, etc. must also be taken into account.[4]

In addition to the chemical effects, the physical effects of sodium, such as good heat transfer properties, must be considered. The effects of reactor trips as well as sodium streams that are at different temperatures mix while impinging upon component surfaces and produce local differences in temperatures. These effects may produce time-dependent or independent fatigue damage depending upon their frequency. Reactor trips (power changes, startups, or shutdowns) are relatively infrequent, perhaps with up to 1000 occurring over the design lifetime of the plant. However, large changes in stress level can occur with resultant low-cycle creep-fatigue damage possible. Impingement of sodium at different temperatures or thermal striping on the surface of above-core structural components or at mixing tees produces a possible high-cycle fatigue (10^9 cycles) problem. In liquid-metal fast breeder reactor systems these combinations of high- or low-cycle time-dependent fatigue interactions are addressed by particular care in design, by proper selection of materials, and by extensive material characterization programs.

Austenitic Stainless Steels

Types 304 and 316 austenitic stainless steels are used extensively as vessel, piping, core support, and heat exchanger materials in fast breeder reactor systems.[5] These components can accrue

creep-fatigue damage as a consequence of exposure to thermal transients, as discussed above. For these reasons a long-term creep-
fatigue program has been and continues to be under way at Argonne
National Laboratory. The objective of this program is to generate
data on types 304 and 316 stainless steel from specimens with
failure times ranging to three years. Figure 1 summarizes much of
the data currently available from this program plotted in a "t-N_f"
or log total test time-vs-log cycles to failure diagram for various
strain ranges and tensile hold times.[6] All the data came from
tests conducted in air and in strain control with the indicated
hold periods at the peak tensile strain point on the hysteresis
loop. The data show a continued degradation in fatigue life with
increasing duration of hold time with no indication of saturation
of the hold-time effect.

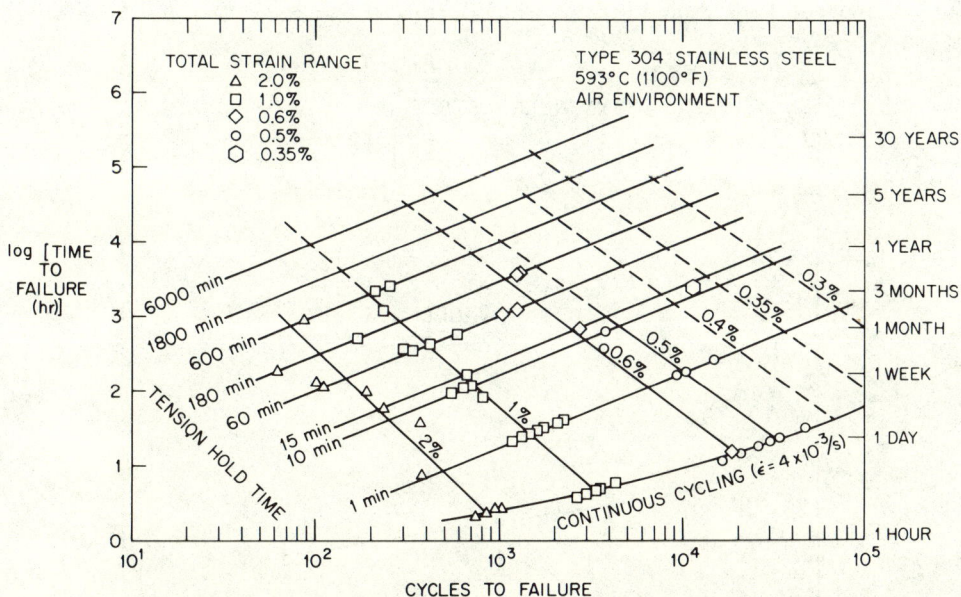

Fig. 1.   Time to failure vs cycles to failure showing the
           influence of tensile hold times on the fatigue life of
           type 304 stainless steel.   Source: C. R. Brinkman,
           V. K. Sikka, M. K. Booker, "An Overview of the U.S.
           Programs on Properties of Primary Circuit Materials,"
           pp. 13—23 in Specialist Meeting on Primary Circuit
           Materials Including Environmental Effects, IWGFR/22,
           International Atomic Energy Agency/International Working
           Group on Fast Reactors, Bergisch Gladbach, Federal
           Republic of Germany, October 17—21, 1977.

Environments that limit or prevent surface oxidation at high
temperatures typically result in marked improvements in the continuous cycle life of types 304 and 316 stainless steel. Comparison
of data[7] generated from specimens tested in high vacuum with air
data indicated an improvement in cycle life with factors ranging
from 3 to 5 at strain ranges from 0.5 to 2.0%. In sodium with 1 to
2 ppm oxygen a similar beneficial effect is noted in comparison
with data generated in air, particularly at low strain ranges. This
beneficial effect of sodium has tentatively been attributed to
improved crack nucleation resistance, as fewer surface cracks are
noted on post-test examination for specimens tested in sodium in
comparison with specimens subjected to identical test parameters
in air.[8] On the other hand, a decrease in crack growth rates is
also observed in compact tension specimens tested in sodium in
comparison with those similarly tested in air.[2]

Figure 2 compares strain-controlled low-cycle fatigue data for
type 304 stainless steel generated both with and without tensile
hold times at 593°C and at a single strain range. The data show

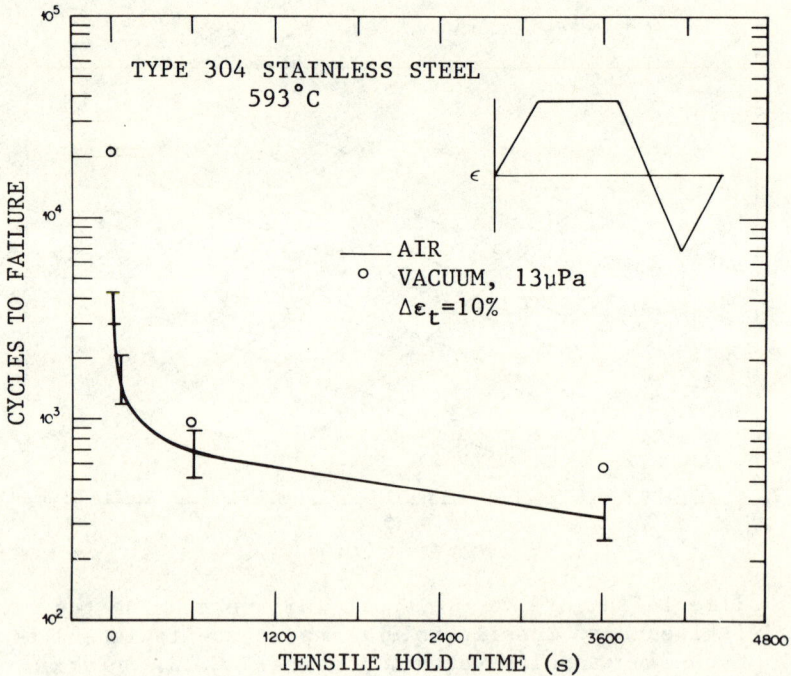

Fig. 2. Degradation of fatigue life resulting from tensile hold
         times becomes nearly the same as the length of the hold
         period increases for high vacuum and air environments.
         Source: P. S. Maiya, "Effects of Waveshape and Ultra-
         High Vacuum on Elevated Temperature Low-Cycle Fatigue in
         Type 304 Stainless Steel," submitted to Materials Science
         and Engineering.

that large differences exist for cycle life between continuous cycle data ($\dot{e}$ = 4 v 10^{-3}/s) generated in air and under high vacuum [1.3 v 10^{-6} Pa (10^{-8} torr)]. However, these differences tend to be minimal as the length of the tensile hold time increases. A similar conclusion has been reached in comparing limited results of strain-controlled tests conducted on type 316 stainless steel tested in high-purity sodium.[8] The significance of waveform in producing creep damage that leads to intergranular fracture in type 304 stainless steel[7] is shown in Fig. 3. Here the ratio of cycle life in vacuum to cycle life in air is plotted against plastic strain range. Tensile strain-controlled hold times or triangular "slow-fast" ramp rates lead to intergranular fracture with the result that high vacuum tends to be less important than when the waveform imposed leads to transgranular fatigue crack propagation. These results tend to demonstrate a true creep-fatigue effect for this material. That is, degradation in cycle life under conditions that produce considerable intergranular cavitation and crack propagation primarily results from creep damage rather than environmental interaction, as has been suggested by several investigators.[9,10] Sadananda et al.[11] similarly have concluded from crack growth studies conducted on several austenitic stainless steels tested in vacuum [1.3 v 10^{-4} Pa (10^{-6} torr)] and in air that enhanced crack growth under hold times primarily results from creep-fatigue interaction.

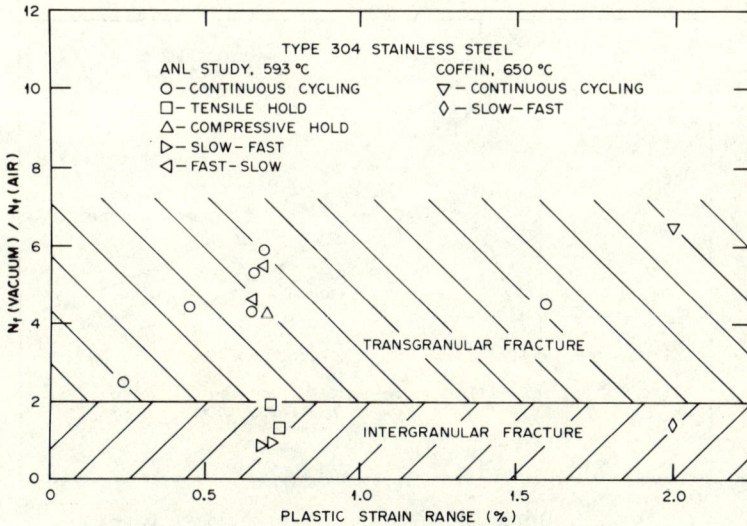

Fig. 3. Effect of waveshape and environment on fracture mode. Source: P. S. Maiya, "Effects of Waveshape and Ultra-High Vacuum on Elevated Temperature Low-Cycle Fatigue in Type 304 Stainless Steel," submitted to Materials Science and Engineering.

Recently investigators conducting strain–controlled exploratory fatigue tests on the austenitic stainless steels in both the
United States and Japan[12] have been imposing hold periods on the
hysteresis loop each cycle at locations other than peak tensile or
compression values. An example of this effort is shown in Fig. 4
for type 304 stainless steel tested at 650°C in air.[12] Here a
strain hold period of 0.17 h (10 min) for the example shown is
introduced at the same point on the hysteresis loop each cycle
until failure. These tests are being conducted to generate a data
base for model substantiation or development with hold periods
imposed at locations such as zero stress, as shown, or at points
on the hysteresis loop where little or no stress relaxation
occurs. Data generated in air of the type shown in Fig. 4 tend to
show the following:

    1. Tensile hold times at peak strain values are more damaging
than compression hold times of equal duration.

    2. Hold periods imposed at other locations on the hysteresis
loops, such as at zero stress or zero stress relaxation points,
degrade fatigue life but not as much as hold periods imposed at
peak tensile strain values.

    3. Hold periods imposed on the tension–going side of the loop
tend to be more deleterious than those imposed on the compression-
going side.

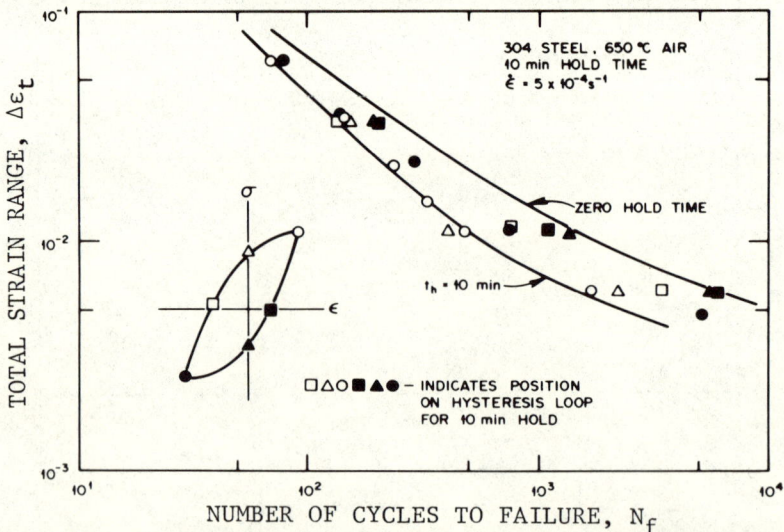

Fig. 4. Results of Japanese exploratory creep–fatigue tests.
         Source: Y. Asda and S. Mitsuhashi, "Creep–Fatigue Inter-
         action of 304 and 316 Stainless Steels in Air and Vacuum,"
         paper presented at Fourth International Conference on Press-
         sure Vessel Technology, London, England, May 19–23, 1980.

Another area of current interest for design of fast breeder reactor systems is fatigue crack propagation. Recent results have shown that a phenomena called "accelerated crack propagation" can occur for particular combinations of waveform, hold-time duration, and perhaps metallurgical state.[13] Figure 5 shows an example for type 316 stainless steel tested in the thermally aged condition before testing at 593°C. The data show that when intergranular

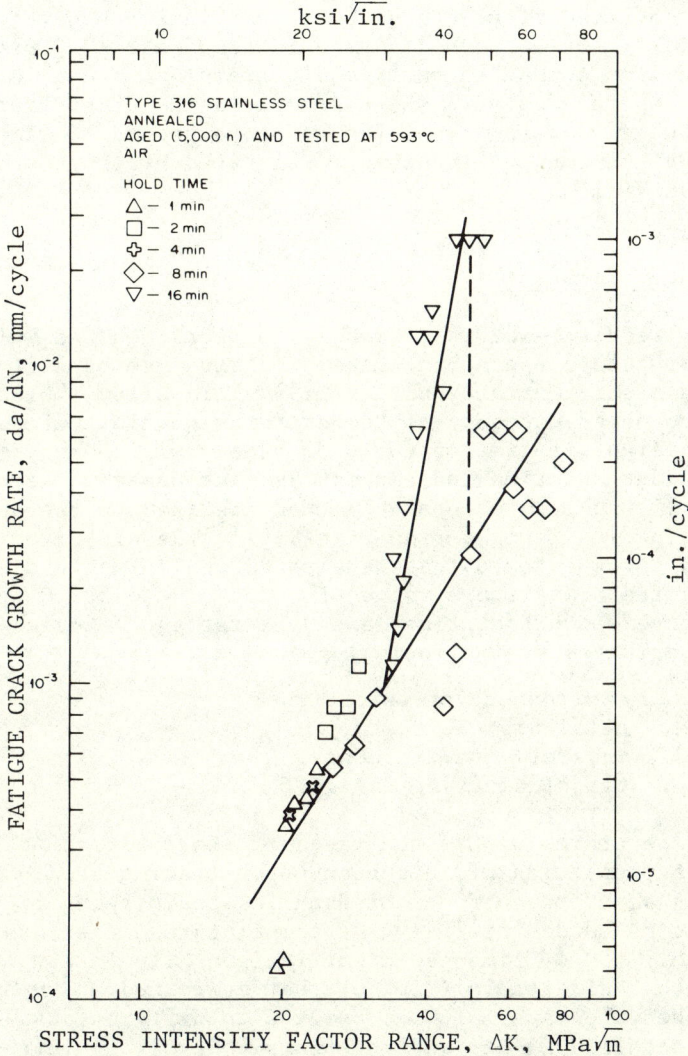

Fig. 5. Accelerated crack propagation in solution–annealed, thermally aged 316 stainless steel tested at 593°C. Source: D. J. Michel and H. H. Smith, "Accelerated Creep–Fatigue Crack Propagation in Thermally Aged Type 316 Stainless Steel," to be published in Acta. Metallurgica.

precipitate particles are present, combined static and dynamic
(zero to tension loading plus load-controlled) hold periods can
lead to higher crack propagation rates. For the example shown,
hold periods equal to or in excess of 0.27 h (16 min.) led to
accelerated crack propagation rates associated with intergranular
crack propagation. It is thought by some [13] that the increase in
crack propagation rate is related to a critical cavity size and
spacing associated with intergranular carbides. However, other
investigators have not noted such a relationship.[14] Further,
similar studies conducted on type 316 stainless steel in air at
elevated temperatures have shown some evidence of environmental
interaction.[14] Ongoing test efforts are expected to clarify the
role of environment in producing accelerated crack growth behavior
in this material.

Fe-2 1/4 Cr-1 Mo Steel

Fast breeder reactors presently in operation, as well as those
planned for future operation, make extensive use of 2 1/4 Cr-1 Mo
steel as a steam generator material.[15] This alloy, which will
undergo prolonged exposure at temperatures within the creep range
during a design lifetime of up to 30 years, will be subject to both
time-dependent and time-independent fatigue damage. Accordingly,
the material has been extensively characterized in the annealed
condition for its fatigue properties.[16,17] Results from strain-
controlled fatigue tests that were conducted in various environments over the temperature range of about 370 to 593°C on annealed
material have shown that time-dependent fatigue lifetime depends
upon the influence of the following:

1. environment or oxidation,
2. metallurgical state,
3. waveform and frequency,
4. classical creep damage.

At temperatures within the range of about 371 to 482°C, tensile
and fatigue properties are dependent upon testing strain rate, primarily because of the effects of dynamic strain aging or interaction
solid-solution hardening.[17,18] At temperatures in excess of
approximately 450°C, time-dependent fatigue life in air is dependent upon the oxide scale (Fig. 6) that is formed and upon its
behavior when the material is subjected to different waveforms.[18,19]
Test data obtained in air and in strain control with either tensile
or compressive hold times or with both tensile and compressive
hold periods introduced each cycle have shown the following trends
over the temperature range 427 to 593°C:

1. Compressive hold times are more damaging than tensile
holds, particularly at low strain ranges where resistance to crack
nucleation governs lifetime (Fig. 7).

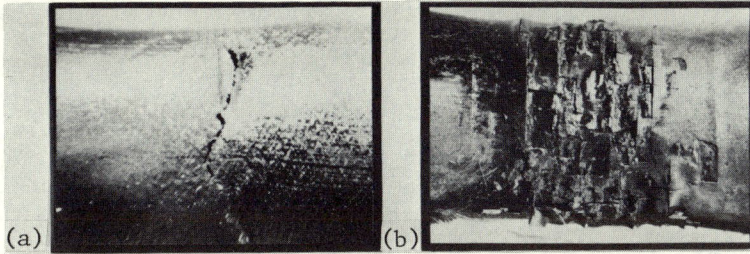

Fig. 6. A low-oxygen environment is a significant factor affecting
         elevated-temperature fatigue resistance. Surfaces of
         specimens tested in air show extensive oxidation compared
         with specimens tested in high-temperature gas-cooled
         reactor (HTGR) helium. (a) Impure Helium. (b) Air.

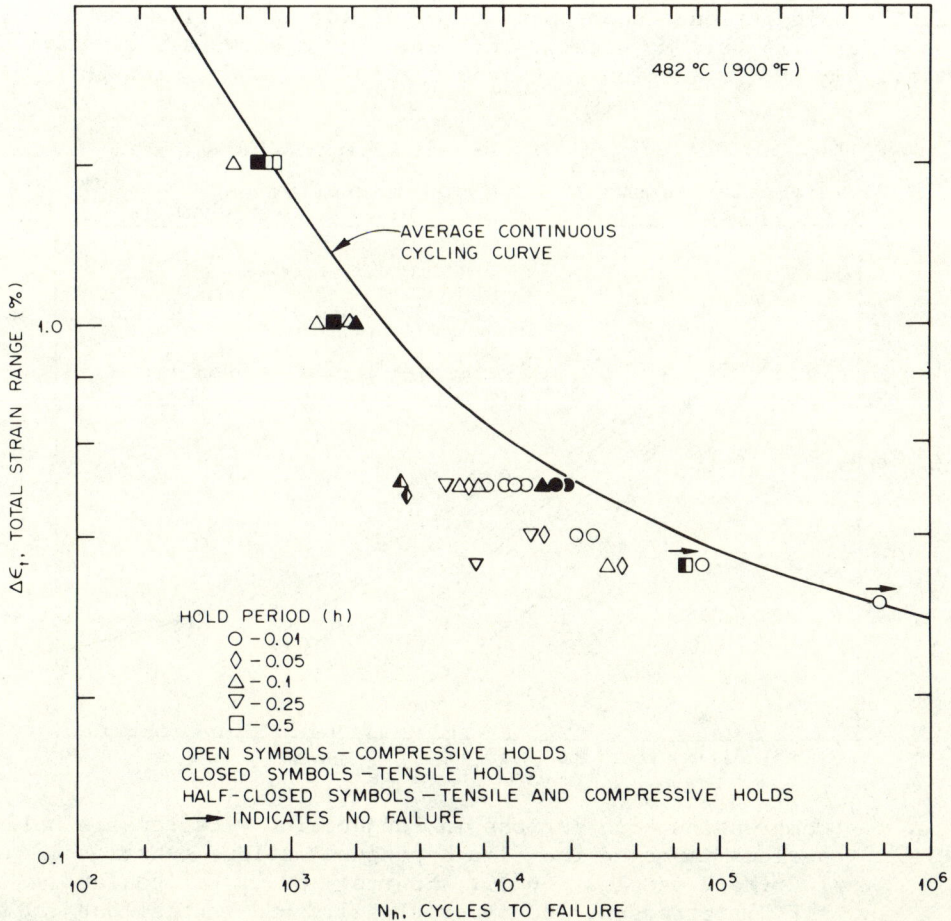

Fig. 7. Results of strain-controlled fatigue tests conducted on
         annealed 2 1/4 Cr-1 Mo steel with various hold periods.

2. A cycle with both a tensile and a compressive hold period
at low strain ranges may be more damaging than the cycle with only
a tensile or compressive hold with all hold periods of the same
duration (Fig. 7).

Not all low-alloy steels that have been tested to date
demonstrate this type behavior under cyclic and time-dependent
loading conditions. For example, a rotor steel (1 Cr-Mo-V) that
was extensively tested[20] indicated that tensile hold times were
most damaging. Furthermore, compression holds when introduced
into cycles that already contained tension holds were beneficial
in that cycle life was improved (Fig. 8).

When tests are conducted at 538°C or higher in environments
that limit or prevent oxidation [e.g., impure or high-temperature
gas-cooled reactor (HTGR) helium (Fig. 9) or sodium[21] (Fig. 10)]
tensile hold periods become more damaging for cycle life for
annealed 2 1/4 Cr-1 Mo steel. Further, when a slow-fast waveform
(i.e., 4×10^{-5}/s tension going and 4×10^{-3}/s compression going)

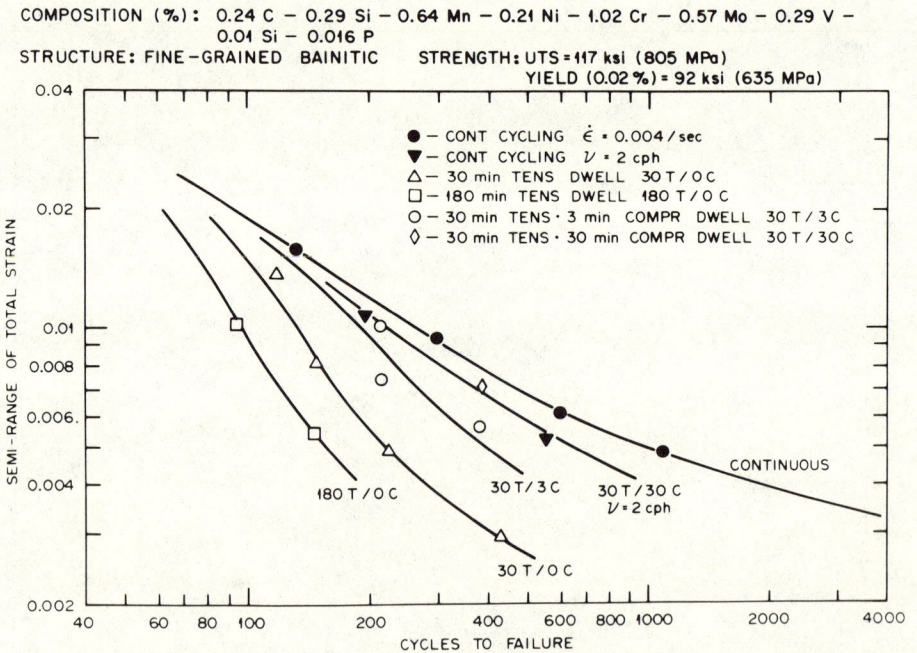

Fig. 8. Compression hold periods in conjunction with tensile hold
periods improved the time-dependent fatigue behavior of a
1 Cr-Mo-V steel at 565°C. Source: E. G. Ellison and
A.F.J. Patterson, "Behavior of a 1 Cr-Mo-V Steel Subject
to Combinations of Fatigue and Creep Under Stain Control,"
Proc. Inst. Mech. Eng. London 190: 333–40 (1976)

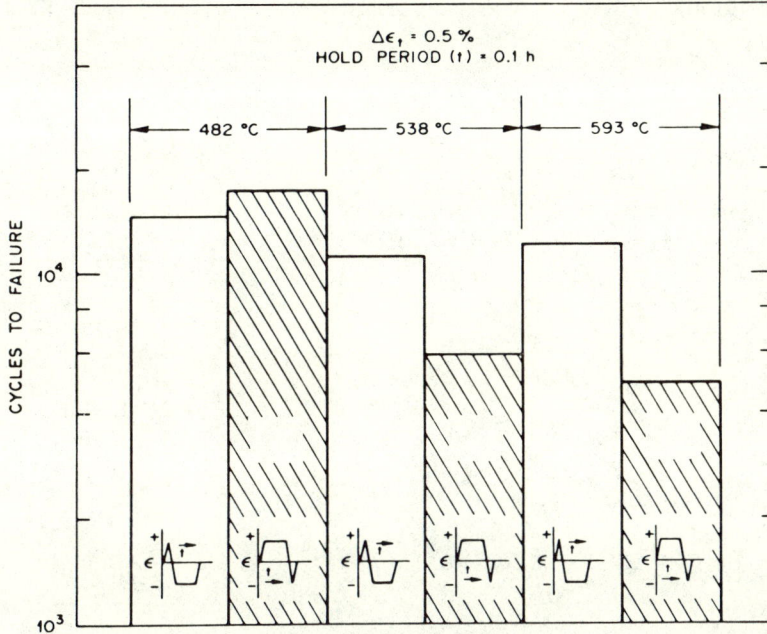

Fig. 9.   Tensile hold times appear to be more damaging than
          compressive hold times in impure helium at temperatures
          equal to or in excess of 538°C.

was employed in low-oxygen sodium environment tests, grain
boundary cavitation was seen on both the circumferential surfaces
(at 482 and 538°C) (Fig. 11) and within the bulk (at 538°C)
(Fig. 12) of the tested specimens, demonstrating classical creep
damage.

Exploratory time-dependent and strain-controlled tests similar
to those previously discussed for type 304 stainless steel have
also been conducted on annealed 2 1/4 Cr-1 Mo steel.   Figure 13
summarizes some of the results from these exploratory tests.   All
the tests were run in strain control at a single strain range and
temperature with a hold period introduced each cycle at a single
point on the hystersis loop, as shown.   The duration of the hold
period was either 0 or 0.1 h, as indicated.   Results of these tests
conducted in air demonstrate the following for the particular
strain range and temperature shown:

1.   Compression holds are more damaging than tensile holds
(6,111 vs 20,147 cycles to failure).

2.   In comparison with zero hold-time or continuous cycle
tests, tests conducted with a hold period introduced at zero stress
points show decreased fatigue life: the average cycle life for

CONTINUOUS CYCLE   FAST-SLOW CYCLE   SLOW-FAST CYCLE
        □   ◨   ■

$\dot{\varepsilon}=4\times10^{-3}s^{-1}$

$\dot{\varepsilon}=4\times10^{-3}s^{-1}$
$\dot{\varepsilon}=4\times10^{-5}s^{-1}$

$\dot{\varepsilon}=4\times10^{-5}s^{-1}$
$\dot{\varepsilon}=4\times10^{-3}s^{-1}$

CONTINUOUS CYCLING IN SODIUM

CONTINUOUS CYCLING IN AIR

TOTAL STRAIN RANGE, $\Delta\varepsilon_t$ (%)

CYCLES TO FAILURE

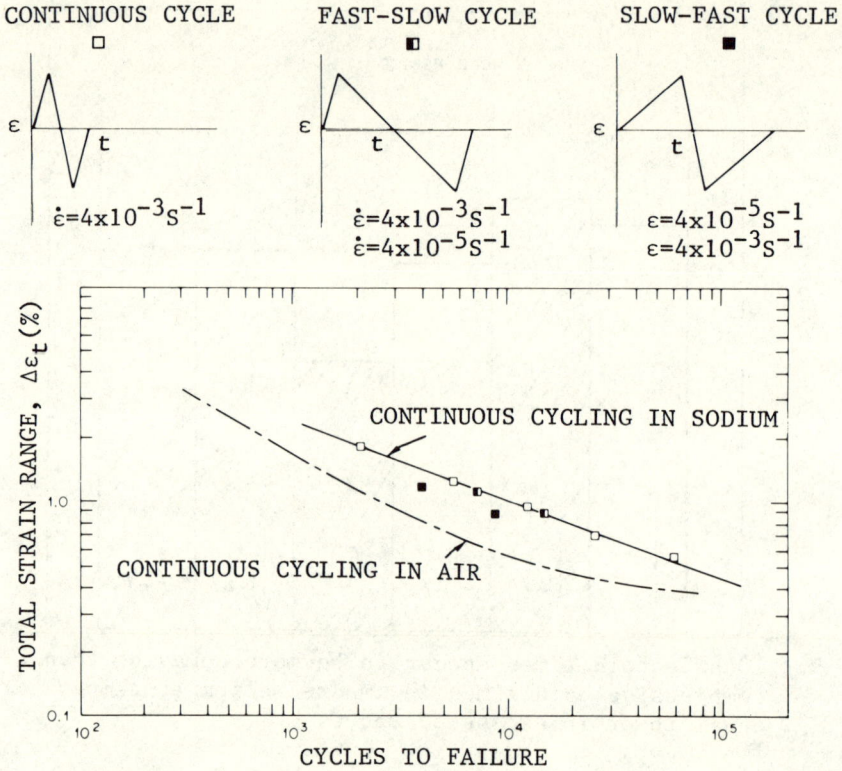

Fig. 10.   Waveform is an important variable only in the slow-fast
           cycle for 2 1/4 Cr-1 Mo steel tested at 538°C in sodium.
           Source: O. K. Chopra, K. Natesan, and T. F. Kassner,
           "Influence of Sodium Environment on the Low Cycle Fatigue
           and Creep-Fatigue Behavior of Fe-2 1/4 Cr-1 Mo Steel,"
           paper presented at Second International Conference on
           Liquid Metal Technology in Energy Production, Richland,
           Washington, April 10–24, 1980.

three specimens subjected to continuous cycling at a strain rate of
4 v 10^{-3}/s was 37,329 vs cycle lives of 6,317 and 15,557 for the
zero stress points shown.

    3. After continuous cycling at the indicated strain range,
it was possible to locate points on the hysteresis loop in both
tension and compression where little or no stress relaxation
occurred (Fig. 14).   A point in compression was at approximately
—190 MPa for the conditions indicated in Fig. 13, and the resultant
fatigue life was 11,667 cycles to failure. This test again shows
degradation in lifetime in comparison with the continuous cycle
life at these conditions (i.e., 11,667 vs 37,329 cycles to
failure).

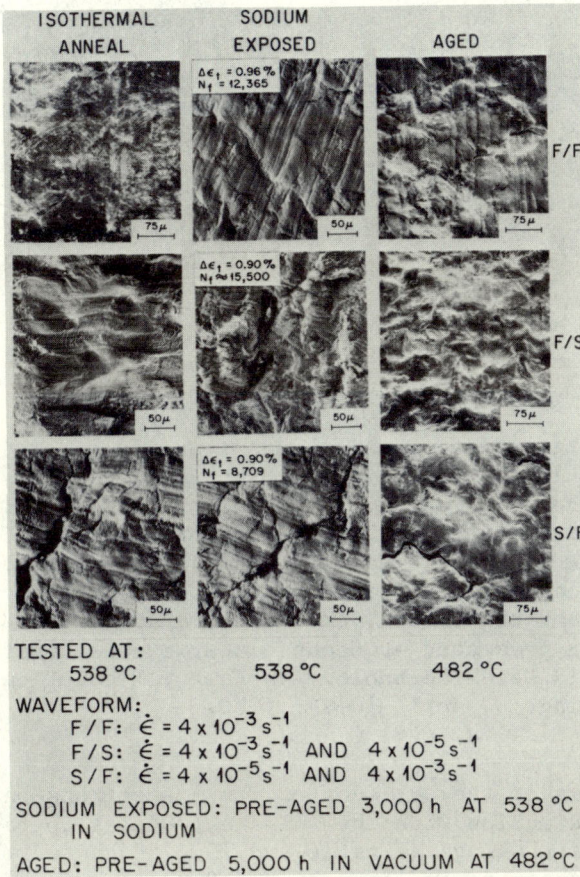

Fig. 11. Results of scanning electron microscope examination
of the circumferential surfaces of specimens of
Fe-2 1/4 Cr-1 Mo steel fatigue tested in sodium.
Note grain boundary cracks at boundaries perpendicular
to direction of applied stress for slow-fast (S/F)
waveform. Source: O. K. Chopra, K. Natesan, and
T. F. Kassner, "Influence of Sodium Environment on
the Low Cycle Fatigue and Creep-Fatigue Behavior of
Fe-2 1/4 Cr-1 Mo Steel," paper presented at Second
International Conference on Liquid Metal Technology
in Energy Production, Richland, Washington,
April 10–24, 1980.

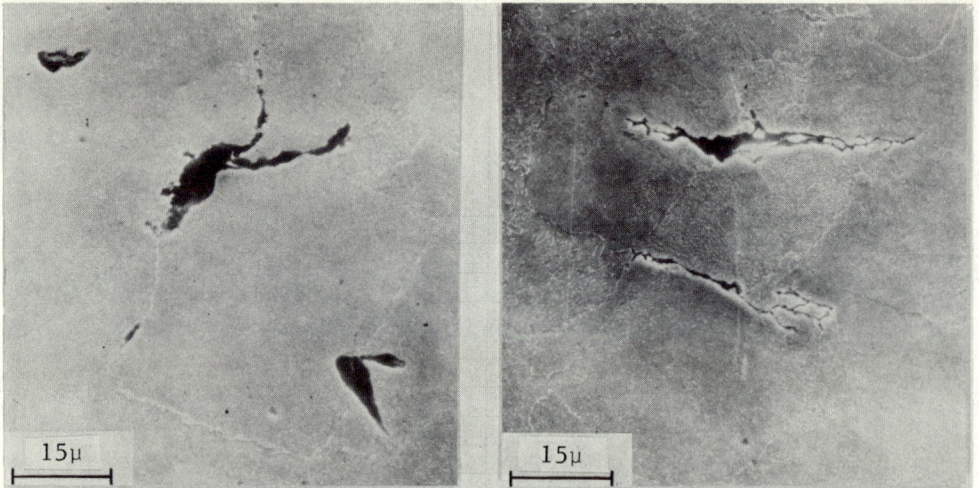

Fig. 12. Intergranular cavities were found in specimens of
          Fe-2 1/4 Cr-1 Mo steel subjected to slow-fast waveform
          at 538°C in sodium. Cavities of this type were not
          found in specimens similarly tested at 482°C. Source:
          O. K. Chopra, K. Natesan, and T. F. Kassner, "Influence
          of Sodium Environment on the Low Cycle Fatigue and
          Creep-Fatigue Behavior of Fe-2 1/4 Cr-1 Mo Steel,"
          paper presented at Second International Conference on
          Liquid Metal Technology in Energy Production, Richland,
          Washington, April 10—24, 1980.

     Examination of the surfaces of these specimens has shown
that oxide interaction with the surface produces characteristic
circumferential markings (Figs. 6 and 15), which are thought to
decrease the number of cycles required for crack initiation with
resultant reduction in cyclic life.[17-19] These markings are
absent on the surface of specimens tested in nonoxidizing environments (Figs. 11 and 16).

     It should also be noted that a protective environment
that limits or prevents oxidation also markedly decreases high-
temperature crack propagation rates in comparison with data
obtained in an air environment for this material.[22]

     Linear damage summation of time and cycle fractions has been
performed on the various tests shown in Fig. 13. The resultant
D_t or total damage summation values are all less than 1.00 and tend
not to sum to a unique value under strain- or load-controlled conditions, making linear damage summation a questionable method for
extrapolation, at least for air environments.

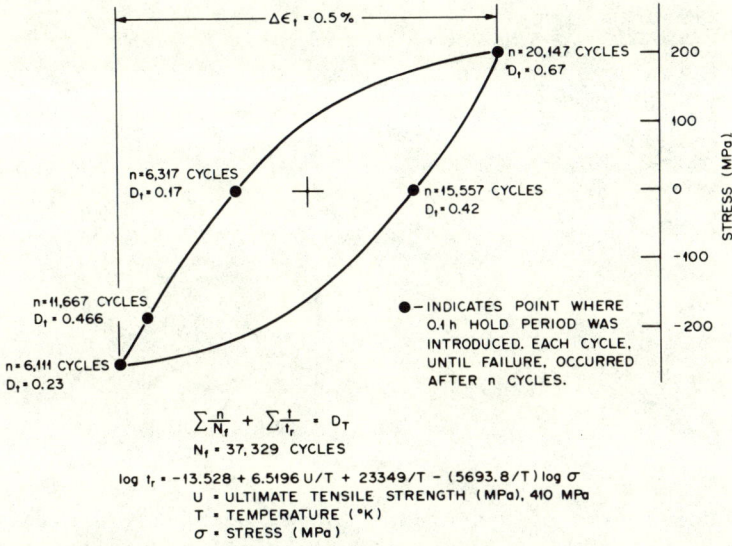

Fig. 13.   Time and cycle fraction damage analysis for annealed
          2 1/4 Cr-1 Mo steel heat 3P5601 tested at 482°C in air.

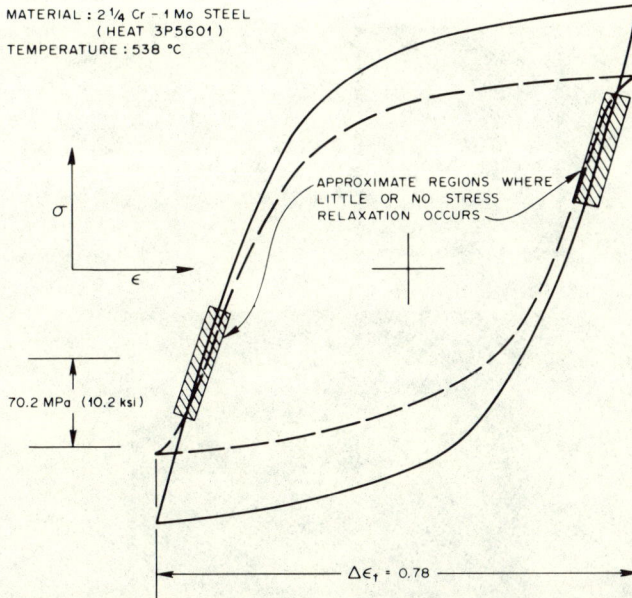

Fig. 14.   Dashed lines represent locus of stress relaxation points
          following 0.1-h strain hold periods from various posi-
          tions on the solid curve.   Intersecting points represent
          positions of zero stress relaxation.

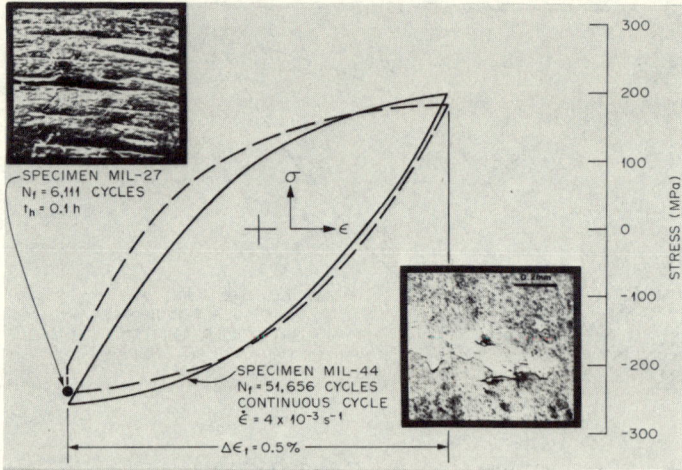

Fig. 15. Comparison of hysteresis loops, surfaces, and fatigue
lives of two specimens subjected to strain control cycling
at 482°C. Note that specimen MIL-36 had a 0.1-h hold
period introduced each cycle at peak compressive strain.

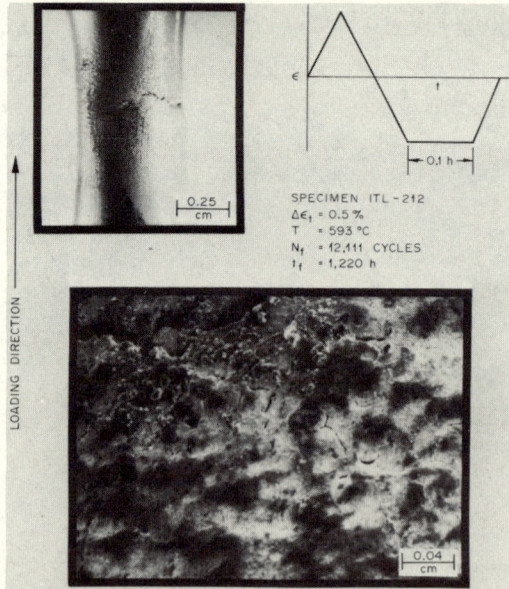

Fig. 16. The surface of a specimen of 2 1/4 Cr-1 Mo steel tested
in impure helium in strain control with a compression
dwell period each cycle.

ENVIRONMENTAL INTERACTION AND TIME-DEPENDENT FATIGUE
IN HIGH-TEMPERATURE GAS-COOLED REACTOR SYSTEMS

Alloy 800H

Gas-cooled nuclear reactors presently under consideration, in which the coolant or heat transfer medium is helium with low levels of impurities, also require design for the prevention of time-dependent fatigue. The helium environment may be oxidizing, reducing, carburizing, or decarburizing, depending on the alloy involved, the temperature, the moisture content, and the carbon potential of the gas relative to the carbon activity of the metal.[23] An example[24] of the effects of impure helium and air on the strain-controlled fatigue properties of alloy 800H tested at 650 and 760°C is given in Table 1. A comparison of the data shown follows.

Table 1. Impure Helium Environments that Can Produce
Carburization Appear to Degrade Time-Dependent
Fatigue Life of Alloy 800H Subjected
to Tensile Hold Times[a]

| Temperature (°C) | Strain Range (%) | Hold | | Cycles to Failure | |
		Mode	Time (min)	Impure Helium[b]	Air
650	0.4	0	0		>10^6
650	0.4	Tension	1	5,465	26,767
650	0.4	Tension	2.5	3,629	14,790
760	0.4	Tension	2.5	2,790	1,053
650	0.4	Compression	1	10,672	13,000
650	0.4	Compression	2.5	10,308	8,836
760	0.4	Compression	2.5	4,785	1,628

[a]Source: D. I. Roberts, S. N. Rosenwasser, and J. F. Watson, "Materials Selection for Gas-Cooled and Fusion Reactor Applications," paper 9 presented at Conference on Alloys for the 80's, Ann Arbor, Michigan, June 17—18, 1980.

[b]Helium composition, ppm: CH_4 = 50; CO = 50; H_2 = 500; H_2O = 1.

1. At low strain ranges compression hold periods are more damaging than tensile holds at 650°C in air, but the reverse appears to be true at 760°C.

2. Impure helium appears to cause a marked decrease in the fatigue life of specimens subjected to tensile hold times in comparison with similar tests conducted in air at 650°C. This is attributed to carburization.

There are indications that the magnitude of the hold-time effect in strain-controlled tests is dependent upon temperature and strain range as well as upon the magnitude and direction of any mean stress that is developed during a given test.

ENVIRONMENTAL INTERACTION AND TIME-DEPENDENT FATIGUE
IN FUSION REACTOR FIRST-WALL SYSTEMS

Cyclic thermal stresses will occur in first-wall and blanket materials of pulsed fusion power generating devices. Because of this, fatigue data are currently being generated on candidate materials, including cold-worked type 316 stainless steel, refractory alloys, and a number of martensitic low-alloy steels with chromium contents in the range 9 to 13%. Environments or possible coolants associated with the first-wall and blanket structure may be liquid metals, gas, molten salts, or water. In addition, the first wall of a fusion reactor will also be subjected to intense high-energy neutron irradiation damage, causing atom displacement damage as well as helium and hydrogen generation.

The influence of high irradiation-induced helium contents and displacement damage on resultant tensile and fatigue properties of 20%-cold-worked (20% reduction in area by swaging) type 316 stainless steel is currently being investigated.[25] A plot showing the effects of high helium content, etc. on the resultant strain-controlled fatigue properties of type 316 stainless steel is given in Fig. 17. Note the high scatter in the irradiated data and the resultant overall degradation in fatigue life, particularly at the low-cycle end of the curve. Tensile ultimate strengths and reduction of area values were approximately 614 MPa and 36%, respectively, in comparison with unirradiated values of 650 MPa and 60% near the indicated temperature and irradiation conditions. Figure 18 is a transmission electron micrograph showing extensive helium bubbles within the microstructure. At higher temperature, helium bubbles at grain boundaries are thought to be particularly deleterious because of the probability of increased intergranular crack propagation rates under cyclic loading conditions.[26] Work dealing with the influence of irradiation on time-dependent fatigue properties, data analysis, and extrapolations has been published elsewhere.[27]

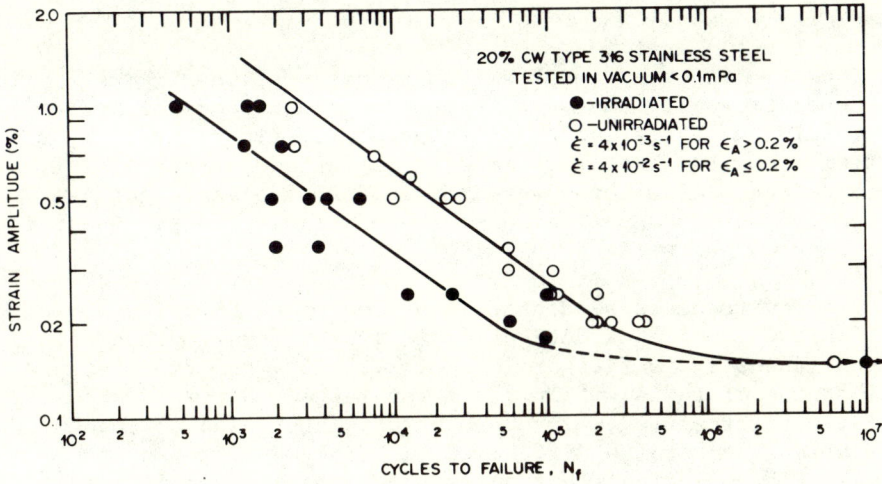

Fig. 17. Low-cycle fatigue life is reduced by factors of 3 to 10 for 200 to 1000 ppm He and 5 to 15 dpa at 430°C, and indicated endurance limit is at a strain amplitude of 0.15%. Source: M. L. Grossbeck and K. C. Liu, "Fatigue Behavior of Type 316 Stainless Steel Following Neutron Irradiation Inducing Helium," paper presented at American Nuclear Society meeting, La Vegas, Nevada, June 8–13, 1980.

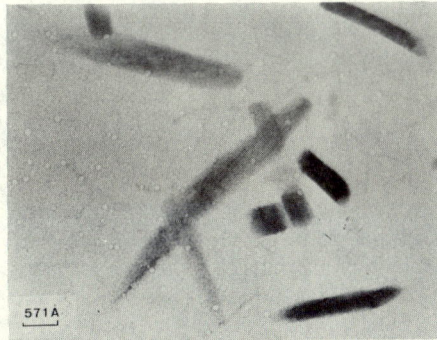

Fig. 18. Microstructure of 20%-cold-worked type 316 stainless steel irradiated in the High Flux Isotope Reactor (HFIR) for 2,770 h at 470°C (12 dpa, 540 at. ppm He). Note the presence of helium bubbles randomly distributed throughout the matrix. Transmission electron micrograph courtesy of P. J. Maziasz, Oak Ridge National Laboratory.

SUMMARY AND CONCLUSIONS

Recently reported results from a number of ongoing materials
data generating programs were reviewed. These programs are aimed
at determining the influence of temperature, time, waveform, and
environment on the elevated-temperature fatigue properties of
several structural alloys presently in use or planned for use in
a number of advanced nuclear power generating systems. Specific
major conclusions were as follows:

1. Environments such as high vacuum or high-purity sodium
that limit or impede oxidation result in a marked improvement in
the continuous cycle fatigue life of types 304 and 316 stainless
steel at temperatures within the creep range. However, when
loading waveforms are employed such that intergranular crack
propagation occurs, the differences in fatigue life tend to be
minimal. This finding, particularly for long-term test results,
supports the concept of a true creep-fatigue effect as a major
contribution to the observed decrease in cycle life.

2. Exploratory strain-controlled fatigue tests conducted on
type 304 stainless steel in air have shown that some degradation
in fatigue life in comparison with continuous cycle tests can
occur when hold periods are introduced at zero stress or zero
relaxation points on the hysteresis loops. Similar tests conducted on annealed 2 1/4 Cr-1 Mo steel in air and at low strain
ranges tend to show significant degradation in cycle life. This
has been attributed to waveform-oxide interaction that facili-
tates crack nucleation and accelerates crack propagation.

3. Cycle lives of specimens of 2 1/4 Cr-1 Mo tested in nonoxidizing environments, such as sodium, do not show the significant waveform dependency effects that are found for similar tests
in air. However, slow-fast triangular waveforms at high temperature reduce cycle lives in sodium probably from the generation of
grain boundary voidage.

4. Limited results of strain-controlled time-dependent fatigue
tests conducted on alloy 800H in air have shown that the fatigue
life is dependent upon cyclic waveform and strain-range interaction and that a carburizing environment can be detrimental,
depending again upon the cyclic waveform imposed.

5. Classical methods, such as time and cycle fraction summation for predicting or extrapolating data, appear questionable,
particularly at temperatures where there is strong environmental
interaction.

6. An example was selected for 20%-cold-worked type 316
stainless steel to show that environments generated within a

structural material, such as the displacement damage and helium bubbles obtained by irradiation, must also be considered in design against fatigue damage when appropriate.

REFERENCES

1. K. Natesan, T. F. Kassner, and C. Y. Li, "Effect of Sodium on Mechanical Properties and Friction-Wear Behavior of LMFBR Materials," React. Technol. 15(4): 244–71 (1972–73).

2. W. Charnock, C. P. Haigh, C.A.P. Horton, and P. Marshall, Underwriting Structural Steels for the Sodium-Cooled Fast Reactor, No. 10, Central Electricity Generating Board Research, United Kingdom, November 1979, pp. 3–14.

3. H. Huthmann, G. Menken, H. U. Borgstedt, and H. Tas, "Influence of Flowing Sodium on the Creep-Rupture and Fatigue Behavior of Type 304 Stainless Steel at 550°C," paper presented at Second International Conference on Liquid Metal Technology in Energy Production, Richland, Washington, April 20–24, 1980.

4. A. R. Keeton and C. Bagnall, "Factors That Affect Corrosion in Sodium," paper presented at Second International Conference on Liquid Metal Technology in Energy Production, Richland, Washington, April 20–24, 1980.

5. C. R. Brinkman, V. K. Sikka, and R. T. King, "Mechanical Properties of LMFBR Primary Pipe Materials," Nucl. Technol. 33(1): 76–95 (April 1977).

6. C. R. Brinkman, V. K. Sikka, and M. K. Booker, "An Overview of the U.S. Programs on Properties of Primary Circuit Materials," pp. 13–23 in Specialist Meeting on Primary Circuit Structural Materials Including Environmental Effects, IWGFR/22, International Atomic Energy Agency/International Working Group on Fast Reactors, Bergisch Gladbach, Federal Republic of Germany, October 17–21, 1977.

7. P. S. Maiya, "Effects of Waveshape and Ultra-High Vacuum on Elevated Temperature Low-Cycle Fatigue in Type 304 Stainless Steel," submitted to Materials Science and Engineering.

8. K. Natesan, O. K. Chopra, and T. F. Kassner, "Creep-Rupture and Low-Cycle Fatigue Behavior of Types 304 and 316 Stainless Steel Exposed to a Sodium Environment," paper presented at Second International Conference on Liquid Metal Technology in Energy Production, Richland, Washington, April 20–24, 1980.

9. L. F. Coffin, Jr., "The Effect of High Vacuum on the Low Cycle Fatigue Law," Metall. Trans. 3: 1777–78 (July 1972).

10. L. A. James, "Some Questions Regarding the Interaction of Creep and Fatigue," J. Eng. Mater. Technol. 235–43 (July 1976).

11. K. Sadananda and P. Shahinian, "Effect of Environment on Crack Growth Behavior in Austenitic Stainless Steels Under Creep and Fatigue Conditions," Metall. Trans. 11A: 267–76 (February 1980).

12. Y. Asda and S. Mitsuhashi, "Creep-Fatigue Interaction of 304
     and 316 Stainless Steels in Air and Vacuum," paper presented
     at Fourth International Conference on Pressure Vessel Technology,
     London, England, May 19—23, 1980.

13. D. J. Michel and H. H. Smith, "Accelerated Creep-Fatigue
     Crack Propagation in Thermally Aged Type 316 Stainless Steel,"
     submitted to Acta Metallurgica.

14. P. Marshall and C. R. Brinkman, "The Influence of Environment
     and Microstructure on Fatigue and Creep Crack Growth in Thick
     Section AISI Type 316 Stainless Steel," paper presented at
     ASME Symposium on Material Environment Interactions in
     Structural and Pressure Containment Service, Chicago, Illinois,
     November 16—21, 1980.

15. C. R. Brinkman and M. Katcher, "Materials Technology for Steam
     Generators in Liquid Metal Fast Breeder Reactor Systems,"
     Met. Prog. 116(2): 54—61 (July 1979).

16. C. R. Brinkman, J. P. Strizak, and M. K. Booker, "Experiences
     in the Use of Strainrange Partitioning for Pedicting
     Time-Dependent Strain-Controlled Cyclic Lifetimes of Uniaxial
     Specimens of 2 1/4 Cr-1 Mo Steel, Type 316 Stainless Steel,
     and Hastelloy X," pp. 15-1—15-18 in Characterization of Low
     Cycle High Temperature Fatigue by the Strainrange Partitioning
     Method, AGARD Conf. Proc. 243, Technical Editing and Repro-
     duction, Ltd., Harford House, London, April 1978.

17. C. R. Brinkman, J. P. Strizak, M. K. Booker, and C. E. Jaske,
     "Time-Dependent Strain-Controlled Fatigue Behavior of Annealed
     2 1/4 Cr-1 Mo Steel for Use in Nuclear Steam Generator Design,"
     J. Nucl. Mater. 62(2,3): 181-204 (November 1976).

18. K. D. Challenger, A. K. Miller, and C. R. Brinkman, "Elevated-
     Temperature Fatigue with Hold Time in a Low Alloy Steel:
     Creep Damage or Environmental Damage? Part I — Physical
     Mechanisms," submitted to Journal of Engineering Materials
     and Technology, 1979.

19. H. Teranishi and A. J. McEvily, "The Effect of Oxidation on
     Hold Time Behavior of 2 1/4 Cr-1 Mo Steel," Metall. Trans.
     10A: 1806—07 (November 1979).

20. E. G. Ellison and A.J.F. Patterson, "Behavior of a 1 Cr-Mo-V
     Steel Subject to Combinations of Fatigue and Creep Under
     Strain Control," Proc. Inst. Mech. Eng. London 190: 333—40
     (1976).

21. O. K. Chopra, K. Natesan, and T. F. Kassner, "Influence of
     Sodium Environment on the Low Cycle Fatigue and Creep-Fatigue
     Behavior of Fe-2 1/4 Cr-1 Mo Steel," paper presented at
     Second International Conference on Liquid Metal Technology in
     Energy Production, Richland, Washington, April 20—24, 1980.

22. C. R. Brinkman, M. K. Booker, J. P. Strizak, and
     T. Weerasooriya, "Fatigue Behavior of 2 1/4 Cr-1 Mo Steel in
     Support of Steam Generator Development," pp. 59—72 in Time

and Load Dependent Degradation of Pressure Boundary Materials, Innsbruck, Austria, November 20–21, 1978, International Atomic Energy Agency, IWG–RRPC–79/2.

23. D. I. Roberts, S. N. Rosenwasser, and J. F. Watson, "Materials Selection for Gas–Cooled and Fusion Reactor Applications," paper presented at Conference on Alloys for the 80's, Ann Arbor, Michigan, June 17–18, 1980.

24. Personal communication, J. L. Kaae, General Atomic Co., La Jolla, Calif., June 1980.

25. M. L. Grossbeck and K. C. Liu, "Fatigue Behavior of Type 316 Stainless Steel Following Neutron Irradiation Inducing Helium," paper presented at the American Nuclear Society meeting, Las Vegas, Nevada, June 8–13, 1980.

26. M. L. Grossbeck and P. J. Maziasz, "Tensile Properties of Type 316 Stainless Steel Irradiated in a Simulated Fusion Reactor Environment," J. Nucl. Mater. 85 and 86(II,B): 883–87 (1979).

27. C. R. Brinkman, K. C. Liu, and M. L. Grossbeck, "Estimates of Time–Dependent Fatigue Behavior of Type 316 Stainless Steel Subject to Irradiation Damage in Fast Breeder and Fusion Power Reactor Systems," pp. 490–510 in Effects of Radiation on Structural Materials, ASTM Spec. Tech. Publ. 683, J. A. Sprague and D. Kramer, Eds., American Society for Testing and Materials, Philadelphia, 1978.

THE EFFECT OF ENVIRONMENT AND TEMPERATURE ON

THE FATIGUE BEHAVIOR OF TITANIUM ALLOYS

J. A. Ruppen, C. L. Hoffmann,
V. M. Radhakrishnan and A. J. McEvily

Department of Metallurgy and
Institute of Materials Science
University of Connecticut
Storrs, Connecticut 06268

INTRODUCTION

The high strength to weight ratio of titanium alloys has made them particularly attractive for a variety of structural applications at moderately elevated temperatures. In recent years there has been an attempt to raise the service temperature of these alloys through the development of near-α alloys which are more creep resistant. For these newer alloys as for titanium alloys in general the microstructure can be significantly varied as a function of thermo-mechanical processing history and it is of interest to determine the fatigue properties in terms of microstructure. In addition, the oxidation characteristics of titanium can be expected to influence the fatigue process particularly since titanium has a high solubility for oxygen, the basis for its excellent diffusion bonding characteristics. Further, as the service temperature is increased the potential for creep-fatigue interaction is also increased. The aim of this paper is to review the current state of knowledge concerning the potential environmental-creep-fatigue interactions in titanium alloys.

PROCESSING, MICROSTRUCTURES AND DEFORMATION MODES

Titanium alloys have particular characteristics which make them different from other alloy types such as aluminum alloys or steels. For example, certain titanium alloys, when properly processed, are remarkably free of inclusions or second phase particles.

In addition, the microstructures are often duplex in nature, containing an intimate mixture of the high temperature β-phase (bcc)
together with the low temperature α-phase. As a result, for large
scale plastic deformation cooperative processes of compatible
shearing must take place in both phases.

Titanium alloys are also interesting in that a wide variety of
microstructural variations can be obtained as a function of alloy
content and thermomechanical processing.[1-4] Through controlled
processing and heat treatment procedures titanium alloys of equivalent strength levels can often be obtained for different microstructures, thus allowing for optimization of various mechanical properties such as fatigue and fracture toughness.[1]

Generally speaking, the alloys fall into three major categories;
α, α/β or β. As indicated in Fig. 1a-c certain elements such as
aluminum and oxygen extend the α field and are referred to as α-
stabilizers. Other elements such as vanadium extend the β-field
and are known as β-stabilizers. Various additional transformation
products such as the omega (ω) phase and the martensitic phase α'
can also occur.[2]

Figure 2 illustrates some of the microstructural features
developed in titanium alloys and Table 1 contains further details
concerning the relationship between processing and microstructure.
Particular points of interest include the fact that primary α (Fig.
2d) is formed by a nucleation and growth process in alloys where the
final hot working operation is completed in the α+β phase field.
The primary α morphology depends on the extent of working performed
below the β-transus and can vary from elongated plates in lightly
worked material to equiaxed grains in heavily worked materials,[1-3],
see Figs. 2c and 2d.

The processing or the solution treatment of α+β and near α titanium alloys above the β transus usually results in microstructures
consisting of packets of crystallographically aligned α platelets
called colonies (Widmanstätten)[1-4,7,8] Figs. 2a, 2b, 2j with the α
phase platelets separated by the bcc β phase and an interface
phase layer.[9-12] The orientation of the α platelets of a particular colony is related to the prior β grain orientation by one of
the twelve variants of the Burgers relation[2,13]

$$<110>_\beta || (0001)_\alpha$$
$$<111>_\beta || <11\bar{2}0>_\alpha$$

The probability of obtaining a large colony microstructure of
similarly aligned α platelets (Figure 2a) is increased by a low content of β stabilizing elements, annealing for long times in the β

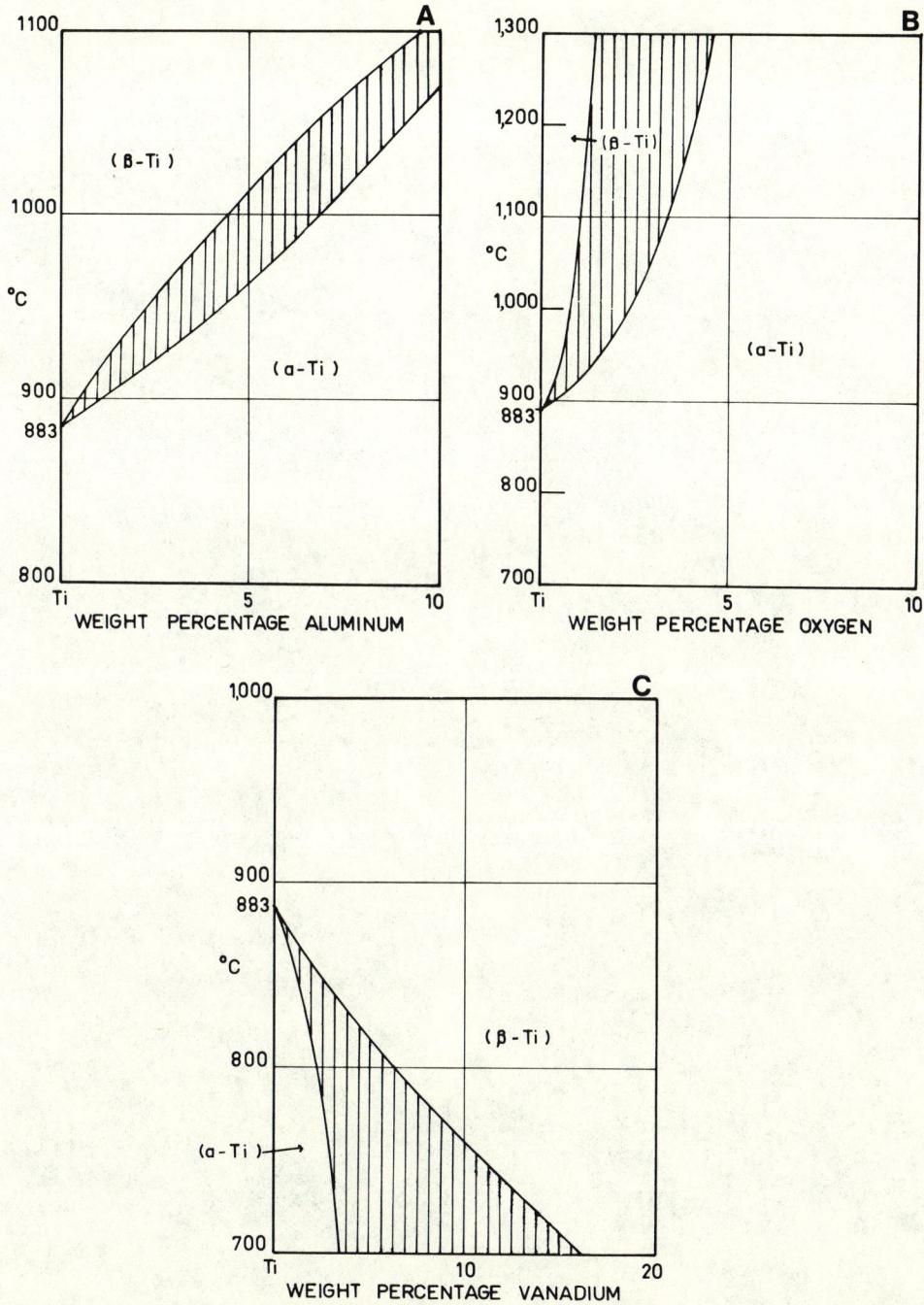

Fig. 1.   The influence of (a) Aluminum, (b) Oxygen and (c) Vanadium
          on the stability of α and β phases. [5]

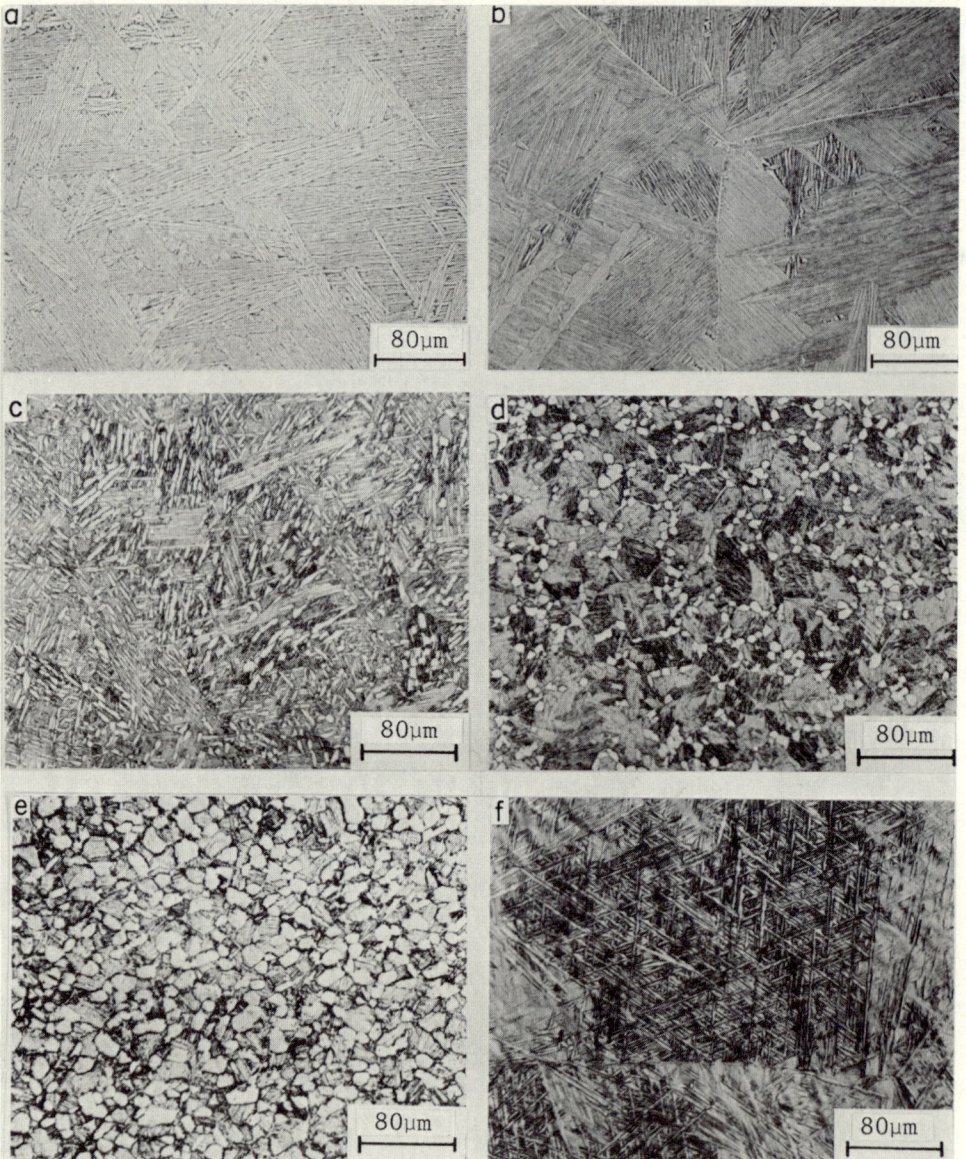

Fig. 2.   (a-f)   Microstructures of titanium alloys.

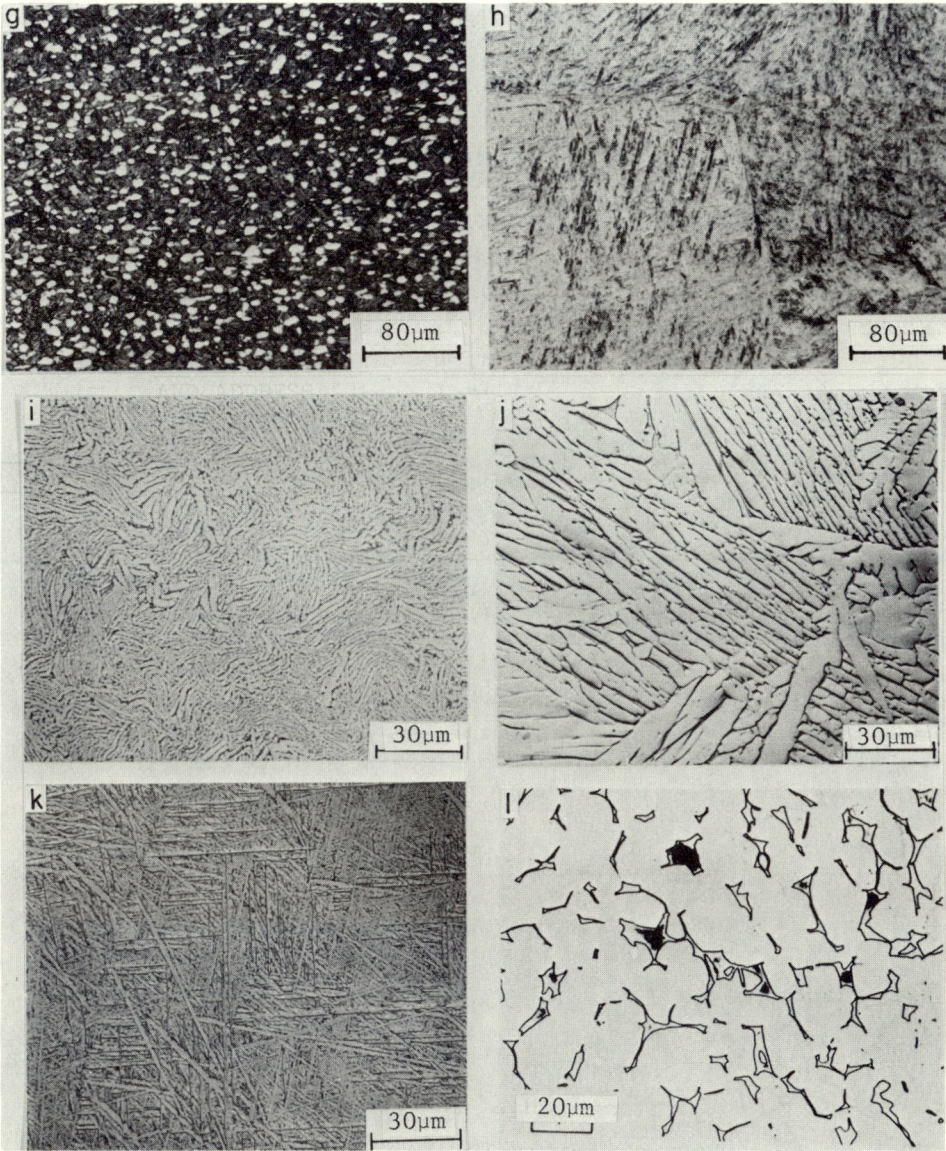

Fig. 2. (g-1) Microstructures of titanium alloys.

Table I
Processing Schedules with Resulting Microstructures
As Shown in Figures 2a thru 1

Material	Micrograph in Fig. 2	Processing	Microstructural Characteristics
Ti-6242	a	β forged α+β heat treated	Widmanstätten colony
(Ti-6Al-2Sn-4Zr-2Mo-0.1Si)	b	α+β forged β heat treated	Large prior β grains, Widmanstätten colony, Grain boundary α phase
	c	β forged α+β heat treated	Large β grains, coarse elongated α platelets
	d	α+β extruded α+β heat treated	Small β grains, equiaxed α and Widmanstätten colonies
IMS 685 (Ti-6Al-5Zr-0.5Mo-0.25Si)	e	α+β forged α+β heat treated	Equiaxed α
	f	β forged α+β heat treated	Acicular α/β platelets
Ti-5524 (Ti-5Al-5Sn-2Zr-4Mo-0.25Si)	g	α+β forged α+β heat treated	Equiaxed α in α+β matrix
	h	β forged and α+β forged β heat treated	Large β grains, tempered α'
Ti-6Al-4V	i	α+β rolled bar Mill annealed (MA)	Discontinuous β
	j	α+β forged β heat treated	
	k	α+β forged β quenched and annealed (BQA)	Fine acicular α needles tempered
	1	α+β forged α+β heat treated Recrystallization Annealed (RA)	

field, low amounts of β work and slow cooling rates. With increasing cooling rates or with increased β stabilizer content, nucleation of additional variants of the Burgers relation[2,13] become more
prevalent and the number of plates in the Widmanstätten α packet
decreases, becoming more acicular in character, until a point is
reached where the transformed regions consist of a random mixture
of α platelets belonging to the different variants of the orientation relation, Fig. 2f, 2k.[3] Quenching from above the β transus
leads to the formation of martensitic structures.

Plastic deformation of hcp α titanium occurs by both slip and
twinning processes, however, slip appears to be the dominant deformation mode in fatigue. The reported slip and twin modes in α-
phase titanium are listed in Table II.[14,15]

Table II. Slip and Twin Modes in α-Phase Titanium

Slip Modes

	Slip Direction	Slip Plane
$\bar{a}$ slip	$\langle 11\bar{2}0 \rangle$	$\{0001\}$ basal
	$\langle 11\bar{2}0 \rangle$	$\{10\bar{1}0\}$ prism
	$\langle 11\bar{2}0 \rangle$	$\{10\bar{1}1\}$ pyramidal
$\bar{c}+\bar{a}$ slip	$\langle 11\bar{2}3 \rangle$	$\{11\bar{2}2\}$ pyramidal - type II
	$\langle 11\bar{2}3 \rangle$	$\{10\bar{1}1\}$ pyramidal

Twin Modes

$\{10\bar{1}2\}$, $\{11\bar{2}1\}$, $\{11\bar{2}2\}$, $\{11\bar{2}3\}$, $\{11\bar{2}4\}$, $\{10\bar{1}1\}$

Slip on $\{10\bar{1}1\}$ and the (0001) planes tend to become relatively
more favorable compared to the $\{10\bar{1}0\}$ planes as the interstial content or the temperature is increased.[16] In addition, $\bar{c}+\bar{a}$ slip is
quite common in Ti-Al alloys.[17]

An important aspect of multiphase alloys is the influence of
crystallographic relationships between phases and slip processes on
fracture behavior. Although the close packed α slip direction
$\langle 11\bar{2}0 \rangle$ and β slip directions $\langle 111 \rangle$ are parallel and the $(11\bar{2})_\beta$ and
$(1\bar{1}00)_\alpha$ planes coincide, the other α and β slip planes are not
parallel. Therefore, slip transference across the interface can
be difficult and result in strain localization at the interface.
The interface phase may also restrict slip transference across the
interface.[18] However, in microstructures containing α/β colonies
oriented by the Burgers relation it is possible for adjacent colonies of common prior β grain origin to have a common basal plane[7]
with shear related slip and crack extension proceeding across
several colonies. [4,7,8,19]

OXIDATION

If titanium alloys are to be cyclically loaded at increasingly higher service temperatures it is important that the nature and effects of the oxidation process be understood, particularly since it is already known that oxide film formation and the absorption of oxygen into the substrate can degrade properties such as ductility and toughness.[20-25] This degradation has been associated with the diffusion of oxygen into the titanium to form a hardened layer whose characteristics relate to the exposure time and temperature. During oxidation above the β-transus the dissolution of oxygen into titanium can produce a transformed layer known as the "α-case". Because oxygen is an α-stabilizer the β-transus temperature is increased with increasing oxygen content. Once the oxygen content reaches a level at which the β-transus temperature is equal to the exposure temperature, a transformation from α to β will spontaneously occur. This α-case layer is extremely hard and brittle and may even crack simply because of the thermal stress induced on cooling.[21] This α-case layer does not form during oxidation at lower temperatures below the allotropic transformation temperature, such as 800°C. However, the diffusion layer still exists as can be shown by hardness measurements.[26-28] The significance of this layer is revealed by tensile tests and impact tests performed on air exposed samples. In these tests microscopic cracks and crevices were observed of a depth which was directly proportional to the depth of the gas saturated layer.[20,29] Below 550°C oxidation (scaling) has generally not been considered to limit the use of titanium alloys, since strength and creep properties were generally low. However, as more creep resistant alloys are developed for use at higher temperatures the oxidation characteristics will be of greater concern. It has already been shown that the tensile ductility of titanium alloys has been reduced significantly after periods of relatively mild oxidation at temperatures where titanium alloys are normally considered to be oxidation resistant.[27] It is to be expected that the formation of an oxide film together with a hardened substrate layer would be detrimental to low cycle fatigue behavior.

In addition to the uniform surface oxidation, it has been suggested that oxidation diffusion rates are enhanced along α/β interfaces[30-33] and prior β grain boundaries.[34] The increased penetration of oxygen into titanium alloys at these locations due to the enhanced diffusion is responsible for accelerated crack initiation and substantial reductions in the high temperature low cycle fatigue and creep lifetimes of these alloys. Additionally, it has been observed in general that plastic zones at the tip of growing fatigue cracks are smaller in air than in vacuum.[35] This decrease in plastic zone size has been associated with the presence of oxygen within the plastic zone. In the case of titanium, Swanson and Marcus[36] have suggested that a mechanism of transport of oxygen by dislocations within the plastic zone is operative. This dislocation sweeping is needed to account for the large depth of penetration of oxygen into titanium at room temperature, and such a mechanism is likely to be important at elevated temperatures as well.

The nature of the oxidation process itself is also dependent upon temperature, and Table III provides a summary of the various temperature dependent kinetic relations which have been proposed. In addition, there is evidence that cyclic straining can also accelerate the oxidation process.[42-44] Alloying additions can influence the rate of oxidation as well as the depth of penetration of oxygen into the substrate as shown in Figs. 3 and 4. Chemical analyses of the oxide scale indicate that it is comprised of rutile, TiO_2, and in alloys containing aluminum, Al_2O_3, in amounts greater than to be expected in proportion to other alloying elements present. The particularly pronounced participation of aluminum in the scale apparently provides a certain protective effect. In heavily oxidized samples the concentration of aluminum in the scale layer is most pronounced immediately below the surface with high Al_2O_3 proportions in the outer zones of the scale acting as a diffusional barrier for the penetration of oxygen.[21,28,45]

Table III. Oxidation Kinetics of Titanium

Film Oxidation Rate	Rate Equation	Temperature Range	
Logarithmic	$x = k_1 \log(t+t_o) + k_2$	300°C	Thin film formation, little or no O_2 diffusion into metal
Cubic	$x^3 = k_3 t + C_3$	300–600°C	Transition stage from logarithmic to parabolic oxidation, due to the onset of significant amounts of oxygen dissolution into metal
Parabolic	$x^2 = k_p t + C_p$	600–900°C	Simultaneous oxide film formation and growth and oxygen dissolution into the metal forming a gas saturated layer.
Linear	$x = k_1 t + C_1$	900°C	Oxygen diffusion predominates at T>900°C. Above 900°C diffusion of titanium ions become increasingly important

[ref. 37-41]

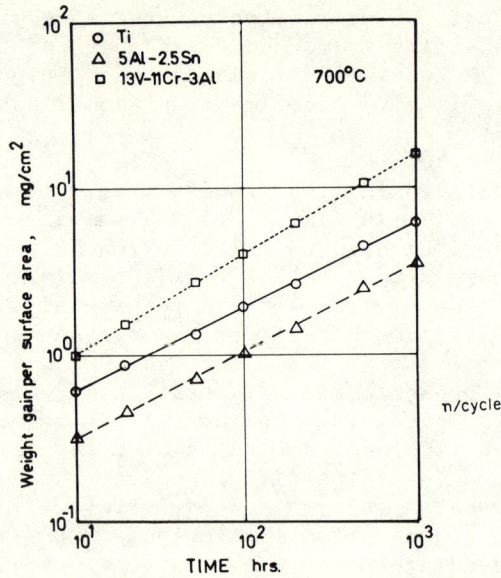

Fig. 3.   Effect of alloying elements on oxidation rate for commercially pure titanium, an alpha alloy (5Al–2.5Sn) and a
beta alloy (13V–11Cr–3Al). [21]

In comparing the resistance to oxidation of various alloy types,
it has been found that the β–alloy Ti–13V–11Cr–3Al and the aluminum-
free α+β alloys are least scale-resistant and that the most favorable behavior is exhibited by the α–alloy Ti–5Al–2.5Sn and the α+β
alloys of higher aluminum contents.[21] If scaling should prove to
be a problem at temperatures where otherwise the mechanical properties are satisfactory, then consideration may have to be given to
protective coatings based perhaps upon silicon, molydisilicides or
nickel–aluminum alloys.[45]

CREEP DEFORMATION AND RUPTURE

The creep strength and stability of a titanium component is
often the limiting parameter in a highly stressed, high temperature
application. Primarily because of the poor oxidation resistance
and phase instability of beta alloys at 500°C, alpha or near alpha
alloys are preferred for long time elevated temperature service.
However, perhaps the most important problems with alpha–Ti alloys
is their tendency to order when containing as little as 5 or 6 wt
% Al and to embrittle when containing more than about 8.5% Al.
Thus, to be conservative, the Al equivalence (= % Al + 1/3 % Sn +
1/6 % Zr) should not exceed about 8.5% in Ti alloys.[46]

The creep strength of commercial Ti–alloys generally depends
on the alloying elements, Al, Zr, Sn, Mo and Si.[47–49] Table IV

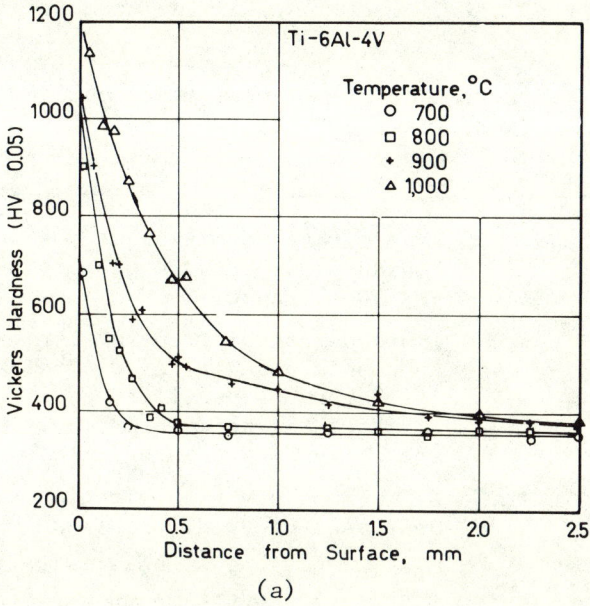

(a)

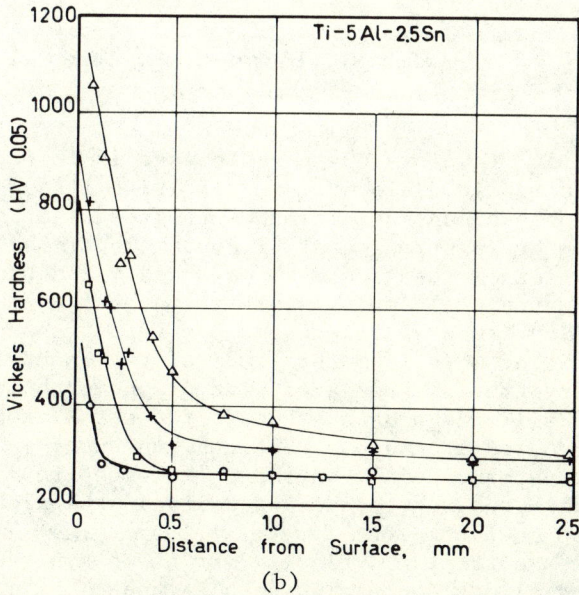

(b)

Fig. 4.   Depth of penetration of oxygen into the substrate for an
annealing time of 200 hrs. detected by microhardness
measurements a) Ti-6Al-4V, and b) Ti-5Al-2.5Sn. [21]

lists the effect of alloying elements on the creep resistance of
titanium.

Table IV. Effect of Alloying Elements on Creep
Resistance of Titanium

Alloying Element

1. Mo (3%-4% max)	Precipitate formation at α-plate grain boundary. Creep strength improved in 400-500°C range.
2. Si, C (0.5%)	Improves creep strength. Forms hard stable compounds with Ti. At equivalent creep strengths the ductility of carbon-containing alloys is significantly lower than that of the silicon-containing alloys.
3. Si + Mo (0.8%)	Grain boundary precipitation occurs which decreases creep deformation.
4. Si (0.5%) + Mo (2-3%)	Optimum balance of tensile, creep, and stability properties obtained. Elevated temperature strength improved.
5. Al (max 8%)	Improves creep resistance. Tendency to form ordering and hence chances of embrittlement.
6. Zr (5%) + Si (11%)	Imparts solid solution strength to α-phase. With Si, improves the creep strength.

In Si bearing Ti alloys creep resistance is improved by Si
precipitation on mobile dislocations, resulting in a pinning of
these dislocations and an inhibition of their further movement.[50]
In addition, the creep strength of Si containing Ti alloys is
highly dependent on micro-structural morphology and hence on heat
treatment. A substantial improvement in creep strength occurs
when the annealing temperature exceeds the beta transus, with the
Si alloys benefiting more from this type of treatment. The improved creep strength is associated with grain morphology in which
alpha, beta and silicide phases assume different microstructural
relations when cooled from above the beta transus.[46] Silicide
precipitation treatment being constant, the creep resistance can
be correlated with the degree of continuity in the alpha matrix,
with an increase in alpha mean free path being characterized by a
decrease in creep strength.[4] Finally, Zr has a synergistic effect
with silicon and alloys containing both Zr and Si have significantly higher creep strengths than those in which Zr is omitted. It
has been suggested that the composition and nature of the silicide
precipitate may be influenced by Zr.

Dislocation mechanisms operative at various temperature ranges
in titanium for a strain rate of 10^{-4}/sec have been fairly well
established.[51,52] At temperatures up to 0.3 T_m plastic deformation
is thermally activated and results from overcoming interstitial
solute atom obstacles, while at 0.3 to 0.45 T_m dynamic strain aging
occurs. Atmosphere drag and dynamic strain aging effects from both
interstitial and substitutional alloy elements, are also responsible
for the "athermal plateau" between 200°C and 400°C. Above 0.45 T_m
grain boundary sliding, dislocation glide and climb appear to be
the controlling mechanisms and creep rates follow the power law as
described by Weertman's dislocation climb model.[53] Fracture at
high temperatures appears to be controlled by the same mechanism
since the activation energy for rupture is similar to that for
creep. Finally, Orr et al.[54] have shown that stress rupture data
for titanium correlates very well with the relation

$$\sigma_f = \text{func } [t_r \ (-60,000/RT] \tag{1}$$

where σ_f is the fracture stress and t_r the rupture time.

In the temperature range of 350°C to 750°C Cuff and Grant[55]
observed sharp breaks in stress rupture plots for commercially pure
Ti and were attributed to a change in fracture mode from transgranular to intergranular. At these temperatures grain boundary
sliding was more predominant which led to wedge type intergranular
cracks.[46]

Creep deformation is often analyzed by parametric methods
such as the Larson-Miller creep parameter.[56] The improvement in
creep resistance and the relative time for deformation at 500°C as
a result of alloying elements are shown in Table V on the basis of
Larson-Miller parameter, computed for a stress of 245 MPa and a
strain of 0.2%.[46]

Table V.   Larson-Miller Parameters for Titanium
Alloys Computed for 0.2% Strain at 245 MPa

	Ti Alloy	L-M Parameter $\sigma_R(20+\log t) \times 10^{-3}$	Temp. 500°C time (hrs)
1.	Ti-6Al-4V	27.8	1
2.	Ti-7Al-4Mo	29.7	16
3.	Ti-8Al-1Mo-1V	30.6	72
4.	Ti-2.5Al-11Sn-5Zr-1Mo-0.2Si	31.4	270
5.	Ti-6Al-2Sn-4Zr-2Mo	31.6	360
6.	Ti-6Al-5Zr-0.5Mo-0.25Si	32.5	1650
7.	Ti-6Al-2Sn-1.5Zr-1Mo-0.35Bi-0.1Si	34.0	19000

The deformation characteristics can be better understood by
deformation maps as originally suggested by Weertman[57] and later

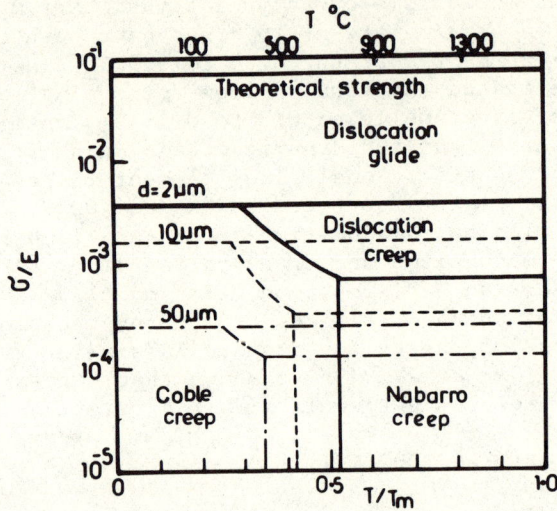

Fig. 5.   Deformation mechanism map for commercially pure titanium.
          d = grain size in microns

developed in detail by Ashby.[58] These maps identify the fields of
stress and temperature in which a particular mechanism of plastic
flow is dominant. Figure 5 shows the deformation mechanism map
for polycrystalline titanium for 2 to 50 μm grain size.[59] The map
for titanium is similar to those for other metals except that the
boundary between Coble creep and Nabarro creep occurs at a relatively lower temperature due to the large bulk diffusion coefficient
of titanium.

     In addition, fracture mechanisms operative over a wide range
of stress and temperature can be easily visualized by fracture maps
as developed by Ashby.[60] In this binary plot, normalized stress is
related with either homologous creep analysis and is used to identify the different modes of failure, namely ductile transgranular
fracture, creep transgranular fracture, creep intergranular fracture
and rupture in the different regions on the σ versus t_r plane.
Above 900°C and at relatively short creep times transgranular fracture was observed.[61] Under these conditions large ductility results
with the specimen drawing down to a point, such a fracture process
is classified as rupture. A fracture map for titanium identifying
these fracture modes is shown in Fig. 6.[62] It can be seen that
large ill-defined areas exist on either side of the boundaries where
the fracture can be of mixed mode.

     With respect to creep fatigue interaction, several aspects
are to be anticipated; one of these is the increase in non-elastic
deformation as cyclic frequencies are reduced and hold times are
increased. In terms of the Coffin-Manson approach, any increase in

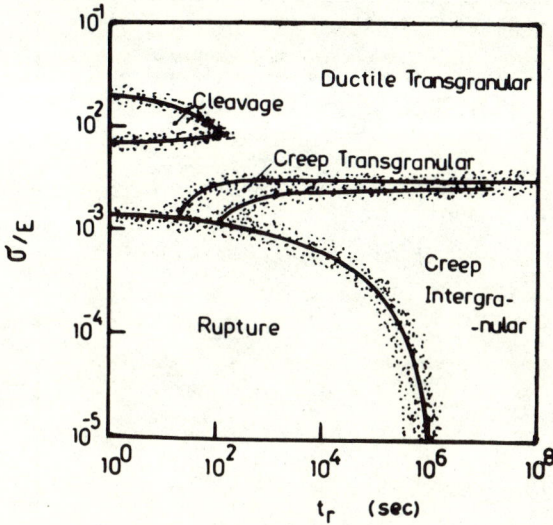

Fig. 6.   Fracture map for commercially pure titanium.

the plastic range per cycle should be damaging. In addition, if
grain-boundary sliding is important in one portion of the cycle and
transgranular slip important in another, this lack of reversibility
in deformation may prove to be deleterious. There may also be an
interaction between a growing fatigue crack and wedge type of cracks
which will accelerate the crack growth process.

It has been shown that the creep resistance of titanium alloys
is greater in vacuum than it is in air.[30,31,34] This is further
evidence of the strong environmental effect on titanium alloys.
Creep tests in air and vacuum have shown that preferential oxidation along prior β grain boundaries[34] and at α/β interfaces[30,31]
is responsible for the reduced creep resistance of titanium alloys
in air. The thermal stability of titanium alloys after prolonged
exposure to elevated temperature under stress is also affected by
the oxidation process. The combined effects of stress and oxidation at high temperature lead to a sharp reduction in ductility.[34]
Creep behavior in air can be improved through the use of coatings,
however coatings which are presently used still do not provide the
creep resistance for titanium alloys that is observed under high
vacuum conditions.

LOW CYCLE FATIGUE

As indicated in Figs. 7 to 9 both processing history and environment can have a marked effect on the low-cycle fatigue properties of titanium alloys at elevated temperatures. The better low
cycle fatigue resistance of the α/β processed material appears to
stem from the presence of the shorter α/β interfaces associated

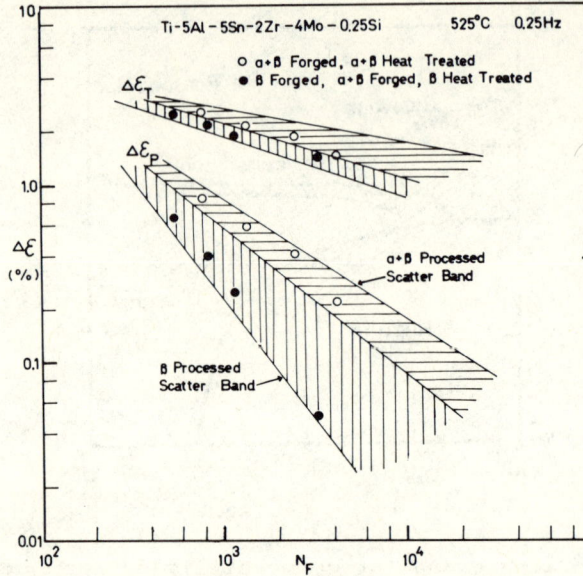

Fig. 7. Total strain range ($\Delta \varepsilon_t$) and plastic strain range ($\Delta \varepsilon_p$)
         versus cycles to failure (N_F) for Ti-5524 alloy. Micro-
         structures of the $\alpha+\beta$ processed and the β processed
         conditions are shown in Figures 2g and 2h respectively.

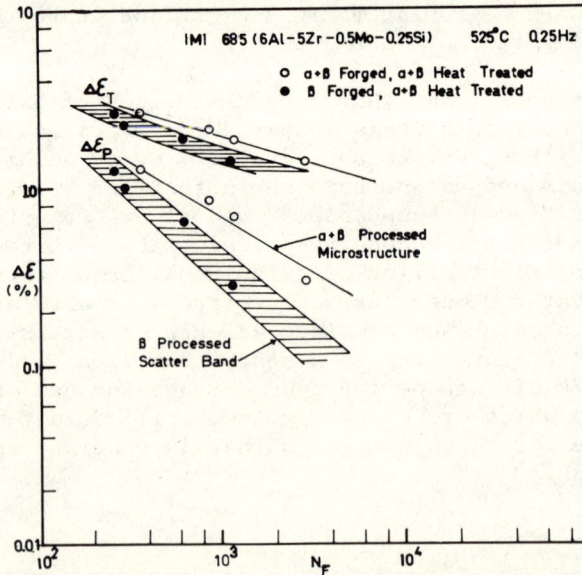

Fig. 8. Total strain range ($\Delta \varepsilon_T$) and plastic strain range ($\Delta \varepsilon_p$)
         versus cycles to failure (N_F) for IMI 685. Microstructures
         of the $\alpha+\beta$ processed and the β processed conditions are
         shown in Figures 2e and 2f respectively.

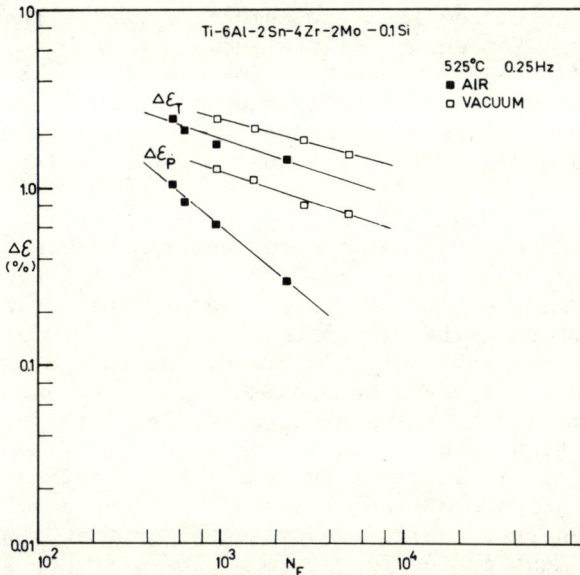

Fig. 9.   Total strain range ($\Delta\varepsilon_T$) and plastic strain range ($\Delta\varepsilon_p$) versus cycles to failure (N_F) for Ti-6242 alloy in air and vacuum at 525°C. (Figure 2c shows the microstructure of the material used for these tests).

with equiaxed primary α grains.[18,63] Similarly, Eylon et al.[32] have found that microstructures with shorter α-plate morphology have better HTLCF life than structures with longer α-plate morphology. The initial cracks were much longer in conditions with longer α/β interfaces.

Structural size is also important. It has been shown by Lucas[64] that the fatigue properties depend upon the α-grain size, the smaller the α grain size the greater the fatigue resistance. In addition, Eylon and Hall[65] have found that by reducing the colony size, the size of shear related initial cracks could be limited resulting in an improved fatigue life. These findings indicate that the smaller the length of the slip band the less the stress intensity at the α-β interface where the slip band is blocked (and crack initiation is likely to occur). Processing which produces equiaxed-primary α results in low cycle fatigue properties which are superior to those of the coarse colony microstructures because blockage of slip at α/β interfaces and prior β grain boundaries is less pronounced. Within this phase plastic deformation can be more readily accomodated and some dispersal of the slip process can take place.[18] As a result crack initiation occurs at a larger number of cycles than for β-processed material. In accord with this view, Sattar et al[63] found that the number of cycles to crack initiation in α/β forged Ti-6Al-4V tested at room temperature was from 2 to 4 times higher than the number of cycles required for β-forged material. Somewhat similar findings have been made for the alloy Ti-17 [Ti-5Al-2Sn-2Zr-4Mo-4Cr].[18] In this case the initial crack growth was more structurally dependent in the β-processed material with macrocracks forming at 70% of the life for both α+β and β-processed material as the result of the linking up of micro-cracks.

At elevated temperatures the fatigue resistance of titanium alloys decreases, the more so the lower the test frequency and the coarser the α-platelet size. Eylon et al[32] have suggested that for tests in air the diffusion of oxygen along α/β interfaces may be primarily responsible for this trend. Diffusion of oxygen along such interfaces would be enhanced and cracks would occur along interfaces due to a lessening of their resistance to the plastic deformation associated with low cycle fatigue. Results of our tests are in accord with this view. For specimens tested in air crack nucleation occurred at the surface and a substantial amount of secondary cracking was present. On the other hand, a number of specimens tested in vacuum showed a tendency for forma-tion of sub-surface sites, Fig. 10, a phenomenon peculiar to titanium alloys,[65-67] with virtually no secondary surface cracking.[68] The subsurface initiation sites bear a resemblance to those ob-served in fatigue at room temperature which relates to shear band formation and do not appear to be a result of creep cavitation. It can also be noted in Fig. 9 that for specimens tested in air a smaller plastic strain range is obtained for a given total strain range as compared with similar tests performed in vacuum (10^{-5} Torr). The stress range is also smaller for a test in vacuum than for a test in air with a similar plastic strain range. The cause for this behavior is believed to be air embrittlement with associa-ted loss in ductility and decrease in amount of plastic strain. Results of such elevated temperature tests indicate that the envi-ronment may be a more dominant factor in lowering resistance to low cycle fatigue than is the creep process.

As a further indication of the importance of the environment, preliminary low cycle fatigue tests show a reduction in fatigue life on comparing room temperature results with elevated tempera-ture (525°C) results in air. However, the fatigue lifetime at elevated temperature in vacuum is greater even than that for room temperature in air. Clearly the ambient environment has a pro-nounced effect on the low cycle fatigue behavior of titanium alloys at room temperature as well as elevated temperatures.

It should also be noted that some of the microstructural fea-tures which enhance resistance to crack initiation may prove to be detrimental to crack propagation and to fracture toughness. This situation is due to the fact that the crack path is more tortuous and crack bifurcation occurs more readily in β-processed material

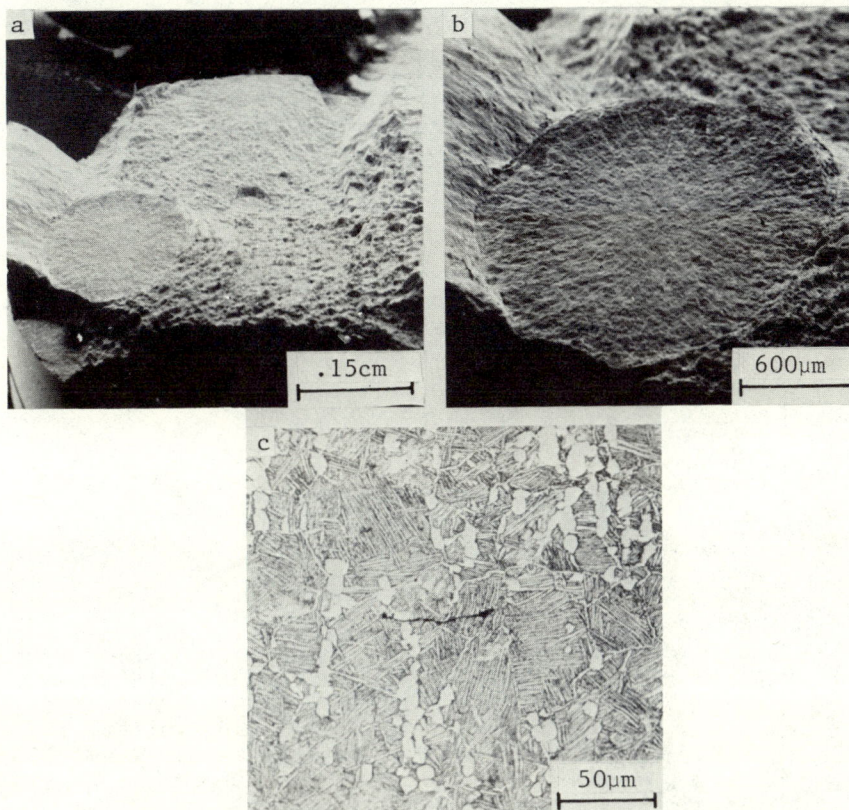

Fig. 10. Subsurface crack initiation sites in Ti-6242 samples tested in vacuum at 525°C. (a) Fracture surface showing several subsurface initiation sites. (b) Higher magnification of large subsurface site in 1(a). (c) Subsurface crack observed in longitudinally sectioned fatigue specimen.

than in α-β processed material. As a result the average crack growth rates can be slower and the apparent fracture toughness higher.[65]

FATIGUE CRACK GROWTH

In recent years considerable attention has been focussed on the fatigue crack propagation characteristics of titanium and its alloys.[1,3,4,6,8,18,65,69-101] In general, crack extension is not directly related to overall ductility, strength level or toughness;[84] however, there is extensive evidence to indicate that crack growth rates are affected significantly by variables such as microstructure, environment, temperature, interstitial content and load variation.

Room Temperature

Figures 11 and 12 show the great range in crack growth rates that can be obtained in titanium alloys. For a given alloy, crack growth rates vary by as much as an order of magnitude in response

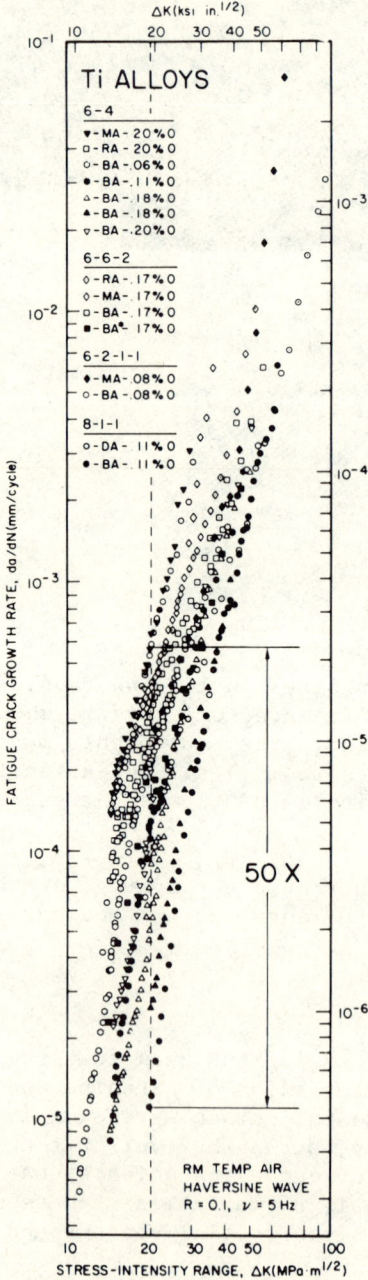

Fig. 11. Fatigue crack growth rates for a wide variety of materials from four α+β titanium alloy systems. Note 50-fold different in growth rates for different materials at ΔK=21 MPa√m. (after ref. 86)

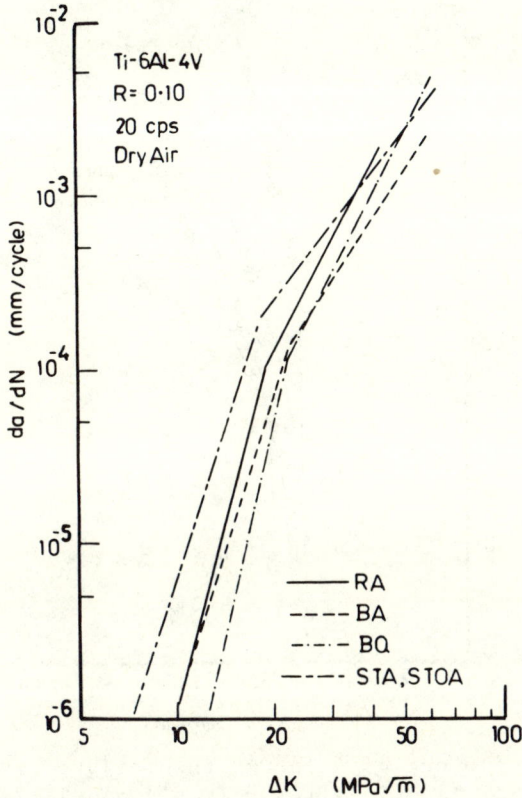

Fig. 12.   The influence of microstructure on fatigue crack propagation rates. (after ref. 84)

to heat treatment, with the usual trend being the larger the microstructural size the slower the crack growth rate.[73,82,86,87,89]
Furthermore, in general, for α/β alloys, the best crack growth
properties result from β-processing or heat treating in which
Widmanstätten microstructures are developed, whereas the poorest
are obtained from α+β processing and α+β heat treating in which an
equiaxed α phase in a Widmanstätten matrix is developed.[1,3,4,6,8,
65,69,70,71].

Crack growth in β-phase alloys has also been investigated.
Figure 13 illustrates the influence of aging on the fatigue propagation behavior of two β-Ti alloys, Ti-30V and Ti-24V.   The most pronounced effect of heat treatment (Ti-30V) on crack propagation
occurred after aging at 300°C in which coherent ω-phase particles
formed.   Aging for longer times at 300°C, increased the ω phase
volume fraction and resulted in planar slip, low energy failure
modes, and faster growth rates.[83]

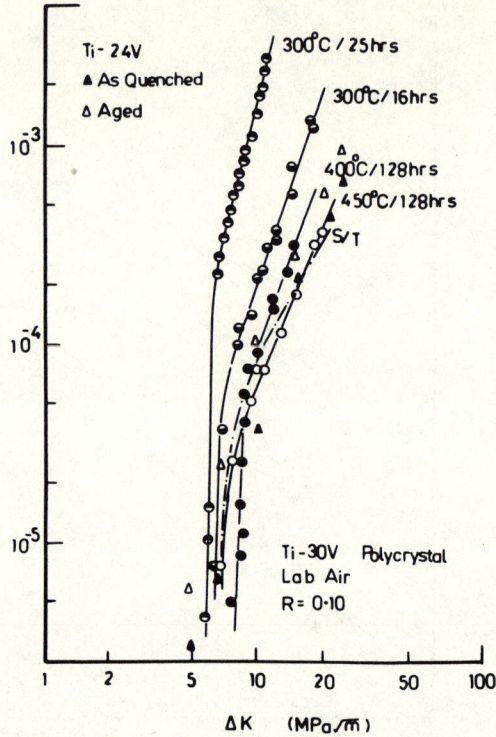

Fig. 13.   The influence of aging on the fatigue crack propagation
           rates of Ti-30V (after ref. 83) and Ti-24V (after ref.
           102).

The above results show that an important aspect of the fatigue
crack growth process in titanium alloys is that it can be more
structure sensitive than in the case of aluminum alloys or steels.
Much of this structure sensitivity is due to a greater tendency
for the crack to deviate from a plane perpendicular to the tensile
stress axis as it follows forward microstructural paths.[1,3,4,6-8,
65,70,73,74,82,85,89] Such an instance is shown in Fig. 14.   Crack
bifurcation is also more common in titanium alloys with branch or
secondary cracks tending to propagate perpendicular to the length
of the α platelets with the influence of this cracking behavior on
crack growth rates being greater for larger α grains and colony
size.   This occurrence of microstructurally sensitive crack growth
has been related to a grain size-colony size/crack tip plasticity
interaction in which structure sensitive growth occurs when the
scale of the microstructure is less than the reversed plastic zone
size.[70,73,80,74,85-87,101] In particular, the sharpness in the
transition from structure sensitive to structure insensitive growth
has been related to the degree of clustering of the colony packet
size[87] indicating that the nature of the packet size distribution
can control the shape of the fatigue crack growth rate plot.

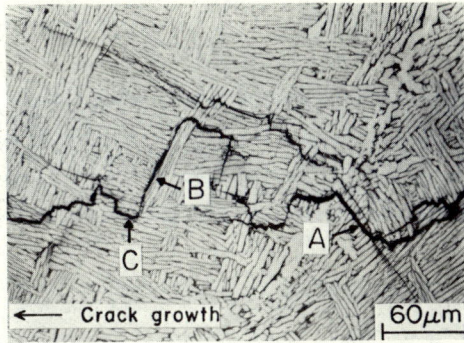

Fig. 14. Micrograph of β forged material illustrating; A,
          cracking perpendicular to α platelets; B, cracking along
          a colony boundary, and C, cracking along an α/β interface.

The improved crack growth resistance, especially in the near-
threshold regime, associated with a β processed or heat treated
condition is due to increased crack path tortuosity, and secondary
cracking occurring both along α/β interfaces and perpendicular to
the length of α platelets.[1,4,6,8,65,69,71,84,101] Multiple
cracking reduces the effective stress intensity by dispersing the
strain field energy of the macroscopic crack among multiple crack
tips[85] and crack tortuosity increases the effective crack path,
thereby leading to lower crack growth rates. In addition, cracks
also tend to grow in a start/stop mode, being arrested at colony
or prior β grain boundaries followed by reinitiation and change in
crack growth direction into the adjacent colony.[8]

The near-threshold regime is also characterized by transgranular facets, Fig. 15, whose size is related to the α-grain, colony
or prior β grain size.[19,74,82,85,86] Both X-ray and electron
channeling techniques have identified basal (0001)α fatigue fracture
facets.[81,103]   The formation of these facets appears to be influenced by environment[19,88] (since facet formation is depressed in vacuum and enhanced at elevated temperature) and slip
behavior at the crack tip. Figure 16 illustrates slip band formation with coincident secondary cracks. Facet formation likely
involves crack propagation (at least in colony microstructures)
along intense shear bands, similar to Stage I cracking, which are
formed ahead of the crack tip and can extend across several colonies on the same plane.[4,7,8,19] Transmission electron microscopy
of areas taken close to the fracture surface illustrating intense
basal and prismatic slip bands support this view.[104] Furthermore,
observations suggest that this crystallographic cracking is an

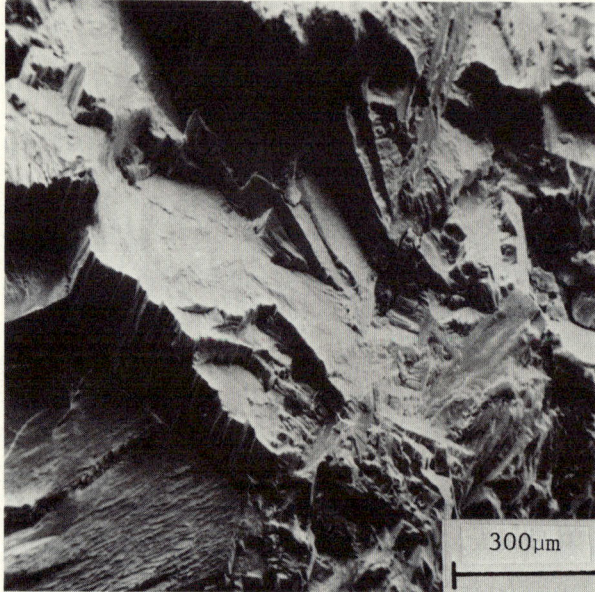

Fig. 15. Fatigue crack propagation fracture topography of microstructure shown in Fig. 2b (da/dN = $5 \cdot 0 \cdot 10^{-6}$ mm/cycle)

Fig. 16. Micrograph illustrating slip bands and slip band cracking
(marked A) produced during fatigue crack growth. Also
note the blockage of slip at prior β grain boundary.
Microstructure shown in Fig. 2b.

easy fracture path, since crack growth rates over the length of
the facet can increase by a factor of two or three compared to
baseline growth rates.[19]

At high stress intensities, the density of shear bands increased, Fig. 17, and a transition to structure insensitive growth
occurs in which striations, and dimpled rupture characterize the
fracture surface. It should be noted, however, that even at higher
growth rates the underlying microstructure is often visible in the
fracture topography.[19] In this regime, secondary cracks often
nucleate in front of the main crack followed by their connection,
Fig. 17.

The effect of interstial elements such as oxygen, silicon and
hydrogen on crack growth behavior is not simply described, since
various results have been found. For the case of oxygen, additions
from 0.06 to 0.18% to Ti-6Al-2Cr-2Zr-1Mo-0.25Si had no effect on
crack growth rates at 25°C or 540°C.[72] Similar results were obtained in a fine grained (.027mm) commercially pure α-Ti alloy.[82]
However, Beevers showed that in large grained α-Ti (0.23 mm) a
marked decrease in growth rates occurred with increasing oxygen
content [89] and Yoder et al [85] showed slower growth rates with higher
oxygen content in the Ti-6-4 alloy in microstructures having similar prior β grain size. On the other hand, an increase in fatigue
crack propagation occurred with increasing oxygen content for
Ti-O alloys [90] and Ti-6Al-4V alloys.[6,91]

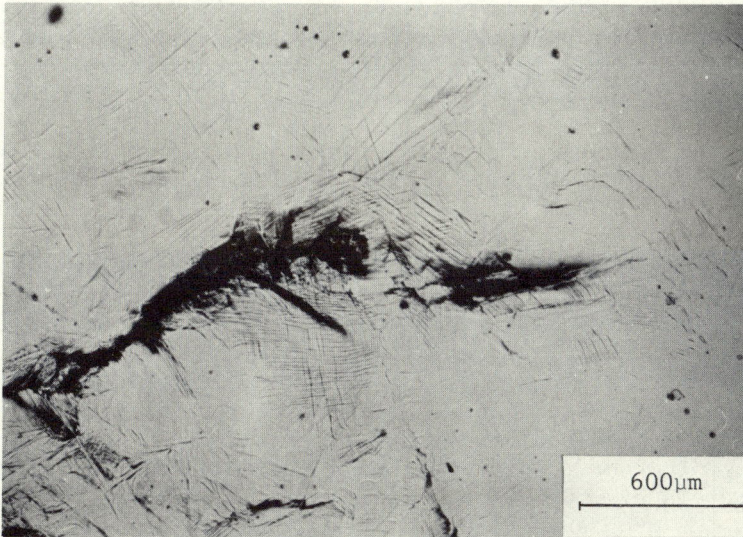

600µm

Fig. 17.   Micrograph illustrating secondary crack nucleation in
          front of main crack. ΔK = 80 MPa$\sqrt{m}$. Room temperature,
          air.

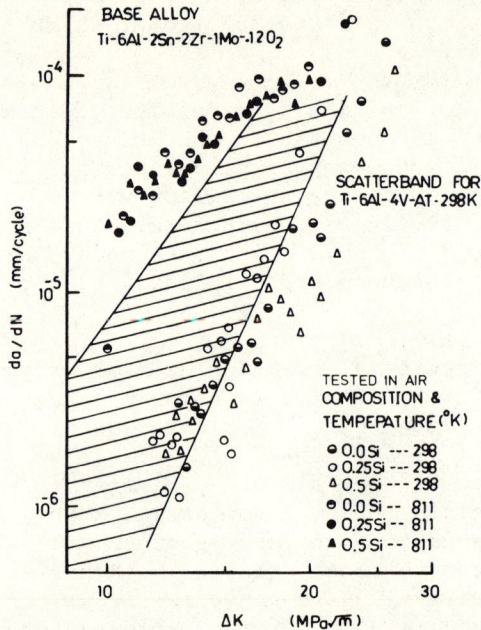

Fig. 18.   Fatigue crack growth rates at 298° and 811°K as a function
           of Si content and stress intensity.

Figure 18 reveals that silicon additions to the creep resistant Ti-6Al-2Sn-2Zr-1Mo-.12O$_2$ alloy did not have any consistent
effect on fatigue crack growth rates. However, it appeared that
the alloy containing 0.5Si had the greatest crack growth
resistance.[72]

With regard to the influence of hydrogen, for the Ti-6Al-4V
alloy tested in vacuum the incidence of cleavage increased with
internal hydrogen concentration. At a hydrogen concentration of
215 ppm, cleavage was prevalent and well defined. At 64 ppm
cleavage markings were less prevalent, while at 8 ppm they were
difficult to detect.[78] Enhanced growth rates in Ti-6Al-6V-2Sn at
0.2 Hz as compared to 20 Hz above a ΔK of 32 MNm$^{-3/2}$ were associated with increased cleavage attributed to hydrogen embrittlement.[79]
For the creep resistant Ti-6Al-2Sn-2Zr-1Mo-.25Si alloy, it appeared
that hydrogen had a small accelerating effect on growth rates,
particularly at intermediate growth rates (10^{-5}–10^{-4} mm/cycle), at
a frequency of 10 Hz. However, when a 5 minute hold was applied,
hydrogen levels of 150-350 ppm significantly increased growth rates
and the cracks propagated parallel to basal hydrides which were
believed to be stress induced.[72] In other investigations the con-
flicting evidence concerning the influence of dwell periods on
crack growth rates in influenced by a combination of hydrogen
content, dwell loading, microstructure and texture.[65,93-97]

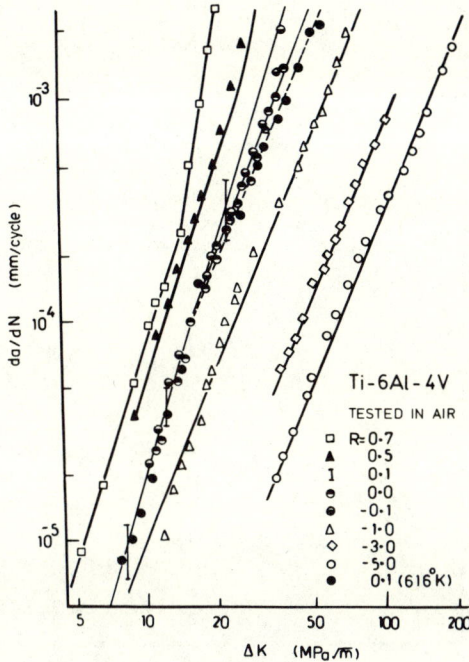

Fig. 19. Fatigue crack growth rate as a function of stress inten-
           sity range for various R ratios (after ref. 77)

     The typical influence of R ratio on crack growth rates in an
air environment is shown in Fig. 19.   For a constant stress intensity range crack growth rates increased as the load ratio increased.
In other investigations involving this alloy (air environment) increasing load ratio increased fatigue propagation rates in the low
and high ΔK regimes but had very little effect at intermediate ΔK
values.   The greatest effect occurred between R = 0.3 to R = 0.5.
Increasing R from 0.5 to 0.7 had little effect on growth rates.[6]
Tests performed at R = 0.12, 0.35 and 0.61 for the same alloy[74]
indicated that above R = 0.35 fatigue crack growth was nearly independent of R while below R = 0.35, a strong element of K_{max} appeared.
Finally, it was observed that the effect of microstructure on
growth rates and fracture topography was enhanced at low R values.

     The influence of vacuum on R ratio dependence is not clear.
Beevers has reported a significant decrease in R dependence in
vacuum.[105]   However, results for the Ti-6242 alloy in vacuum
$(2 \cdot 10^{-5}$ Torr) for R = 0.05 and R = 0.5 do not support these results,
Fig. 20.

     In comparing fatigue crack growth rates in air with those
obtained in vacuum it has been observed that air environments in-

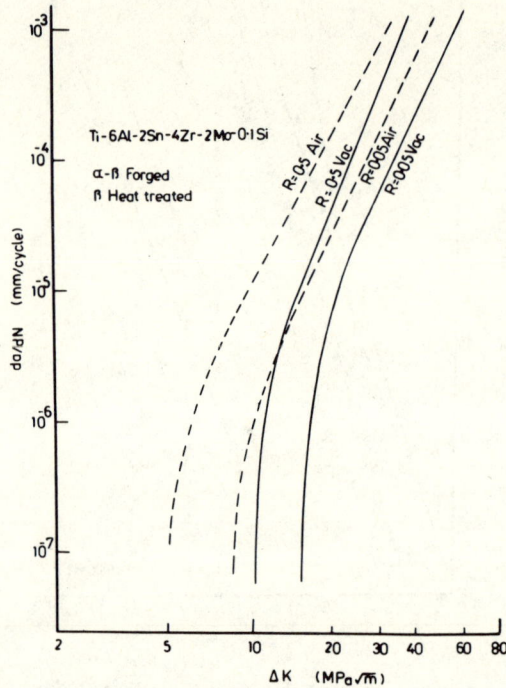

Fig. 20.   Fatigue crack growth rate as a function of stress inten-
           sity range and R ratio in air and vacuum.

crease crack growth rates compared to vacuum. Figure 20 illustrates
this trend for the Ti-6242 Si alloy. The greatest difference in
growth rates occurred at lower stress intensity levels while at
higher growth rates the air and vacuum results are within a factor
of two. Similar results were obtained for Ti-130 [89] and Ti-6-4.
[80,81,88,100]

     These differences in growth rates (air vs vacuum) have been
explained on the basis of factors which include vacuum crack tip
rewelding,[106] crack closure,[107] and the interaction of interstial
oxygen with the base metal, which may modify the local plastic behavior and fracture strain in the vicinity of the crack tip.[36]
Beevers has also suggested that facet linkage, which is K_{max} controlled, is environmentally assisted and may enhance air growth
rates.[80]

     A comparison of the fatigue fracture surfaces in air and
vacuum is also revealing. In general, the vacuum fracture surfaces
are more ductile appearing. This is illustrated in Fig. 21, where
the vacuum facet surface, Fig. 21a, exhibits small dimples whereas
the air facet, Fig. 21b, surface is very crystalline and cleavage-
like.

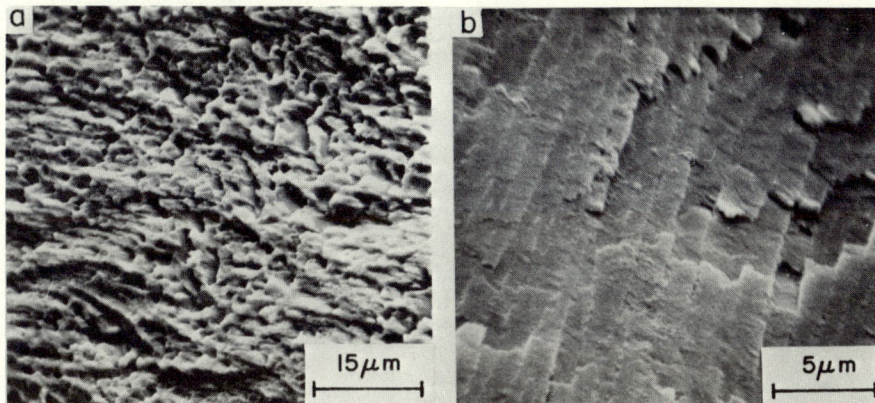

Fig. 21.   Micrographs illustrating facet appearance in (a) vacuum, $\Delta K = 16.5$ MPa$\sqrt{m}$ (b) air, $\Delta K = 9$ MPa$\sqrt{m}$.

Crack Growth at Elevated Temperature

The number of investigations into the fatigue crack growth characteristics of Ti alloys at elevated temperature has been very limited, especially at temperatures in the neighborhood of 540°C. Results for Ti-6-4, Ti-6242 and Ti-6221 are shown in Figs. 18, 22 and 23.   These results indicate a pronounced environmental effect at elevated temperature, particularly at low ΔK levels where time dependent effects such as oxidation, would be expected to predominate.   In the case of Ti-6242 the fracture surface is almost 100% faceted at low ΔK levels.

For elevated temperature tests conducted in vacuum there was no intrinsic temperature effect up to 350°C for the Ti-6-4 alloy.[88] However, testing to 540°C, Ti-6242 Fig. 23 revealed an increase in crack growth rates compared to the room temperature vacuum data.

In other investigations into the effect of temperature on crack growth rates in Ti-6-4, Chesnutt et al[6] found a significant increase in growth rates with temperature (up to 600°F) especially at low ΔK values and high R ratios.   In the work of Hudson[109] the crack growth characteristics of Ti-8Al-1Mo-1V did not change appreciably over the temperature range – 430°C to 290°C, while the crack growth curves for Ti-6Al-4V (annealed) and Ti-4Al-3Mo-1V (aged) appeared slightly more resistant to crack growth at elevated temperature (290°C) than at room temperature.

Crack Growth Modeling

It is of interest to examine results in terms of a crack-opening displacement model for fatigue crack growth which leads to the following equation[81]

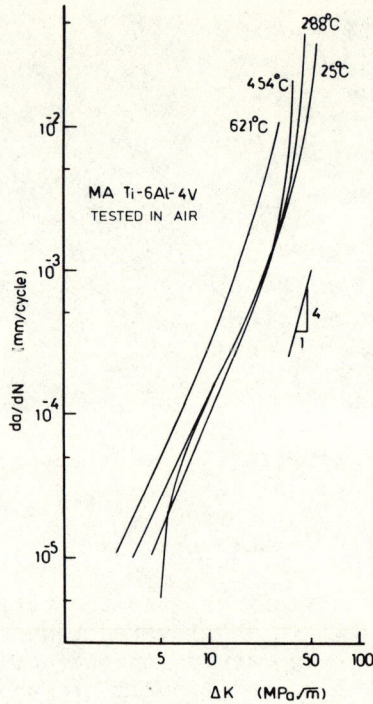

Fig. 22.   Fatigue crack growth rate as a function of stress inten-
           sity level and temperature.[108]

$$\frac{\Delta a}{\Delta N} = \frac{A}{E^2}(\Delta K - \Delta K_{TH})^2 [1 + \frac{\Delta K}{K_c - K_{max}}] \qquad (2)$$

where A is a dimensionless constant and E is the elastic modulus.
The solid curves in Fig. 23 were generated using equation (2) and
it is seen that good correlation to the actual data is obtained.
In this analysis it is found that insofar as the vacuum tests at
25 and 540°C are concerned the increased growth rate at the higher
temperature can be accounted for largely on the basis of the decrease in dynamic elastic modulus with temperature.[101]   Such a
correlation further indicates that time dependent processes, i.e.
creep, are not important under these test conditions. Because of
a strong influence of environment on the crack growth rates in
air at 540°C, no attempt was made to utilize equation (2), since
it is based upon a purely mechanical model.

CONCLUDING REMARKS

     The study of fatigue of creep-resistant titanium alloys at
elevated temperatures is as yet an area of research that has not
drawn extensive attention. On the basis of test programs carried

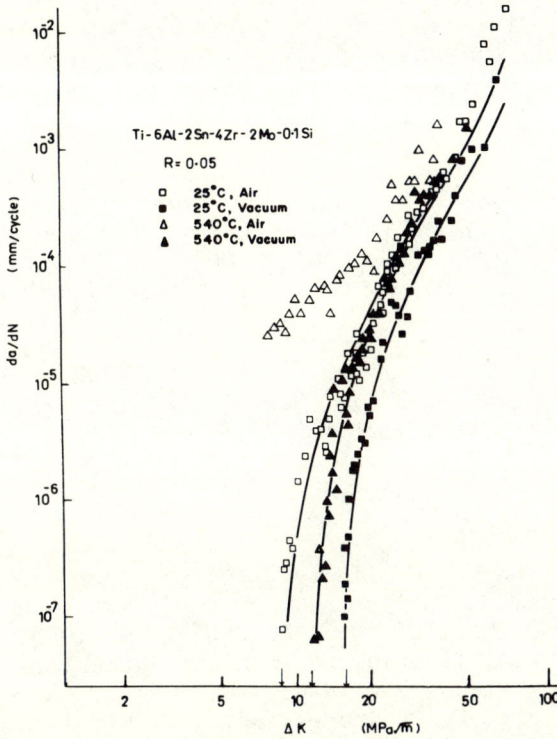

Fig. 23.   Fatigue crack growth rates as a function of environment,
           temperature and stress intensity level.

out, the dominant process affecting both fatigue crack initiation
as well as propagation at elevated temperatures is oxidation.   However, it is to be expected that where longer load time and slower
frequencies are used creep interaction may also become important.
However, the process may be self limiting in actual design, for it
may be necessary to keep the stresses, temperature and times below
the range where creep is significant.

     If oxidation should limit the application of titanium alloys
subjected to cyclic loading at elevated temperatures then more
attention may have to be given to protective coatings as is currently being done with nickel-base superalloys, for example.

     The influence of microstructure on the fatigue process is
understood in a general way.   Titanium alloys display a high sensitivity to microstructure both in crack initiation as well as in
crack propagation.   Unfortunately, those microstructures most resistant to crack initiation under strain controlled low cycle
fatigue conditions may be the least resistant to crack propagation
and vice versa.   In determining thermomechanical processing treat-

ments the materials engineer may have to decide whether crack
initiation or propagation is of greater concern in a particular
application and prescribe the processing procedure accordingly.
This conflict in microstructural requirements is stimulating a new
approach in material design in the form of dual microstructure
components.

By combining processing and heat treating techniques components can be produced with a microstructure at the surface which
is resistant to fatigue crack initiation and an interior microstructure with improved resistance to fatigue crack propagation.

ACKNOWLEDGEMENT

The financial support provided by the Air Force Office of
Scientific Research (Contract No. F49620-79-C-004) is gratefully
acknowledged.

REFERENCES

1. N. E. Paton, J. C. Williams, J. C. Chesnutt and A. W. Thompson,
    Alloy Design for Fatigue and Fracture Resistance, AGARD Con-
    ference Proc. No. 185, Brussels, Belgium, 1975.
2. J. C. Williams, Titanium Science and Technology, Vol. 3, p.
    1433, Plenum Press, New York, N.Y., 1973.
3. J. C. Chesnutt, L. G. Rhodes and J. C. Williams, ASTM STP 600,
    p. 99 ASTM Publication, Philadelphia, PA, 1976.
4. D. Eylon, J. A. Hall, C. M. Pierce and D. L. Ruckle, Met.
    Trans., Vol. 7A, p. 1817, 1976.
5. ASM Handbook, Vol.8: Metallography, Structures & Phase Diagrams
6. J. C. Chesnutt, A. W. Thompson and J. C. Williams, Technical
    Report AFML-TR-78-68, 1978.
7. D. Eylon, Met. Trans., Vol. 10A, p. 311, 1978.
8. D. Eylon and P. J. Bania, Met. Trans., Vol. 9A, p. 1273, 1978.
9. H. Margolin, E. Levine and M. Young, Met. Trans., Vol. 8A, p.
    373, 1977.
10. C. G. Rhodes and J. C. Williams, Met. Trans., Vol. 6A, p. 1670,
    1975.
11. C. G. Rhodes and N. E. Paton, Met. Trans., Vol. 10A, 1979, p.
    1753.
12. C. G. Rhodes and N. E. Paton, Met. Trans., Vol. 10A, 1979, p.
    209.
13. W. G. Burgers, Physica, 1934, Vol. 1, pp. 561-86.
14. P. G. Partridge, Metals and Materials, 1968, pp. 169-194.
15. N. E. Paton, J. C. Williams, and G. P. Rauscher, Titanium
    Science and Technology, Vol. 2, p. 1049, Plenum Press, New
    York, N.Y., 1973.
16. H. Conrad, M. Doner and B. de Meester, ibid., p. 969.
17. T. R. Cass, The Science, Technology and Application of Titanium
    and its Alloys, p. 459, Pergamon Press, London, England, 1970.

18. A. W. Funkenbusch and L. F. Coffin, Met. Trans., Vol. 9A,
     1978, p. 1159.
19. J. A. Ruppen and A. J. McEvily, ASTM, Symposium on Fractogra-
     phy, Williamsburg, VA, 1979, to be published.
20. L. A. Glikman, V. I. Deryabina, N. N. Kolgatin, I. A. Bytenskii
     V. P. Teodorovich and N. S. Teplov, Titanium and its Alloys,
     Publication #10 Investigation of Titanium Alloys, I. I.
     Kornilov Ed., 1963, p. 122.
21. K. Rudinger and H. H. Weigand, Titanium Science and Technology,
     Vol. 4, pp. 2555, Plenum Press, New York, N.Y., 1973.
22. L. Bendersky and A. Rosen, Engineering Fracture Mechanics,
     Vol. 13, 1980, pp. 111-118.
23. M. Kh. Shorshorov, V. N. Mescheryakov and V. A. Matyushkin,
     Titanium Science and Technology, Vol. 4, pp. 2679-2692,
     Plenum Press, New York, N.Y., 1973.
24. H. Kellerer and L. Wingert, Met. Trans., Vol. 2, 1973, p.
     113.
25. A. Rosen and A. Rottem, Materials Sci. and Engr., Vol. 22,
     1976, p. 23.
26. J. E. Reynolds, H. R. Ogden, and R. I. Jaffee, Trans. ASM,
     Vol. 49, 1957, p. 280.
27. K. C. Antony, J. of Materials, Vol. 1, No. 2, 1966, p. 456.
28. S. A. Gorbunov, G. P. Nadutenko, and V. P. Teodorovich,
     Titanium and its Alloys, Publication #10 Investigation of
     Titanium Alloys, I. I. Kornilov Ed., 1963, p. 113.
29. C. E. Shamblen and T. K. Redden, The Science, Technology and
     Application of Titanium, Pergamon Press, N.Y., 1970, p. 199.
30. S. Fujishiro and D. Eylon, Scripta Met., Vol. 11, 1977, p.
     1011.
31. S. Fujishiro and D. Eylon, Scripta Met., Vol. 13, 1979, p.
     201.
32. D. Eylon, M. E. Rosenblum and S. Fujishiro, Proc. Fourth
     Int. Conf. on Titanium, Kyoto, Japan, May 1980, to be pub-
     lished.
33. D. Eylon, T. L. Bartel, and M. E. Rosenblum, Met. Trans.,
     Vol. 11A, 1980, p. 1361.
34. C. Quesne, C. Duong, F. Charpentier, J. F. Fries, and P.
     LaCombe, J. of the Less-Common Metals, Vol. 68, 1979, p. 133.
35. M. C. LaFarie-Frenot, J. Petit and C. Gasc, Fatigue of Engr.
     Mat. and Structures, Vol. 1, 1979, p. 431.
36. J. W. Swanson and H. L. Marcus, Met. Trans., Vol. 9A, 1978,
     p. 291.
37. P. Kofstad, High Temperature Oxidation of Metals, John Wiley
     and Sons, 1966.
38. P. Kofstad, K. Hauffe and H. Kjollesdal, Acta Chem. Scan.,
     12 (1958) 2, p. 239.
39. T. Hurlen, J. Inst. of Metals, Vol. 89, 1960, p. 128.
40. W. A. Alexander and L. M. Pidgon, Canadian J. of Research,
     Vol. 28, Sec. B #1, 1950, p. 60.

41. D. V. Ignatov, Titanium and its Alloys, Publication #10 Inves-
     tigation of Titanium Alloys, I. I. Kornilov Ed., 1963, p.
     261.
42. R. P. Skelton and J. I. Bucklow, Metal Science, Feb. 1978,
     p. 64.
43. L. F. Coffin, Trans. ASM, Vol. 56, 1963, p. 339.
44. G. Ward, B. S. Hockenholl and P. Hancock, Met. Trans., Vol.
     5, 1974, p. 1451.
45. D. V. Ignatov, Z. I. Kornilova and E. M. Lazarev, Titanium
     Science and Technology, Vol. 4, p. 2545, Plenum Press, New
     York, N.Y., 1973.
46. H. W. Rosenberg, Titanium Sci. and Tech., Eds., R. I. Jaffee
     and H. M. Burte, 1973, Plenum Press, N.Y., p. 2127.
47. M. Kehoe and R. W. Broomfield, "Titanium Sci. and Tech." Eds.
     R. I. Jaffee and H. M. Burte, Vol. 4, 1972 , p. 2167.
48. G. S. Hall, S. R. Seagle and H. B. Bomberger, ibid, p. 2141.
49. S. R. Seagle and H. B. Bomberger, The Sci. and Tech. and
     Application of Ti. Eds. R. I. Jaffee and N. E. Promisel,
     Pergamon Press, 1970, p. 1001.
50. N. E. Paton and M. W. Mahoney, Met. Trans., 7A, 1976, p. 1685.
51. H. Conrad, M. Doner and B. deMeester, Titanium Sci. and Tech.,
     Plenum Press, N.Y., 1973, p. 969.
52. M. Doner and H. Conrad, Met. Trans., 4, 1973, p. 2809.
53. J. Weertman, J. Appl. Phys., Vol. 26, pp. 1213, 1955.
54. R. L. Orr, O. D. Sherby and J. E. Dorn, Trans. ASM, Vol. 45,
     1954, p. 113.
55. F. B. Cuff and N. J. Grant, Iron Age, 1952, p. 134.
56. F. R. Larson and J. Miller, Trans. ASME, Vol. 74, p. 765, 1952.
57. J. Weertman, Trans. ASM, 61, 1968, p. 681.
58. M. F. Ashby, Acta. Met., 20, 1972, p. 887.
59. K. Okazaki, T. Odaware and H. Conrad, Scrip. Met., 11, 1977,
     p. 437.
60. M. F. Ashby, "Fracture-1977", Ed. D.M.R. Taplin, Vol. 1,
     Pergamon Press, 1978.
61   L. S. Richardson and N. J. Grant, Trans. AIME, 215, 1959, p.
     18.
62. Y. K. Mohan Rao, V. K. Rao and P. R. Rao, Scripta Met., 13,
     1979, p. 851.
63. S. A. Sattar, D. H. Kellogg, K. J. Oberle and G. N. Green,
     ASM Technical Report, No. D-8-24, 4, 1968.
64. J. J. Lucas, Titanium Science and Technology, Vol. 3, p. 2081,
     Plenum Press, New York, N.Y., 1973.
65. D. Eylon and J. A. Hall, Met. Trans., Vol. 8A, 1977, p. 981.
66. D. F. Neal and P. A. Blenkinsop, Acta Met., Vol. 24, 1976,
     p. 59.
67. J. Ruppen, P. Bhowal, D. Eylon and A. J. McEvily, ASTM STP
     675, p. 47, ASTM Publication, Philadelphia, PA, 1979.
68. C. Hoffmann, D. Eylon and A. J. McEvily, Int. Symp. on Low
     Cycle Fatigue and Life Prediction, Firminy, France, Sept.
     1980, ASTM, to be published.

69. D. Eylon and C. M. Pierce, Met. Trans., 7A, 1976, p. 111.
70. G. R. Yoder, L. A. Cooley and T. W. Crooker, Met. Trans., A, 1977, Vol. 8A, p. 1737.
71. W. R. Kerr, D. Eylon and J. A. Hall, Met. Trans., A, 1976, Vol. 7A, p. 1477.
72. M. W. Mahoney and N. E. Paton, Met. Trans., Vol. 9A, 1978, p. 1497.
73. J. R. Robinson and C. J. Beevers, 2nd Intern. Conf. on Ti, Cambridge, MA, 1972.
74. P. E. Irving and C. J. Beevers, Mater. Sci. and Eng., 14, 1974, p. 229-238.
75. R. J. Bucci, P. C. Paris, R. W. Hertzberg, R. A. Schmidt and A. F. Anderson, ASTM STP 513, p. 125, ASTM Publication, Philadelphia, PA, 1972.
76. R.J.H. Wanhil, Met. Trans., Vol. 7A, 1976, p. 1365.
77. A. Yuen, S. W. Hopkins, G. R. Leverant and C. A. Rau, Met. Trans., Vol. 5A, 1974, p. 1833.
78. D. A. Meyn, ASTM STP 600, 1976, p. 853.
79. M. R. Moody and W. W. Gerberich, Metal Science, March, 1980, p. 95.
80. C. J. Beevers, Met. Sci., 1977, 11, p. 362.
81. R. Ebara, K. Inoue, S. Crosby, J. Groeger and A. J. McEvily, Proc. 2nd Int. Conf. on Mech. Beh. of Mater., ASM, Cleveland, Ohio, 1976, p. 685.
82. J. L. Robinson and C. J. Beevers, Proc.2nd International Conf. on Titan. Sci. and Technology, Titanium Science and Technology, Vol. 2, 1973, p. 1245.
83. G. W. Salgat and D. A. Koss, Materials Science and Eng., 35 1978, p. 263-272.
84. Anthony W. Thompson, James C. Williams, J. D. Fransden and J. C. Chesnutt, Third Int. Conf. on Ti., Moscow, May, 1976.
85. G. R. Yoder, L. A. Cooley and T. W. Crooker, NRL Report 8166, 1977.
86. G. R. Yoder, L. A. Cooley and T. W. Crooker, Jnl. of Eng. Maters. and Techno., Vol. 101, 1979, p. 87.
87. G. R. Yoder, L. A. Cooley and T. W. Crooker, Vol. 11, No. 4, Eng. Frac. Mech., 1979, p. 805.
88. C. J. Beevers and P. E. Irving, ICF3, Munich, Germany, 1973, p. 174.
89. J. L. Robinson and C. J. Beevers, Metal Sci. J., 1973, Vol. 7, p. 153.
90. A. W. Thompson, J. D. Fransden, J. C. Williams, Metal Science, 1975, Vol. 9, p. 46.
91. M. J. Harrigan, M. P. Kaplan and A. W. Sommer, Fracture Prevention and Control, D. W. Hoeppner, ed., Mater./Metal Work, Technical Series, No. 3, p. 225, ASM, Metals Park, OH, 1974.
92. J. C. Chesnutt, Report SG584.40IR, Science Center, Rockwell International, Thousand Oaks, CA, 1977.
93. P. J. Bania and D. Eylon, Met. Trans., Vol. 9A, 1978, p. 848.

94. A. W. Thompson, J. C. Chesnutt and J. D. Frandsen, Interim Report on Air Force Contract No. F33615-74-6-5067, Rockwell Int., June 30, 1976.

95. A. W. Thompson, J. C. Chesnutt and J. D. Frandsen, Interim Report on Air Force Contract No. F33615-74-6-5067, Rockwell Int., Sept. 30, 1976.

96. J. T. Ryder, D. E. Pettit, W. E. Krupp and D. W. Hoeppner, Final Report on Evaluation of Mechanical Property Character. of IMI 685, Lockheed-Calif. Company, Rye Canyon Laboratory, Oct. 1973.

97. C. A. Stubbington and S. Pearson, Eng. Frac. Mech., Vol. 10, pp. 723, 1978.

98. G. R. Yoder, L. A. Cooley and T. W. Crooker, Jnl. of Eng. Mater. Sci. and Tech., Vol. 99, 1977, p. 313.

99. J. A. Carlson and D. A. Koss, Acta Met., 26, 1978, p. 123.

100. P. E. Irving and C. J. Beevers, Met. Trans., Vol. 5A, 1974, p. 391.

101. J. Ruppen and A. J. McEvily, Fatigue of Eng. Mats. and Structures, Vol. 2, 1979, p. 63.

102. S. G. Chakrabortty, Fatigue of Engineering Materials and Structures, Vol. 2, No. 3, 1979, p. 331.

103. D. L. Davidson and D. Eylon, Met. Trans., Vol. 11A, 1980, p. 837.

104. D. Schechtman and D. Eylon, Met. Trans., A, 1978, Vol. 9A, p. 1018-1020.

105. J. L. Robinson, P. E. Irving and C. J. Beevers, Third Int. Congress on Fracture, Munich, 1973.

106. R. M. N. Pelloux, Trans. ASM, 62, 1969, p. 281.

107. W. Elber, Eng. Frac. Mech., 2, 1970, p. 37.

108. P. L. Sallade, Masters Thesis, University of Connecticut, Storrs, CT, 1970.

109. C. M. Hudson, NASA-Langley Report L-3784, 1964.

CREEP-FATIGUE-EFFECTS IN COMPOSITES

N. S. Stoloff

Materials Engineering Department
Rensselaer Polytechnic Institute
Troy, NY 12181

ABSTRACT

The fatigue behavior at elevated temperatures, mostly in vacuum, of several cobalt and nickel-base eutectic composites is described. Decreased fatigue life and increased crack propagation rates are noted as test frequency is reduced. A shift from surface to internal crack initiation occurs at low frequencies and high test temperatures. Fatigue life data at various frequencies for two alloys were correlated by an equation of the form $N_f^{-\beta}\nu = K(\Delta\sigma)$. Creep-fatigue interactions appear to be responsible for the observed behavior, with void growth around cracked fiber ends the mechanism of damage accumulation.

INTRODUCTION

Many polycrystalline alloys display, at elevated temperatures, low cycle fatigue lives or crack growth rates that are susceptible to change as frequency[1,2] or waveform[3-5] of the imposed fatigue cycle is changed. In general, lives are lowered as frequency is lowered and crack propagation rates are increased by hold times or, in some cases, by unbalanced load or strain cycles. In the latter type of test, when the tensile-going strain rate is higher than the compression going strain rate, for a constant test frequency, lives are lengthened, relative to those achieved with a balanced cycle.[6] However, upon reversing the sequence of loading to achieve "slow-fast" cycling, lives usually are reduced. Coffin and co-workers have demonstrated that for A-286 and 304 stainless steel frequency and wave shape effects may be accounted for by interaction between fatigue cracks and the environment.[1,2,6] Thus, at low frequencies, or for long hold-times in tension, chemical processes lead to acceleration of micro and macrocrack growth rates, and fracture is by an inter-

301

granular path. Creep damage, on the other hand, may be produced with
unbalanced waveforms, due to the accumulation of damage by cavity
nucleation and growth at grain boundaries, again with a tendency for
grain boundary fracture. Only at high frequencies in vacuum is transgranular cracking expected to predominate.

In the case of eutectic composites, several instances of frequency or hold-time effects on fatigue behavior have been reported.[7-11]
These reports are noteworthy in that eutectic alloy specimens usually
consist of very few large grains, so that damage accumulation by
environmental attack or cavity nucleation and growth must occur by
some process not associated with the presence of grain boundaries.
It is the purpose of this paper to review frequency, hold-time and
wave shape experiments on eutectic composites, to contrast their
behavior with that of conventional polycrystalline alloys, and to
propose a mechanism by which accelerated crack growth may be achieved.

MATERIALS AND TEST CONDITIONS

Four aligned eutectics, described in Table I, have provided data
which can be interpreted as arising from a creep-fatigue or fatigue-
environment interaction. Three of these alloys, Cotac, Nitac and
$\gamma/\gamma'-\alpha$, are fibrous; the one lamellar alloy, $\gamma/\gamma'-\delta$, is also the only
one which has been tested under low cycle (strain-controlled) condi-
tions.[10] All other alloys have been tested under axially loaded
stress-controlled conditions, with a small positive minimum stress,
i.e., R, the ratio of minimum to maximum stress, is about 0.1.
Smooth bar σ-N data as well as the results of crack propagation
experiments on precracked samples are reviewed. Most experiments
were conducted in vacuum of less than 7×10^{-4} MPa at room temperature and less than 4×10^{-3} MPa at elevated temperatures. For
$\gamma/\gamma'-\delta$, the tests reviewed here were performed in air.[10]

Table I.   Eutectic Alloy Microstructure, Compositions, w%

	Type	Vol. Frac.	Ni	Co	Cr	Al	Nb	Mo	Ta	C
Nitac	Fib.	0.05	69	--	10	5	--	--	14.9	1.1
Cotac	Fib.	0.10	10	65	10	--	--	--	14	1
$\gamma/\gamma'-\delta$	Lam.	0.37	71.5	--	6	2.5	20	--	--	--
$\gamma/\gamma'-\alpha$	Fib.	0.26	65.5	--	--	8.1	--	26.4	--	--

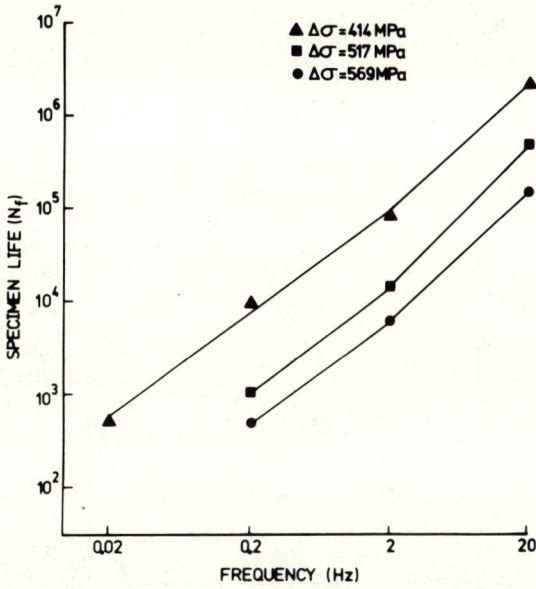

(a)

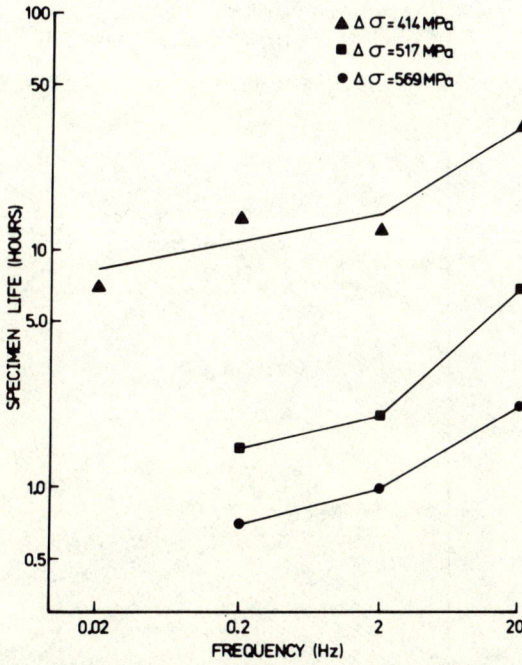

(b)

Fig. 1.   Effects of frequency on high cycle fatigue life of Nitac
at 825°C in vacuum.[7]
(a) number of cycles; (b) duration.

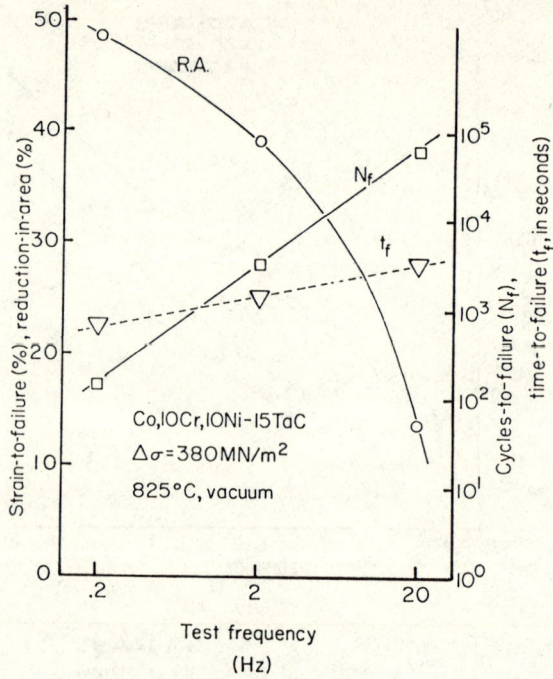

Fig. 2.   Changes in lifetime and ductility of Cotac with test fre-
          quency.[8]

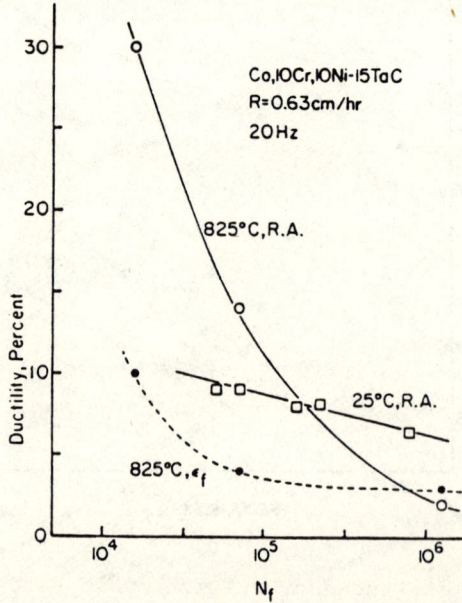

Fig. 3.   Cyclic ductility of Cotac at 25 and 825°C.[8]

EFFECTS OF TEST FREQUENCY AND ENVIRONMENT

Fatigue Life

The effects of frequency on the fatigue behavior of Nitac and
Cotac in vacuum, shown in Figs. 1[7] and 2,[8] are very similar; fatigue
lives, measured in elapsed time, were reduced about ten times at any
stress level for each ten-fold reduction in frequency. Similarly,
the number of cycles to failure decreased significantly. Additional
strong evidence of a time-dependent fatigue process in this class of
alloys is provided by the large increase in percent reduction in area
of Cotac, tested at 825°C, from 13% at 20 Hz to 48% at 0.2 Hz, also
shown in Fig. 2. Also, the cyclic ductility at 825°C was considerably higher than at 25°C (ν = 20 Hz), see Fig. 3,[8,11] for Cotac
tested at relatively high stresses. At 25°C, % RA decreased only
slightly, from 9 to 7%, as N_f increased from 5 x 10^4 to 9 x 10^5
cycles to failure. Lower stress, associated with higher N_f, permits
a longer fatigue crack to exist prior to final overload fracture.
The measured % RA at 25°C is assumed to arise predominantly from the
overload portion of the failure. At 825°C, on the other hand, % RA
increases rapidly as N_f is reduced (higher stress levels), indicative
of an axial creep component that is very sensitive to applied stress
(a stress exponent of 15.1 relating creep rate to stress has been
noted in creep tests on Nitac at 825°C).[7] High stress exponents
generally are noted for other fibrous eutectics.[12]

Similar fatigue data as a function of frequency have been
obtained at 725°C in vacuum for a Ni-Al-Mo, γ/γ'-α alloy designated
AG-15, as shown in Fig. 4 (a).[13] Below $\Delta\sigma$ = 925 MPa, there is a
pronounced decrease in life as frequency is reduced from 20 to 0.2
Hz. Material tested at 20 Hz but at higher stress levels exhibits
a sharp departure in slope of the σ-N curve relative to that noted
at lower stress levels. The data of Fig. 4 (a) are suggestive of a
strong creep effect which may occur at any stress level above the
endurance limit as frequency decreases, but also at stresses above
a critical level for tests run at 20 Hz. A puzzling feature of these
data, however, is the lack of any change in slope of the σ-N curves
for 2 Hz or 0.2 Hz at stress levels below 925 MPa, since one could
expect a pronounced stress effect on creep rate to occur at progressively lower stress levels as test frequency is lowered. Another
unusual observation was the increased elapsed time to failure associated with lower frequencies, see Fig. 4 (b).[13] These data contrast
sharply with previously described results for Nitac, Fig. 1, in which
both cycles to failure and elapsed time decreased with decreasing ν.

Fatigue life data in both air and vacuum have been reported for
Cotac as a function of test temperature, see Fig. 5. At 25°C, vacuum
had a beneficial effect on life, particularly at low stress levels
while at 825°C, no effect of atmosphere was seen at any stress level.

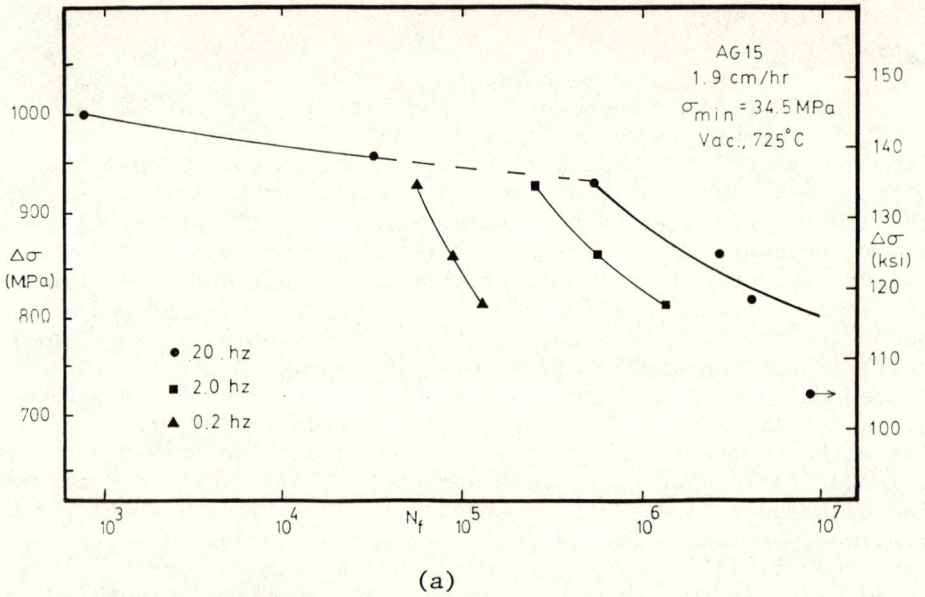

AG 15
1.9 cm/hr
σ_{min} = 34.5 MPa
Vac., 725°C

● 20 hz
■ 2.0 hz
▲ 0.2 hz

(a)

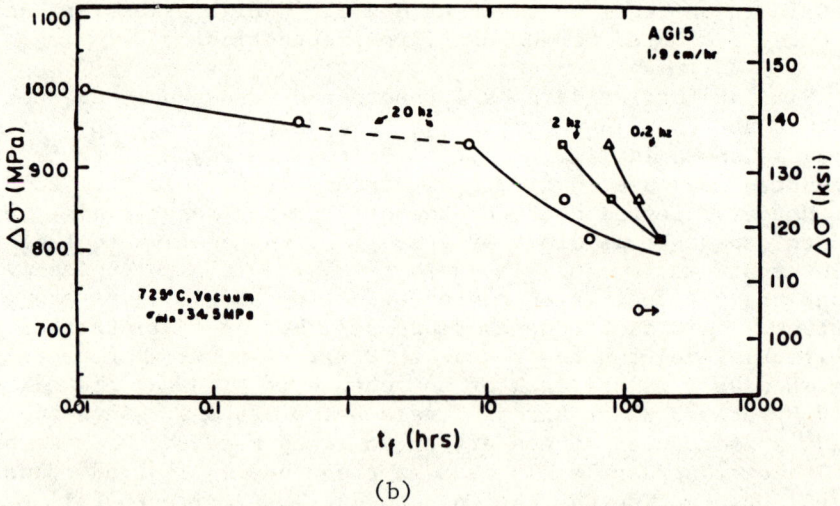

(b)

Fig. 4. Effects of frequency on high cycle fatigue life of γ/γ'-α
(AG–15) at 725°C in vacuum. [15]
(a) number of cycles
(b) duration.

At 1000°C in air, oxidation was readily apparent, and fatigue lives
were much lower than at 825°C.

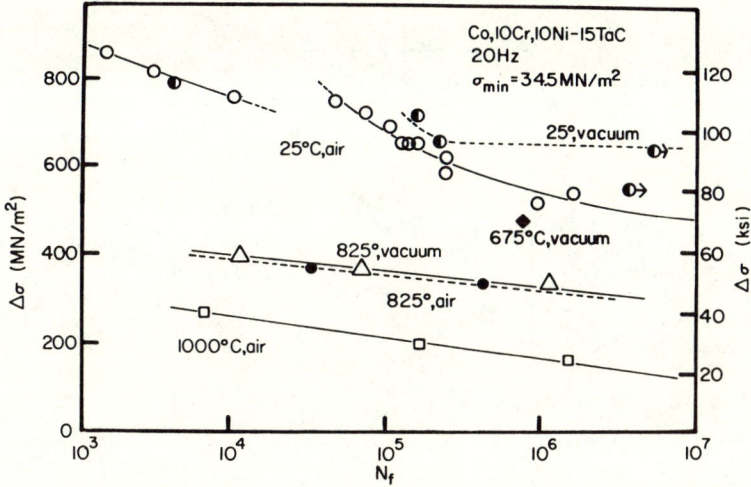

Fig. 5.   Effect of environment on high cycle fatigue life of Cotac at various temperatures.[8]

Crack Propagation

Although test frequency has been shown to exert a profound effect on fatigue _life_ of several eutectic alloys, only crack propagation tests can distinguish between frequency dependent initiation or propagation. Figs. 6 (a) and (b) show fatigue crack growth rates for Cotac at 725°C and 900°C, respectively.[9]   At 750°C there is a significant increase in crack growth rate at the lower frequency, 2 Hz, but only for stress-intensity ranges, ΔK, $\geq$ 20 MPa. At 900°C, there is also an apparent increase in the _slope_ of the da/dN vs. ΔK curve, provided that fibers were intact prior to test. In prestrained material, on the other hand, the limited data suggest that there is no significant effect of frequency upon the slope.

CORRELATION OF CREEP AND FATIGUE DATA

Creep rupture data for Nitac have been obtained at 825°C and 900°C.[7] A modified Larson-Miller parameter,[14] utilizing a stress-dependent constant, was used to correlate rupture data:

$$P_w = T[\frac{2000}{\sigma} + \log t_R] \tag{1}$$

where P_w is the modified parameter, T is temperature (°K), σ is stress and t_R is rupture time, see Fig. 7. The stress-dependent first term was introduced by Woodford[14] to eliminate environmental effects on creep damage in Nitac for low stress, long time tests. Fatigue data obtained on Nitac at 825°C correlated well on the same graph when

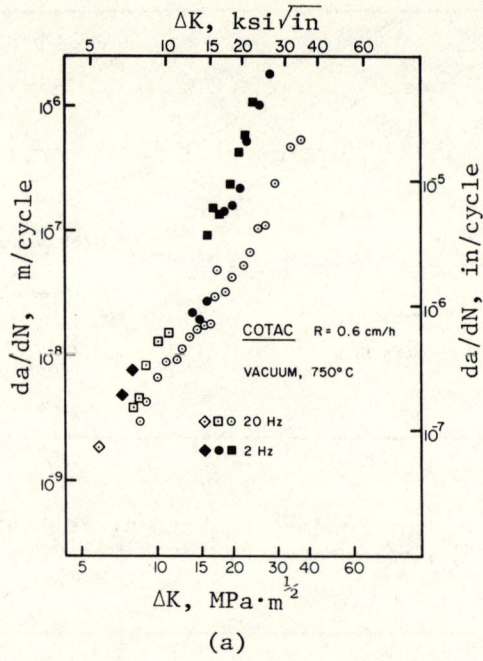

(a)

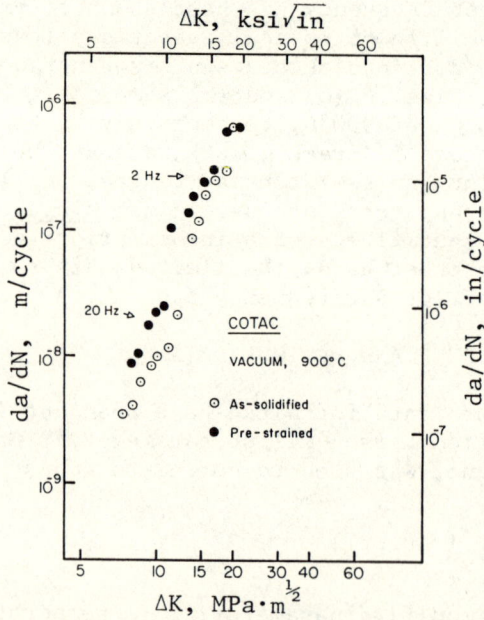

(b)

Fig. 6. Crack propagation rate vs. stress intensity, Cotac.[9]
         (a) 750°C; (b) 900°C.

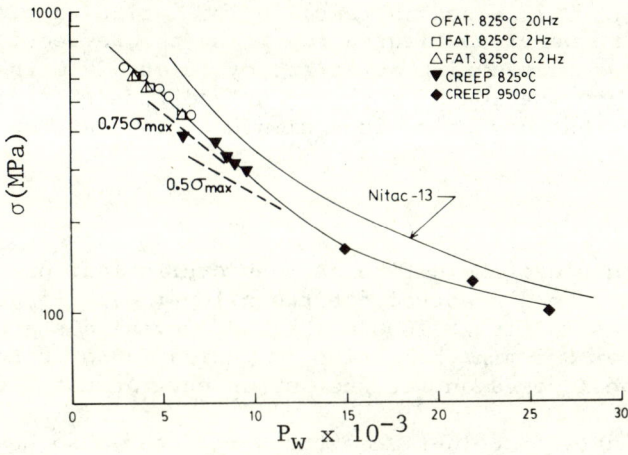

Fig. 7.   Correlation of high cycle fatigue data with creep data
          according to Eq. (1).[14]   For fatigue samples, σ is maximum
          stress or indicated fraction of maximum stress, t_R.

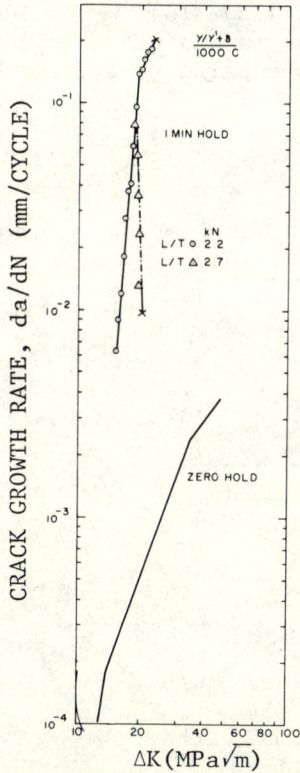

Fig. 8.   Effect of hold-time on crack growth rate of $\gamma/\gamma'-\delta$ at
          1000°C in air.[10]   L/T refers to orientation of sample.

peak stress (or 0.75 x peak stress) in the fatigue cycle and elapsed
time to failure were substituted for σ and t_R, respectively. These
results suggest that during a fatigue cycle at 825°C the damage produced at any combination of peak stress and time to fail the specimen
is similar to that occurring in a simple creep test for the same time
at temperature.

HOLD-TIME EFFECTS

    The influence of hold-time at peak cyclic load on fatigue crack
growth rate has been reported for two alloys: $\gamma/\gamma'-\delta$, and Cotac.
Tests on $\gamma/\gamma'-\delta$ in air at 1000°C, Fig. 8, reveal a sharp increase in
growth rate for one min. hold-time,[10] indicative of either a creep
component superimposed on fatigue, or an environmental effect.

    Limited data on hold-time effects in Cotac also have been
obtained.[9] Two types of load waveform, see Fig. 9, were used in
vacuum in an attempt to determine the roles of creep and strain rate.
The first waveform type increased the applied load linearly to a maximum in 0.1 s, remained at maximum for 0.8 s and then decreased in
0.1 s to minimum load. Maximum load was adjusted often to keep ΔK
constant; minimum load was maintained near zero. An alternate form
of this waveform also was applied, i.e., the hold period occurring

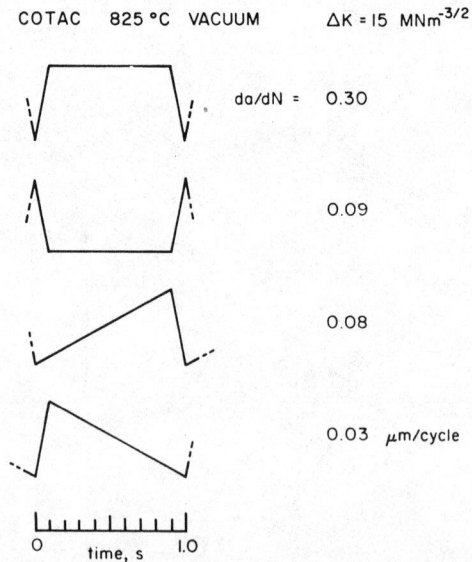

COTAC   825 °C   VACUUM $\Delta K = 15$ MNm$^{-3/2}$

da/dN =   0.30

0.09

0.08

0.03 μm/cycle

O   time, s   1.0

Fig. 9.   Load waveforms used in FCP experiments on Cotac at 825°C
           in vacuum.[9]

at the minimum load, and rise and fall times as before.   Therefore,
the only difference was whether a hold period was applied at maximum
or minimum load.

The second waveform type, also repeated at 1.0 Hz, features a
0.1 s rise time to maximum load followed immediately by a 0.9 s fall
to minimum load.   The alternative of this waveform (0.9 s rise and
0.1 s fall) also was applied.   The sequence of tests and the resulting data are detailed in Table II.   Each value of da/dN in Fig. 8 is
the average of two measurements, corresponding to propagation produced
by each of the four waveforms, on a single test specimen.

Table II.   Sequence of Application of Load Waveforms, shown in Fig. 9.

Waveform	ΔN	a_o, mm	Δa, mm	$\Delta a/\Delta N$, μm/cycle
slow-fast	2275	4.65	0.14	0.060
fast-slow	3175	4.79	0.08	0.026
slow-fast	2080	4.87	0.23	0.11
fast-slow	3850	5.10	0.15	0.039
low hold	1120	5.39	0.15	0.013
high hold	930	5.54	0.34	0.36
low hold	1550	5.97	0.08	0.054
high hold	1025	6.05	0.25	0.25

Three conclusions arise from these results: (1) A maximum load
hold-time causes faster cracking than a low load hold-time.   (2) More
rapid cracking results from a slow loading/fast unloading rate waveform than vice versa.   (3) The waveform with the low hold-time caused
cracking at about the same rate as the slow loading/fast unloading
rate waveform.

The relationship between waveform and frequency effects in Cotac
is not straightforward.   A ten-fold decrease in frequency also produces a ten-fold decrease in average strain rate, which itself can
influence the rate of crack growth by lowering the flow stress of
the material in the plastic zone ahead of the crack tip.   Increasing

frequency does not affect average stress in terms of time or cycles,
but the stress-time product per cycle, $\int_{t_o}^{t_{2\pi}} \sigma dt$, is increased
proportionally. This quantity also is higher for the waveform with

(a)

(b)

Fig. 10. Surface crack initiation sites, $25°C$, $\nu = 20$ Hz.
(a) $\gamma/\gamma'-\alpha$ (AG-15)[15]
(b) Nitac.[7]

the high-stress hold-time relative to the low-stress hold-time. For
both situations the increase in the stress-time product per cycle
causes an increase in strain range at the crack tip. The waveform
testing indicated that average stress was somewhat more important
than strain rate in determining da/dN. On this basis, the observed
decrease in crack growth rate with increasing frequency can be
explained. The changes in frequency dependency with ΔK, Fig. 6 (a),
may be due to changes in the relative importance of strain rate and
stress-time product.

METALLOGRAPHY AND FRACTOGRAPHY

    Fatigue fractures at 25°C in γ/γ'-α, Nitac and Cotac always were
surface initiated. Fig. 10 (a) shows such a fracture in γ/γ'-α with
river patterns emanating from the origin.[15] Stage I cracking occurred
in the fatigue zone, A; the remainder of the fracture surface, B, is
due to overload. A similar surface crack origin for Nitac is shown
in Fig. 10 (b).[7] Fibers are cracked readily during fatigue at 25°C,
but most fiber cracks do not extend into the matrix. The mechanism
of link-up of large cracks is by fracture along slip planes of the
matrix, Fig. 11 (a), leading to a very faceted, stage I-type fracture
surface, Fig. 11 (b).[7] Similar observations have been made on Cotac[9]
and γ/γ'-α.[15]

    At 725°C, the fatigue fractures in γ/γ'-α remained surface initi-
ated at all test frequencies. Fig. 12 (a) depicts a surface initiation site for a 2 Hz test, which was identical in appearance to the
origin in a 20 Hz test. A small amount of stage I cracking was noted
on these samples, but stage II (fibrous) cracking predominated at all
frequencies, as shown in Fig. 12 (b). Outside of the thumbnail
fatigue zone the crack path became very terraced, Fig. 12 (c), before
a final overload zone, very similar to those in room temperature
specimens, appeared, as shown in Fig. 12 (d). The fractographic
features of the two specimens comprising the upper branch of the 20
Hz fatigue curve of Fig. 4 are somewhat different than those described
above. The specimen tested at $\Delta\sigma$ = 1000 MPa (N_f = 780) had no fatigue
zone at all, and in fact the fracture resembled very closely that of
a tensile specimen broken at the same temperature; the sample tested
at $\Delta\sigma$ = 960 MPa revealed a small, surface initiated fatigue zone.

    At 825°C, all tests on γ/γ'-α were run at 20 Hz and the fatigue
fractures were centrally initiated, see Fig. 13, in sharp contrast
to the surface initiation sites noted at 25°C and 725°C. In the
fatigue zone stage II fracture predominated.

    The fracture morphologies noted in Nitac and Cotac at 825°C in
vacuum differed somewhat from those noted in γ/γ'-α. For example,
the appearance of fractographic features in Nitac depended upon
stress level. High stresses induced failure by internal crack
nucleation; no distinct fatigue zone could be distinguished at low

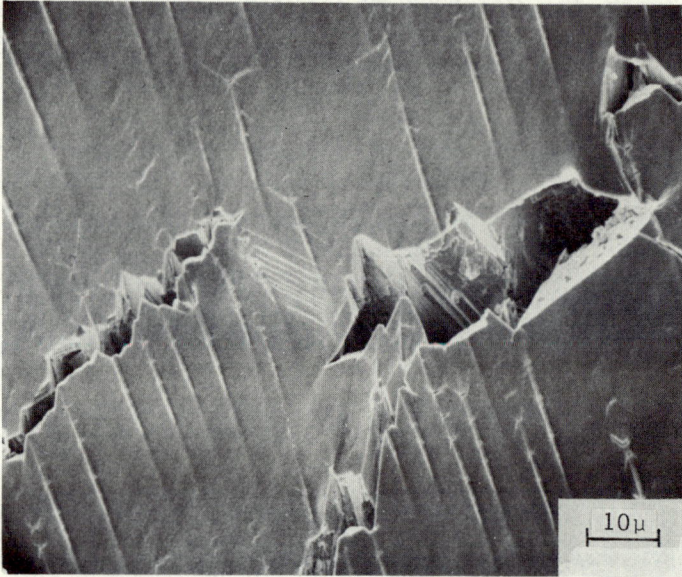

(a)

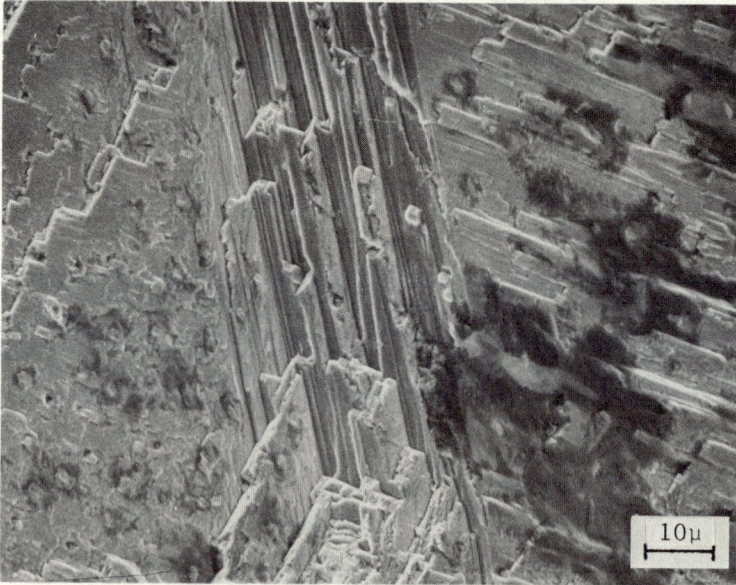

(b)

Fig. 11. Stage I crack growth in Nitac cycled in air at 25°C,
          ν = 20 Hz.[7]
          (a) cracks along slip bands, longitudinal surface
          (b) fractograph.

(a)

(b)

Fig. 12.   SEM fractographs of γ/γ'-α (AG-15) cycled at 725°C in
          vacuum, $\nu = 2$ Hz.
          (a) surface crack origin
          (b) stage II (fibrous) cracking

(c)

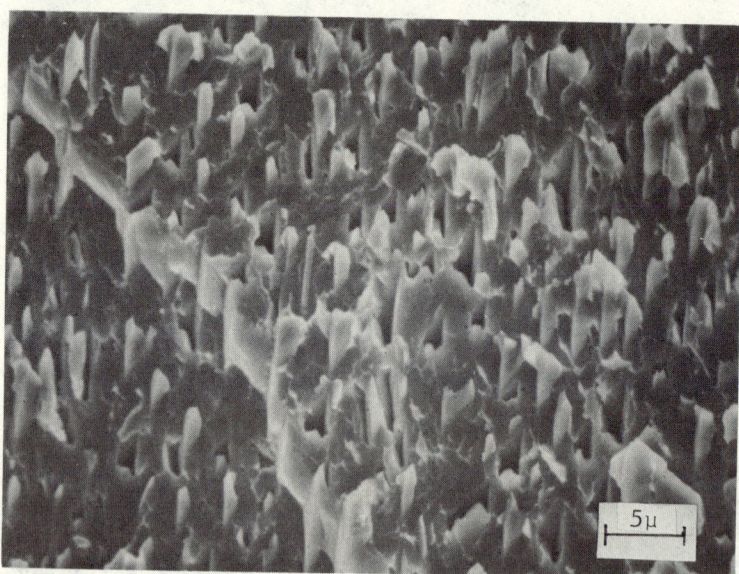

(d)

Fig. 12. SEM fractographs of $\gamma/\gamma'-\alpha$ (AG-15) cycled at 725°C in
          vacuum, ν = 2 Hz.
          (c) terraced region at end of fatigue zone
          (d) overload zone.

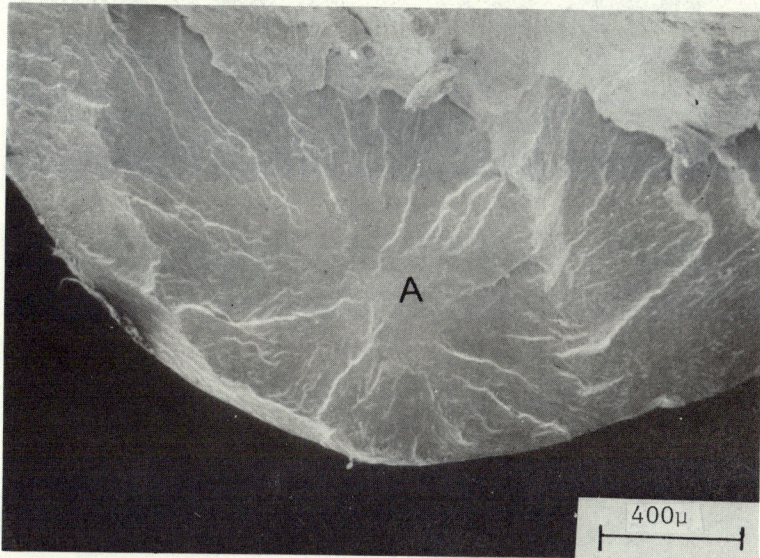

Fig. 13.   SEM Fractograph of $\gamma/\alpha'-\alpha$ (AG-15) cycled at 825°C in
vacuum, ν = 20 Hz, showing internal crack initiation
site (A).

stresses, and cracks were surface initiated, see Fig. 14 (a). Lower
frequencies favored shear-lip type fractures of the type noted only
at high stress levels for 20 Hz tests. Cracks initiated internally,
Fig. 14 (b), and propagated outward by a hole growth and coalescence
mechanism until final separation by a shear mechanism.[16] The tensile
overload zone formed a 100–150 μm ring around the central slow growth
region. Finally, samples tested at low σ and low frequency exhibited
surface voids centered on broken fibers; the voids coalesced along
circumferential deformation bands, see Fig. 15.[16] By way of comparison, samples tested under static load (creep-rupture) exhibited fracture surfaces with two distinct regions, one of slow growth at the
center, surrounded by a tensile overload zone. That is, the fracture
surface of creep-rupture samples resembled those of fatigue specimens
tested at low frequency and low to intermediate stress levels.

Fracture surfaces of Cotac tested at 825°C in vacuum did not
change significantly with test frequency.[8] Dimples were noted for
all frequencies, Figs. 16 (a) and 16 (b), in marked contrast to the
crystallographic, stage I-type fractures noted at 25°C in this and
the other eutectic alloys. When Cotac was prestrained several percent in tension at 25°C, prior to testing at 825°C in vacuum, an
extensive series of microcracks, showing link-up of voids around
individual fibers, could be seen, Fig. 17. Thus both Nitac, Fig. 15,
and Cotac, Fig. 17, display void formation at cracked fiber ends when
tested in fatigue at 825°C.

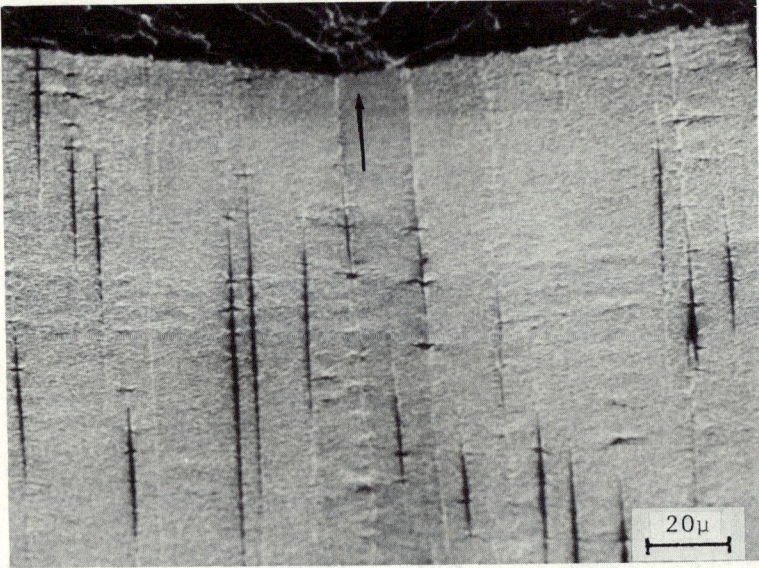

(a)

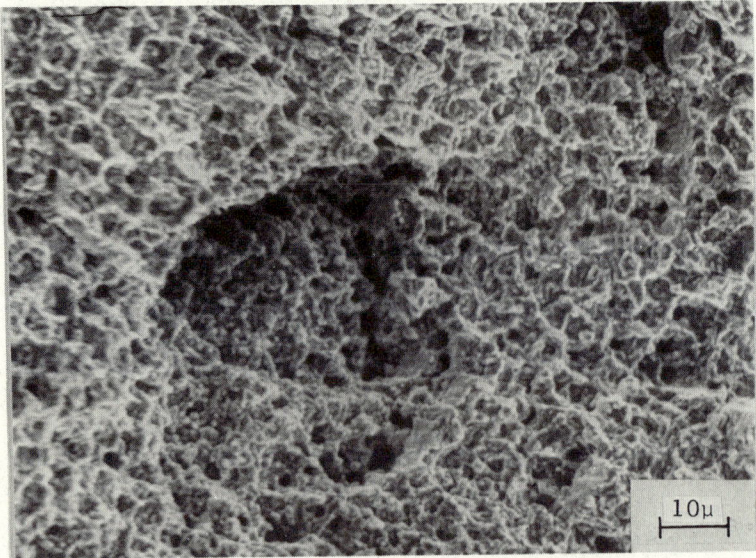

(b)

Fig. 14.   SEM fractographs of Nitac cycled at $825°C$ in vacuum,[18]
           $\Delta\sigma = 414$ MPa.
           (a) 20 Hz, surface initiation
           (b) 0.2 Hz, internal initiation.

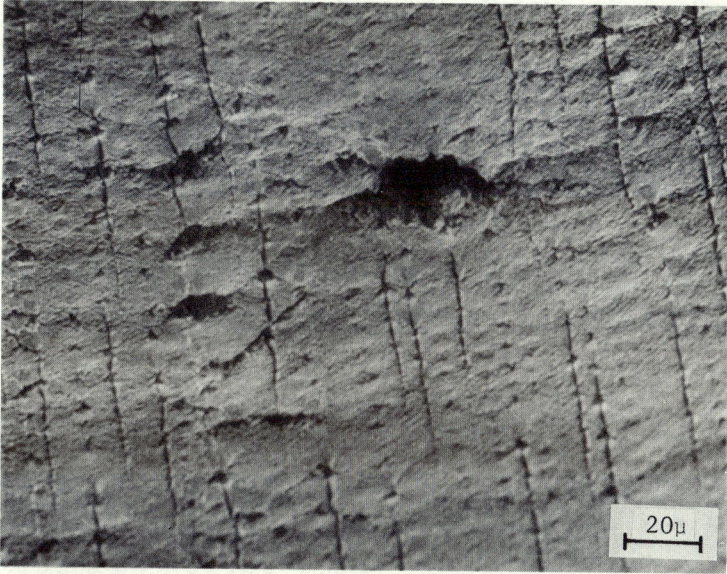

Fig. 15.   Longitudinal surface of Nitac cycled at 825°C in vacuum, $\Delta\sigma$ = 414 MPa, ν = 0.2 Hz, illustrating void coalescence.[18]

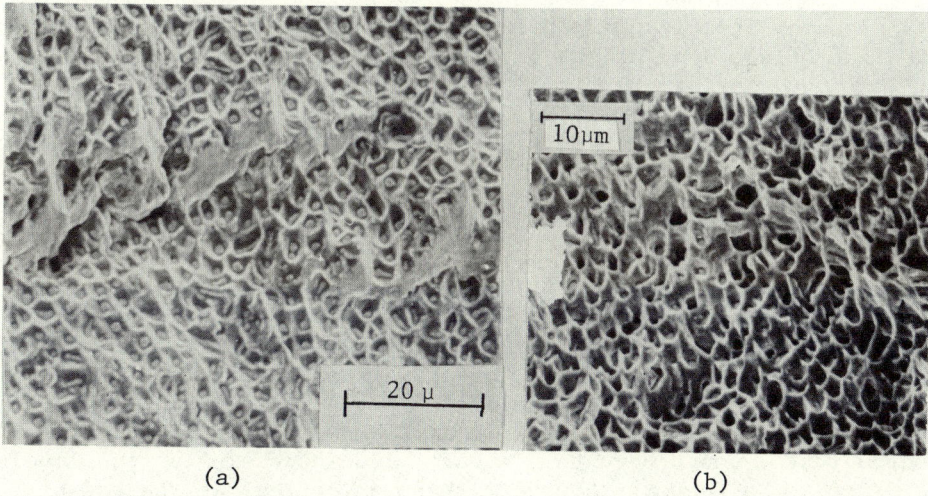

(a)   (b)

Fig. 16.   SEM fractographs of Cotac cycled in vacuum at 825°C, $\Delta\sigma$ = 380 MPa.[16]
   (a) ν = 20 Hz, N_f = 69,000
   (b) ν = 0.2 Hz, N_f = 140.

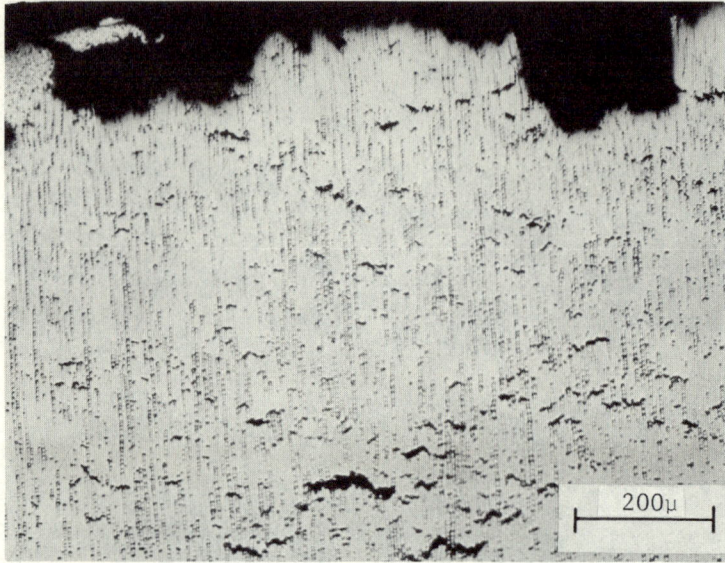

Fig. 17.   Precracked Cotac, cycled at 825°C in vacuum, $\Delta\sigma$ = 175 MPa,
           N_f =   3.36 x 10^5, illustrating void coalescence.

DISCUSSION

     Frequency-induced changes in fatigue life in vacuum are a strong
indication of a creep-fatigue interaction in eutectic alloys.   An
empirical equation of the form:

$$\log \ (\nu \, N_f^{-\beta}) = f(\Delta\sigma) = m(\Delta\sigma) + b \qquad\qquad (2)$$

was used to correlate the 725°C high cycle fatigue data of Nitac at
three cyclic stress levels, as shown in Fig. 18 (a).   Using data from
the four test frequencies (ν = 0.02 to 20 Hz), the constant β was
calculated to be $\simeq$ 0.8.   Utilizing the graph of Fig. 18 (a) to back-
calculate fatigue lifetimes for various levels of $\Delta\sigma$ and ν provided
excellent agreement with the original test data for Nitac, Fig. 18 (b).
The data for γ/γ'-α at the three cyclic stress levels of $\Delta\sigma$ = 813,
862, and 931 MPa also were correlated by Eq. (2), see Fig. 19.   The
scatter was much higher, with β = 1.3, than was shown for Nitac.   It
appeared that β would have to increase with increasing stress levels
to give better agreement; however, this would preclude the use of
Eq. (2) for predictive purposes with γ/γ'-α.

     Equation (2) is a high cycle analogue to the frequency-modified
Manson-Coffin equation for low cycle fatigue[1]

$$(N_f \, \nu^{k-1}) \Delta\epsilon_p^{\,\beta} = C_2 \qquad\qquad (3)$$

where k and β are constants and Δε is the plastic strain range. The term $N_f \nu^{k-1}$ has been termed the frequency modified fatigue life; k is independent of plastic strain but dependent on temperature.

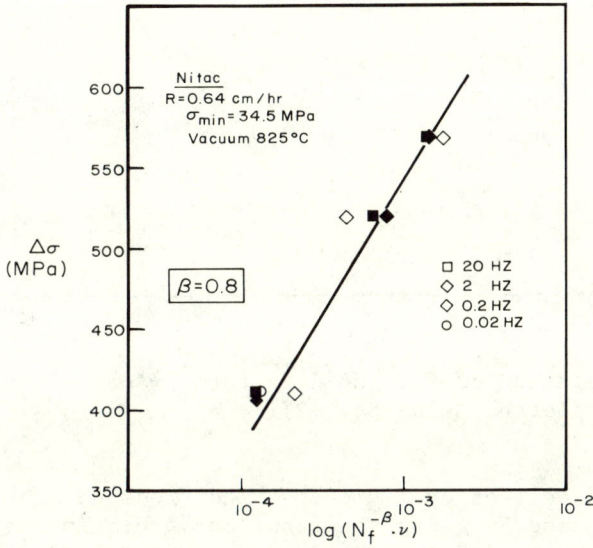

(a)

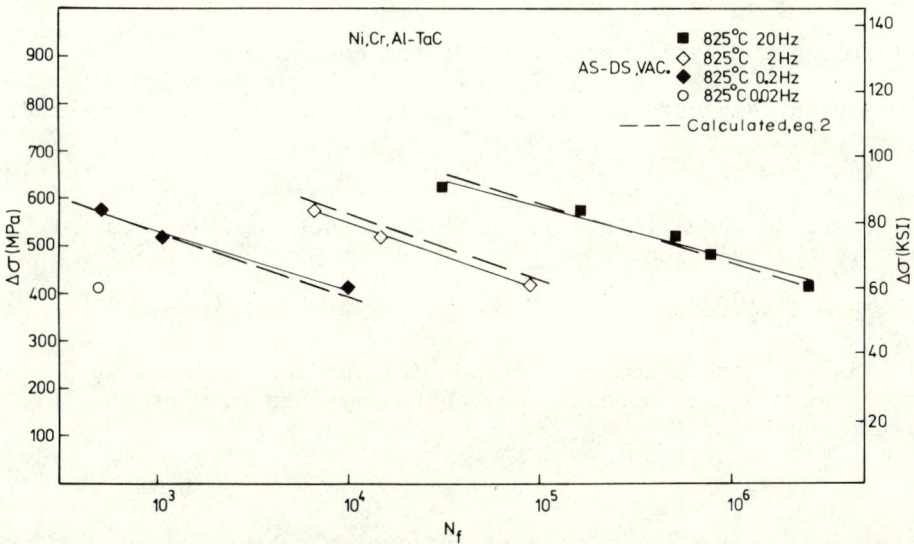

(b)

Fig. 18.   Correlation of fatigue data for Nitac at 825°C (Fig. 1).
(a) data plotted using Eq. (2)
(b) σ–N curve (calculated) compared to actual test data.

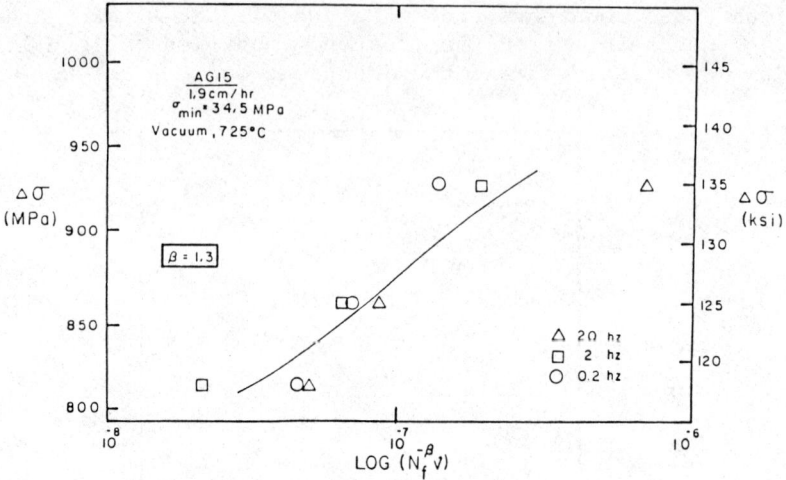

Fig. 19. Correlation of fatigue data for $\gamma/\gamma'-\alpha$ (AG-15), (Fig. 4);
data plotted using Eq. (2).

A relation between cyclic stress and frequency in low cycle
fatigue is obtained by using the functional relation between cyclic
plastic strain range and the steady-state stress range:

$$\Delta\sigma = A(\Delta\varepsilon_p)^{n'} \tag{4}$$

where n' is the cyclic strain hardening coefficient.

A high temperature form of this equation has been suggested:[17]

$$\Delta\sigma = A(\Delta\varepsilon_p)^{n'} \nu^{k_1} \tag{5}$$

Combining Equations (3) and (5), and setting the constant k = 1
leads to the relation:

$$\Delta\sigma = A'N_f^{-\beta} \nu^{k_1'} \tag{6}$$

which is the direct analogue of Eq. (2) for low cycle fatigue. By
setting $k_1' = 0$, one obtains the low temperature form of Eq. (6), the
Basquin equation:

$$\Delta\sigma = A'N_f^{-\beta'} . \tag{7}$$

It appears, then, that similar techniques may be utilized to correlate low cycle and high cycle fatigue data as a function of test
frequency.

Attention is now directed towards the significance of decreased
N_f (and increased or decreased t_F) as ν is decreased. If the only

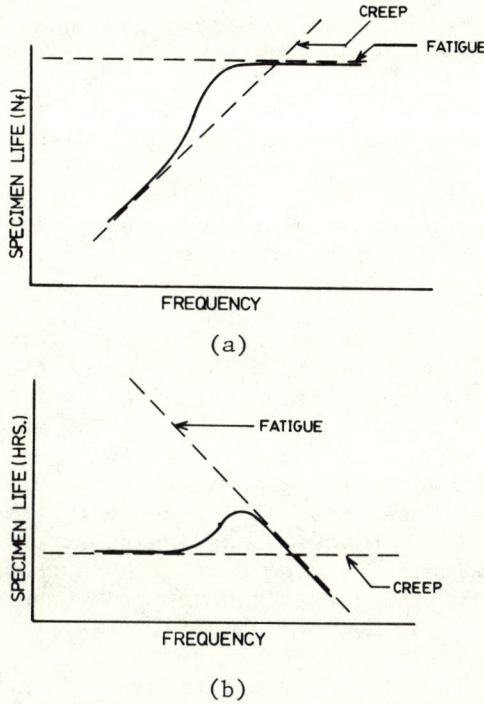

Fig. 20.   Schematic diagrams of specimen life vs. frequency for creep
           and fatigue-dominated processes.
           (a) number of cycles
           (b) duration.

deformation mechanism in operation were fatigue, the number of cycles
to failure at a given stress level would remain constant, as shown
schematically in Fig. 20 (a).   In a purely creep-type test, on the
other hand, time to failure would be independent of frequency, see
Fig. 20 (b).

Low frequencies are equivalent to low imposed stress (or strain)
rates. However, it is apparent that a lower applied stress rate leads
to more time spent by a specimen at high stresses for a given number
of cycles. Accordingly, it is expected that fatigue life, measured
in elapsed time, should increase with decreasing frequency at high
frequencies, when fatigue processes dominate but then could actually
decrease with decreasing frequency for a limited range of frequencies,
as creep processes become more dominant. At the lowest frequencies,
there should be little effect of frequency on fatigue life when creep
processes dominate. These predictions are shown by the solid line of
Fig. 20 (b).

On the basis of this qualitative discussion, it is suggested

that Nitac, tested at 825°C in the frequency range 20 - 0.02 Hz, is
behaving in a manner dictated by creep processes, i.e. both t_F and
N_f decrease as ν decreases. However, $\gamma/\gamma'-\alpha$, tested at 725°C, shows
decreasing N_f and increasing t_F as ν is lowered; the former suggesting
a creep (or environmental) component in the fatigue process, the
latter suggesting neither a creep nor an environmental effect. Since
all fatigue fractures at 725°C were surface initiated, the likelihood
of a creep-dominated process at $\Delta\sigma \leq 950$ MPa is small. Further work
is required, however, to determine whether surface crack initiation
(or propagation) may be accelerated by the presence of small quantities of harmful impurities in the vacuum system.

A large % RA also can indicate a large creep component; the trend
in % RA averaged over three samples of $\gamma/\gamma'-\alpha$ tested at different
stress levels was as follows: 5% at 20 Hz, 11% at 2 Hz, and 15% at
0.2 Hz.   The two specimens, tested at $\Delta\sigma = 958$ and 1000 MPa with $\nu =$
20 Hz, showed still higher ductilities, 16.1 and 33.7%, respectively.
The similarity between fractographic features of the $\Delta\sigma = 1000$ MPa
fatigue tests and those of a 725°C tensile test suggest that this
test ($N_f = 780$) could be modeled as a high strain rate creep or tensile test.   Additional low frequency, high stress testing at 725°C is
required to resolve the question of why two-branched curves were not
noted in tests at 2 Hz and 0.2 Hz, Fig. 4 (a).

In Nitac, a transition from surface to central initiation was
observed fractographically at 825°C when the frequency was lowered.[7]
This additional evidence for a creep-fatigue interaction was not
observed in the fatigue testing of $\gamma/\gamma'-\alpha$ at 725°C.[15]   All nine specimens of the latter used in the correlation by Equation (2) exhibited
identical fractographic features, with surface initiation sites.
Since fracture origins in $\gamma/\gamma'-\alpha$ were internal at 825°C, independent
of frequency, it seems clear that a frequency-induced change in fracture origin must occur at some temperature between 725°C and 825°C.

The role of environment in the frequency and waveform experiments
on Cotac and Nitac seems to be minimal.   For one thing, cracks did not
readily propagate at all in Cotac in air due to the blunting of crack
tips by oxide.[8]   However, when vacuum was imposed, cracks not only
propagated, but did so with no surface oxide visible.   Moreover, while
vacuum environment produced a significant improvement in fatigue life
at 25°C in Cotac, no effect of vacuum was noted at 825°C (see Fig. 5);
this observation was consistent with the internal crack initiation
sites in this material at all levels of applied $\Delta\sigma$, in that no effect
of environment is expected when the environment has no access to the
crack tip.

The results of the unequal ramp rate tests on Cotac in vacuum
may be contrasted with the behavior of 304 stainless steel.[6]   For
both alloys the slow-fast sequence is considerably more damaging than
the fast-slow.   Slow-fast cycling is associated with intergranular

cracking in 304 stainless steel, both in air and in vacuum.[6] In the tests on Cotac, no significant shift in fractographic features was noted in any of the tests summarized in Fig. 9, and, of course, all fractures were transgranular.[9]

The waveform experiments on Cotac can be interpreted on the basis of creep-damage as follows: the slow-fast waveform initially imposes stresses high enough to cause creep for a relatively long period; the rapid unload portion of the cycle allows little time for load relaxation to occur, before reimposition of the slow component of the next cycle; crack growth under these conditions is rapid. The fast-slow waveform, on the other hand, permits creep to occur for a relatively short time, followed by a long period of unloading during which appreciable relaxation of strain may occur; this leads to relatively slow crack growth.

Similarly, the hold-time experiments are subject to straightforward interpretation on a creep-damage model. A hold-time imposed at peak tensile stresses caused much more rapid crack growth than did either the same hold-time at minimum tensile stress or the slow-fast and fast-slow waveforms. These results suggest that little or no crack blunting by strain relaxation is possible during the hold-time at the highest stress level, and that creep-assisted damage is significant. When a hold-time is imposed at the minimum tensile stress level, little or no additional creep occurs, and the crack growth rate is governed by the frequency of the load and unload portions of the cycle. However, the crack growth rate of 0.09 μm/cycle is approximately the same as that of the slow-fast waveform, suggesting that strain relaxation in the latter condition is minimal.

There are two pieces of evidence suggesting the nature of creep-damage that may occur during one of the above fatigue cycles. We have already shown that voids develop at cracked fiber ends in Nitac (Fig. 15) and Cotac (Fig. 17) which are tested at low frequencies, 0.2 Hz, but not in the same alloys when tested at 20 Hz. Also, fracture surfaces of Nitac and Cotac at 825°C (e.g. Fig. 16) reveal dimples in the fatigue zone. Therefore, the mechanism of more rapid crack advance at low test frequencies or under other experimental conditions favoring a creep-damage interaction seems to be associated with the link-up of voids around cracked fibers. Consequently, grain boundaries play no role in creep-fatigue interactions in eutectic composites, unlike the situation for conventional polycrystalline alloys.

The fact that frequency and wave shape effects on fatigue processes in aligned eutectics tested in vacuum is so striking arises in part from the tension-tension nature of the fatigue loading. Unlike the case in fully reversed strain controlled cycling, there is a positive mean stress, approximately equal to $\Delta\sigma/2$, which is imposed upon all test specimens, resulting in a situation very favorable for creep. Moreover, there is a possibility that the lower imposed stress rates in low frequency tests lead to a lowered resistance to plastic flow and thereby increased crack growth rates. However, the eutectic alloys are very high strength materials, with flow stresses of the order of 700 MPa even at 725°C (or 825°C), due to the anomolous strengthening of ordered γ' (a major constituent of the matrix in Nitac and γ/γ'-α alloys) as temperature increases. Therefore, it is considered unlikely that stress or strain rate effects contribute significantly to the observed modifications in fatigue behavior with changing frequency.

SUMMARY AND CONCLUSIONS

The elevated temperature fatigue behavior of several eutectic alloys has been shown to depend upon test frequency and the imposition of hold-times. Frequency effects have been noted in three alloys tested in vacuum, (γ/γ'-α, Nitac and Cotac). In all cases each ten-fold reduction in frequency during stress-controlled cycling leads to an approximately ten-fold reduction in number of cycles to failure, N_f. The relation among σ, ν, and N_f has been expressed for Nitac and γ/γ'-α in the form:

$$\log (N_f^{-\beta} \cdot \nu) = m(\Delta\sigma) + b \tag{8}$$

β was determined to be 0.8 for Nitac and approximately 1.3 for γ/γ'-α; m is 140 MPa for γ/γ'-α and 165 MPa for Nitac.

Additional evidence for a creep-fatigue interaction is provided by experiments conducted in air: imposition of a hold-time increases crack propagation rates in γ/γ'-δ. The fact that the general trend of results in these experiments in air parallels those obtained on the other alloys in vacuum, together with the film-free dimpled fracture surfaces displayed after fatigue at 725°C (or 825°C), strongly infers that environmental effects play a secondary role in high temperature fatigue behavior of aligned eutectics. However, tests on γ/γ'-α at 725°C do not support unambiguously a creep-fatigue model. In Cotac, on the other hand, the results of waveform experiments can be interpreted on the basis of creep damage occurring during peak tensile load hold-times and during slow load-fast unload cycles. These preliminary results need to be amplified by additional tests in which the magnitude of the hold-time is changed, together with tests at constant frequency with no hold-time, at the same test temperature.

Damage accumulation at low frequencies has been shown to occur by void formation and growth at cracked fiber ends. There is a marked tendency for interior crack initiation at low test frequencies and high temperatures.

ACKNOWLEDGEMENTS

The author is grateful for helpful discussions with Professor
D. J. Duquette, who collaborated on much of the work reported here,
and to the Air Force Office of Scientific Research for financial support under Grant No. AFOSR-80-0015.

REFERENCES

1. L. F. Coffin, Jr., Met. Trans., V. 3, 1972, pp. 1777-1788.
2. D. A. Woodford and L. F. Coffin, Jr., Fourth Bolton Landing
    Conf., Claitor's Pub. Div., Baton Rouge, LA, 421, 1974.
3. E. Krempl and B. M. Wundt, in ASTM STP489, Amer. Soc. for Test.
    and Mat., 1971.
4. P. Shahinian and K. Sadananda, 1976 ASME-MPC Symposium on Creep-
    Fatigue Interactions, ASME, New York, 1977, p. 365-390.
5. P. Shahinian and K. Sadananda, J. Eng. Mat. and Tech., Trans
    ASME, V. 101, 1979, p. 224.
6. L. F. Coffin, Jr., in Symp. on Creep-Fatigue-Environment Inter-
    actions, Fall Meeting of AIME, Oct. 1979, Milwaukee, WI, to
    be published.
7. W. A. Johnson and N. S. Stoloff, Met. Trans. A, V. 18A, 1980,
    p. 307.
8. C. Koburger, D. J. Duquette and N. S. Stoloff, Met. Trans. A,
    V. 11A, 1980, p. 1107.
9. C. M. Austin, Ph.D. Thesis, Rensselaer Polytechnic Institute,
    Troy, NY, 1979.
10. K. Sadananda and P. Shahinian, Mat. Sci. and Eng., V. 38, 1979,
    p. 81.
11. N. S. Stoloff and D. J. Duquette, in Symp. on Creep-Fatigue-
    Environment Interactions, AIME Fall Meeting, Milwaukee, WI,
    to be published.
12. N. S. Stoloff, in Adv. in Composite Materials, Applied Sci.
    Publ., Reading, England, 1978, p. 247.
13. J. Tartaglia and N. S. Stoloff, submitted to Met. Trans. A,
    1980.
14. D. A. Woodford, Met. Trans. A, V. 8A, 1977, p. 639.
15. J. Tartaglia, Ph.D. Thesis, Rensselaer Polytechnic Institute,
    Troy, NY, 1980.
16. N. S. Stoloff, D. J. Duquette, C. Koburger and W. A. Johnson,
    in Strength of Metal and Alloys, Pergamon, Oxford, 1979,
    p. 1201.
17. L. F. Coffin, Jr., Proc. Air Force Conf. on Fatigue and Fracture
    of Aircraft Materials, AFFDL TR70-144, 1970.
18. L. F. Coffin, Jr., in Ann. Rev. of Mat. Sci., V. 2, 1972,
    p. 313.

THERMAL FATIGUE ANALYSIS

D. F. Mowbray and G. G. Trantina

Corporate Research and Development
General Electric Company
Schenectady, New York 12301

INTRODUCTION

Many high temperature turbine components are subjected to rapid temperature changes during start-up and shutdown, leading to nonuniform heating and cooling. Repetition of such thermal loadings leads to thermal fatigue in critical hot section components such as blades, vanes, shrouds and combustors. This type of thermal fatigue is one of the primary failure modes that must be considered in turbine design and analysis.

A number of approaches to thermal fatigue analysis have appeared over the years.[1,2,3] For the most part the emphasis in these is on assessing the damage to produce cracking. Little attention has been given specifically to the treatment of crack propagation, despite the fact it would appear that this is the primary damage process. Thermal fatigue loadings generally produce high strains and cracking commences in very few cycles. Further, the thermal stress fields generally involve strong gradients, so that the subsequent crack propagation rates may decay as the crack extends. Hence, in thermal stress problems it would appear important to develop tools for assessing crack propagation, as well as crack initiation.

Quantification of the thermal fatigue damage process requires highly complex analysis when temperature and loading conditions induce nonlinear deformations. The analysis must consider the nonuniform, 3-dimensional geometry in the thermal and stress analysis, factor in material property variations with temperature, and account for the accumulation of time independent and time dependent nonlinear deformation.

The present paper explores some analytical methods for treating the overall problem, including the separation of the crack initiation and crack propagation phases of life. Emphasis is placed on simplifying the analyses so that they can be carried out economically. The analysis methods are used to examine the results of thermal shock tests on tapered disks.

TAPERED DISK TESTS

Thermal shock tests were conducted on the tapered disk specimen design shown in Fig. 1. Many of the test results have been reported in earlier papers.[4,5,6,7] The specimens are subjected to alternate heating and cooling shocks by immersion in fluidized baths. The fatigue process in tests of this type is expected to be similar to turbine hot-section thermal fatigue, because of the similar strain-time-temperature histories.

The fourfold variation in peripheral radius (R_p) indicated in Fig. 1 was incorporated as a means for achieving a variation in cyclic stress range for fixed minimum and maximum bath temperatures. For the tests analyzed herein, the cold bath or minimum temperature was always held at $70°F$ for an exposure time (t_c) of 4 minutes.

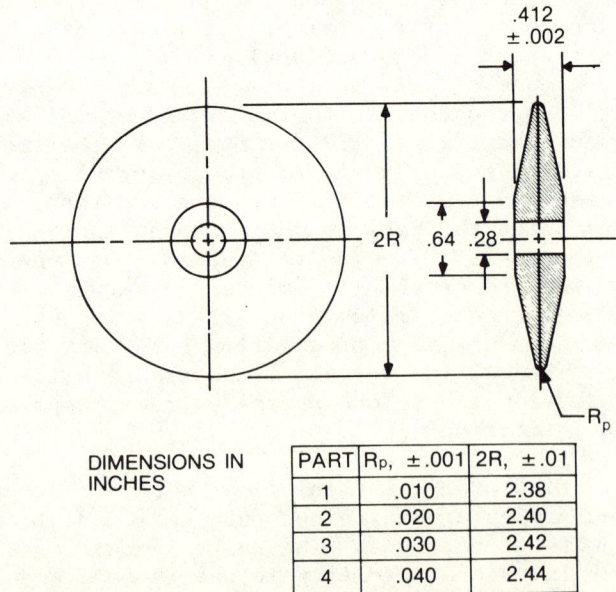

PART	R_p, ±.001	2R, ±.01
1	.010	2.38
2	.020	2.40
3	.030	2.42
4	.040	2.44

DIMENSIONS IN INCHES

Fig. 1. Tapered disk thermal fatigue specimen.

The hot bath or maximum temperature (T_{max}) was varied between 1500 and 1900°F, with exposure times (t_h) of 1, 4, 16 and 60 minutes. These variations in maximum temperature and exposure also gave rise to differing strain ranges and strain-temperature cycles. Some additional variation was achieved in a few tests by varying the heating and cooling rates and by cooling the specimen bore with high pressure air.

The number of cycles to produce a visible crack (crack initiation) and crack growth rates of macrocracks were determined. Crack initiation was defined as a 0.002- to 0.005-inch crack on the crown of the specimen periphery. Crack growth rates were determined by testing the specimen configuration with the diametrally opposed shallow notches. The notches induced immediate crack initiation and permitted observation of cracks to be confined to the two locations.

The test material was the cobalt-base alloy FSX 414.[8] It was procured in the form of cast-to-size specimen blanks and tested after a standard heat treatment.[6]

A major portion of the testing was carried out at T_{max} = 1688°F and an exposure time in the hot bath of 4 minutes. Examples of crack initiation and crack propagation data for this test condition are shown in Fig. 2. The figure illustrates the effect of peripheral radius, or cyclic strain range, on the fatigue response.

The effect of other exposure time and maximum temperature conditions on thermal fatigue resistance are shown for a single peripheral radius in Figs. 3 - 5. Figure 3 illustrates the effect of exposure time and maximum temperature on crack initiation life. Figure 4 illustrates crack growth rate response as a function of maximum temperature; and Fig. 5, crack growth rate response as a function of exposure time.

Several trends in thermal fatigue response are indicated in Figs. 2 - 5. One can see that in general for a smaller peripheral radius (higher cyclic strain), higher maximum temperature and longer exposure time, the fatigue resistance is lowered. There is some exception in this general behavior for the longest exposure times and at some of the higher maximum temperatures. These exceptions or reversals in behavior can be explained by time and temperature dependent changes in microstructure, which alter the strength and ductility of the alloy.[6]

ANALYSIS METHODS

The analysis methods are based on using linear-elastic calculations to obtain total strain-time-temperature histories.

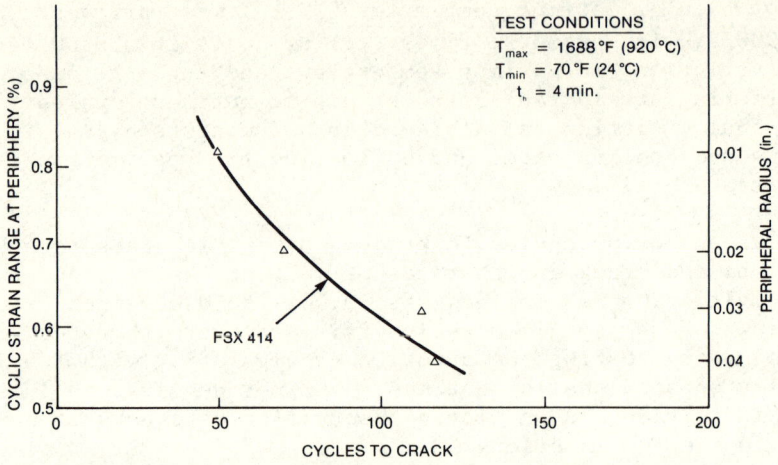

Fig. 2(a).   Example crack initiation data.

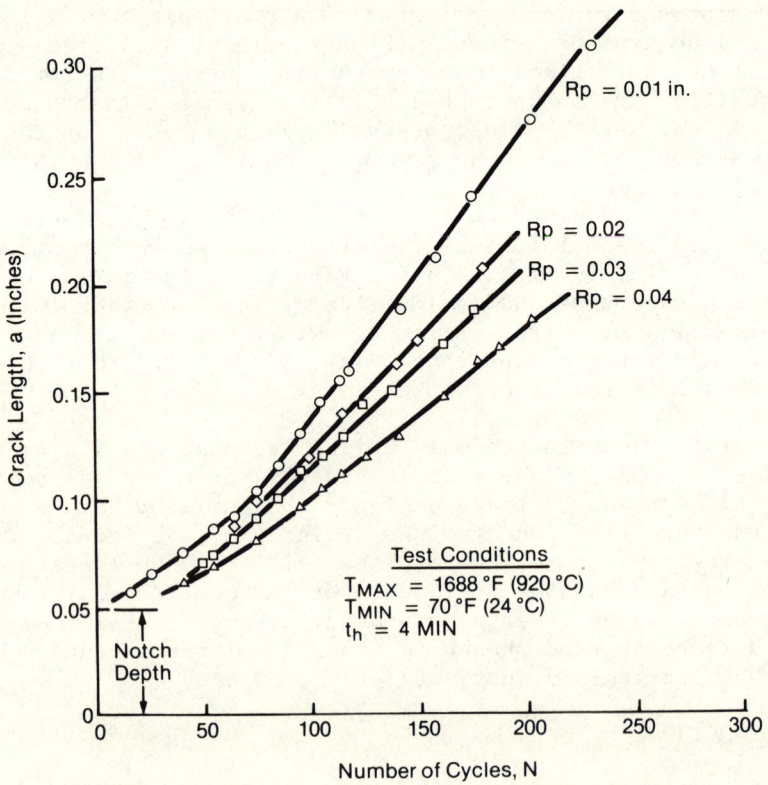

Fig. 2(b).   Example crack growth data.

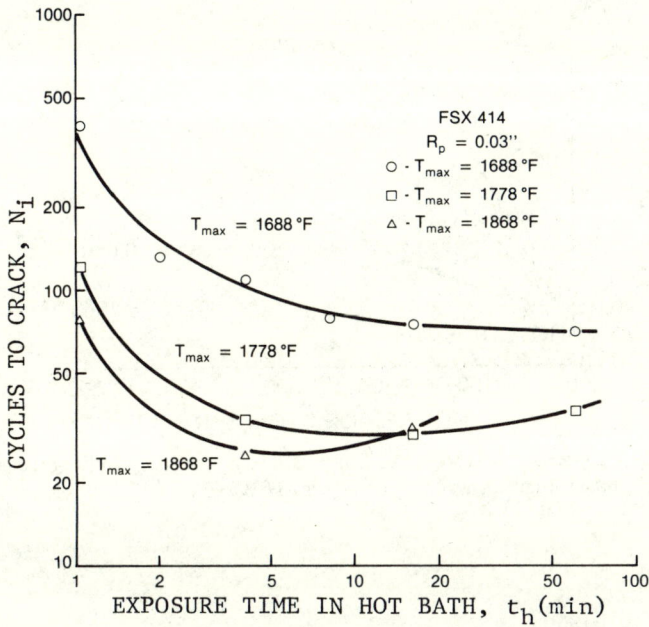

Fig. 3.   Effect of exposure time and maximum temperature on cycles to crack; FSX 414.

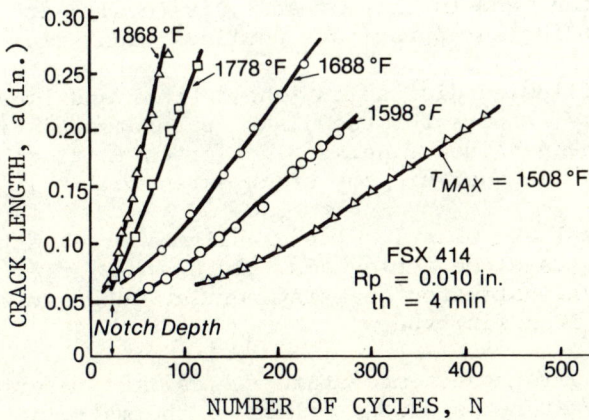

Fig. 4.   Crack growth curves illustrating effect of maximum temperature at an exposure time of 4 minutes.

Linear analysis is known to be satisfactory for obtaining the total strains when there is strictly thermal loading present,[1] as is the case here. However, it does not yield information on the actual stresses and the division of strain into elastic and plastic

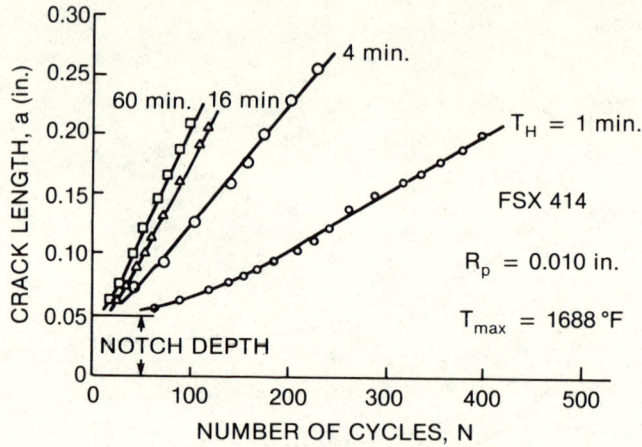

Fig. 5.   Crack growth curves illustrating effect of exposure time
at a maximum temperature of 1688°F.

components. The latter is assumed unimportant in analyzing crack
growth on the basis of a strain intensity factor. In analyzing
crack initiation, separating out the nonlinear components is
assumed important and the linear analysis results are used as a
starting point for an approximate nonlinear analysis.

    In the following discussion, the thermal and linear stress/
strain analysis results for the disks are presented first, followed
in order by the nonlinear analysis for crack initiation and the
fracture mechanics analysis for crack extension.

    The thermal and elastic stress analyses were performed using
compatible finite element programs. An example grid is shown in
Fig. 6. The elements are the axisymmetric, constant strain quadri-
lateral and triangular type.

    In solving for the temperature fields, the materials were
ascribed temperature dependent values of thermal conductivity and
specific heat. The heating and cooling shocks were simulated by
specifying the hot and cold bath temperatures and values of con-
vective heat transfer coefficient at all exterior surfaces. The
latter were measured and controlled during the testing. Comparisons of the calculated time-temperature response of the disks with
the thermocouple measurements at several locations within the disk
are shown in Fig. 7. In general, excellent correspondence was
achieved.

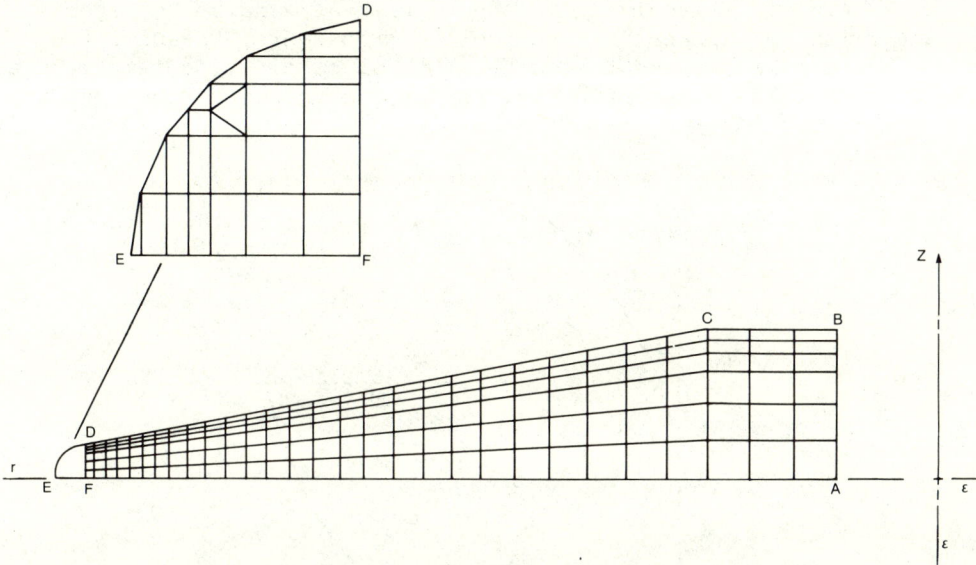

Fig. 6. Finite element grid for thermal and stress/strain analyses.

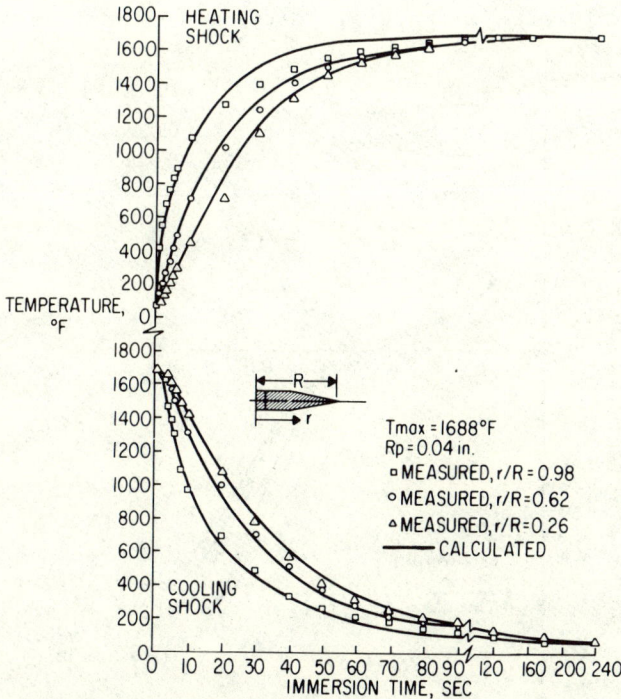

Fig. 7. Comparison of measured and calculated time-temperature response of a disk specimen.

Elastic stress analyses were performed at a number of appropriate times in the heating and cooling transients. Results from
these analyses were used to construct total strain-time-temperature
histories for the disk periphery, and hoop strain distributions in

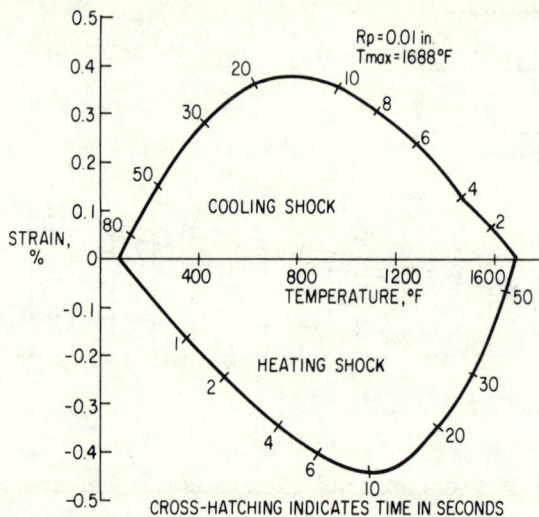

Fig. 8. Example strain-temperature cycle for peripheral fiber.

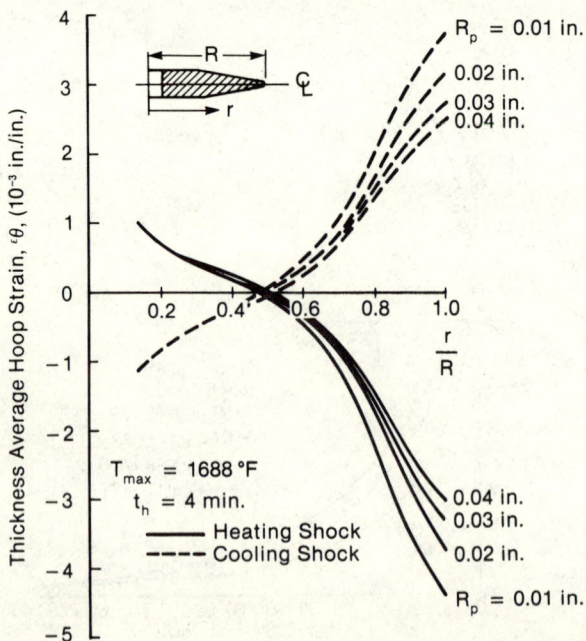

Fig. 9. Calculated hoop strain distributions in radial direction
at peak values of peripheral strain.

the disk radial direction at the times in each transient where the
peripheral strain peaks. Figures 8 and 9 illustrate examples. The
strain-time-temperature calcualtions were used as input for the
analysis of crack initiation damage and the strain distributions
for the analysis of the crack propagation driving force.

A. Crack Initiation

Fatigue damage leading to crack initiation at elevated
temperatures is known to be a complex function of some stress
and/or strain variable, and the degree of environment-metal
reaction.[9] The common approach for representing this interaction is to treat the environmental contribution in parametric
form (hold time, frequency, etc.) and to prescribe the controlling
mechanical variable. Many design approaches incorporate the total
strain range as the controlling variable because it is relatively
easy to calculate. Its use in thermal fatigue would be a gross
simplification, however, because the damage induced in the heating
shock is far more damaging than that in the cooling shock.[2] Other
variables commonly considered to control thermal fatigue damage
include the time dependent plastic and creep strains,[10] total
inelastic strain,[9] and creep damage.[3] The present computational
approach permits each of these variables to be calculated and
utilized in damage predictions.

The nonlinear cyclic plasticity analysis treats the peripheral
fiber as a uniaxial bar subjected to the calculated strain-time-
temperature history such as shown in Fig. 8. The computational
procedure is patterned after that proposed by Spera,[3] wherein the
total strain is broken up into elastic, plastic and creep
components,

$$\varepsilon_t = \varepsilon_e + \varepsilon_p + \varepsilon_c \tag{1}$$

For solution of Eq. (1), ε_e, ε_p and ε_c are expressed as functions of σ. The function for elastic strain, ε_e, follows simply
from Hooke's law:

$$\varepsilon_e = \sigma/E \tag{2}$$

For the time independent plastic strain, ε_p, the function is
kept in open form represented by temperature dependent stress-
strain curves. In the solution procedure, several stress-strain
curves are specified over the temperature range of interest, and
interpolation is used to identify curves at intermediate temperatures. Transfer from one stress-strain curve to another (or one
temperature to another) is facilitated at constant total strain.
Figure 10 shows stress-strain curves for FSX 414.

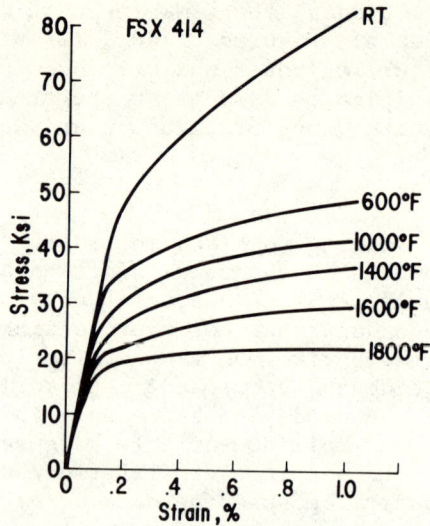

Fig. 10.   Stress-strain and creep properties for FSX 414.

Creep strain, ε_c, is expressed in the form of Eq. (3):

$$\varepsilon_c = 2At^{1/2} \sinh(G\sigma).$$
(3)

In this relationship A and G are functions of temperature. They were evaluated by nonlinear regression analysis of available creep data. For cast superalloys, the following functions of T were found to provide good fits to the data:

$$A = P - Q/(T + 460)$$
(4)
$$G = 1/(R - ST)$$

where P, Q, R and S are regression parameters.

Two additional assumptions are required to carry out the cyclic plasticity and creep analysis. One is concerned with the Bauschinger effect and the second with a method for stress transfer in creep analysis. Evaluations[7] showed the assumptions of equal tension and compression yield points and the strain hardening rule to be most satisfactory for the type of loadings considered.

Calculation of the nonlinear stress-strain behavior using the above equations and assumptions was carried out numerically. A FORTRAN computer program was written based on a time incremental scheme incorporating the Runge-Kutta solution technique.

The ability of the uniaxial modeling approach to predict local stress-strain behavior was evaluated through two preliminary calibration analyses. The first compared experimental results from a uniaxial thermal-mechanical fatigue test with the model calculated response; the second compared the model calculated stress-strain response at the periphery of a disk specimen with results from a nonlinear, 3-dimensional finite element solution of the disk specimen.

The uniaxial thermal-mechanical test result was obtained for a nickel-base alloy, Rene' 77. The test parameters and comparison of results are shown in Figs. 11 and 12, respectively. A reasonably good prediction of the experimental result is achieved, particularly with regard to total inelastic strain and stress level in the tensile or high temperature, creep dominated regime.

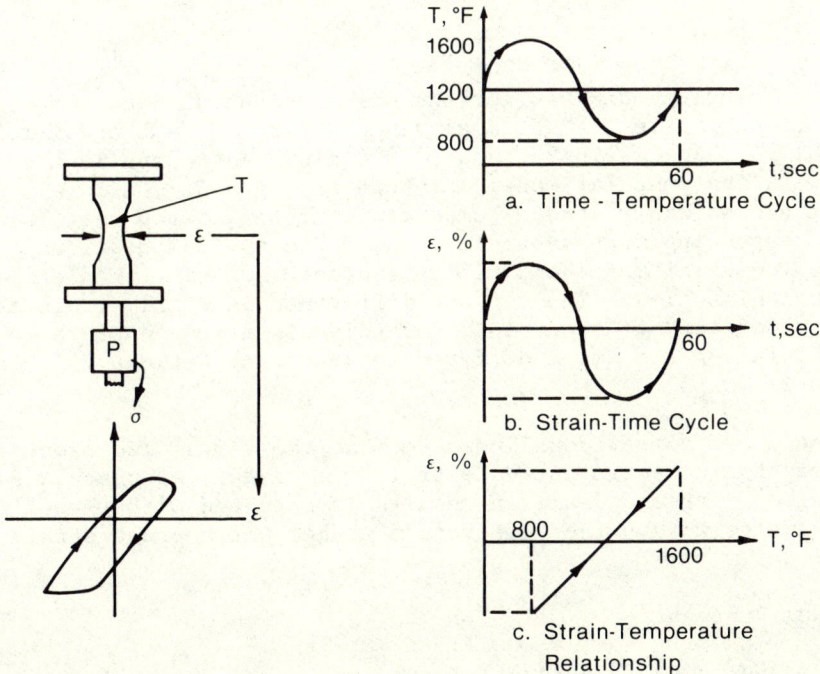

Fig. 11. Description of uniaxial thermal-mechanical fatigue test.

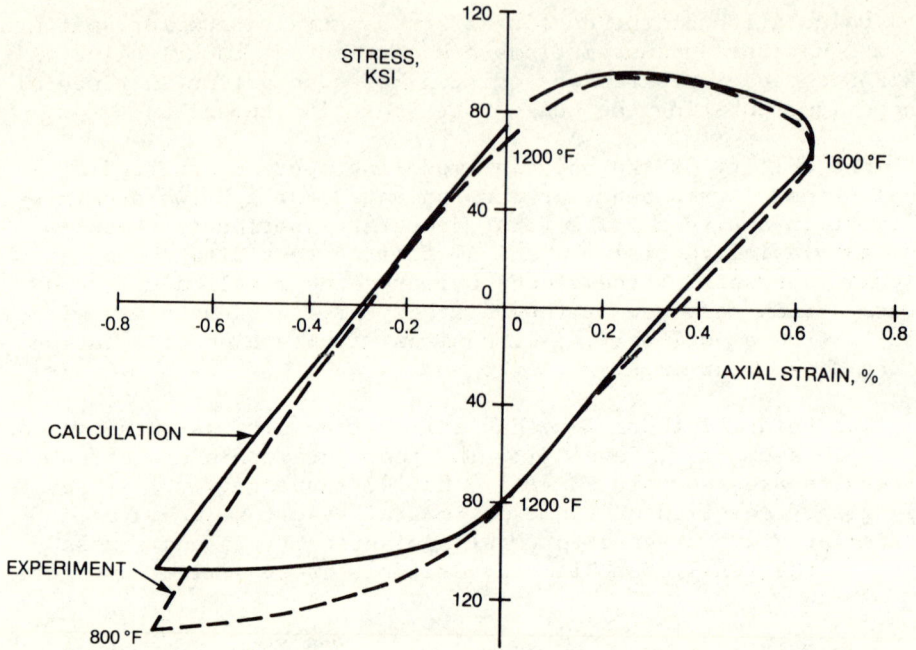

Fig. 12.   Comparison of calculated and experimentally determined
          hysteretic loop; Rene' 77.

       The solution obtained for the disk peripheral behavior ob-
tained using finite element analysis was for the test conditions
R_p = .01 in., T_{max} = 1688°F and t_h = 4 min.   The comparison of results with the uniaxial model is shown in Fig. 13.   Part (a) shows
the peripheral fiber strain-temperature histories and part (b),
the corresponding stress-strain loops.   The two computational
methods (uniaxial model and finite element analysis) predict nearly
identical histories.   The maximum difference in total strain at a
given temperature is about 10%, and this difference appears to be
responsible for the small differences appearing between calculated
hysteretic loops.

       The above comparisons indicate that the simplified procedure
developed for computing stress-strain response is adequate.   Especially good correspondence in results is obtained with the calibration examples within the temperature ranges where creep strain
accumulates.

B.   Crack Propagation

       The analysis of thermal fatigue crack propagation in disk
specimens has been pursued in the past using the linear elastic

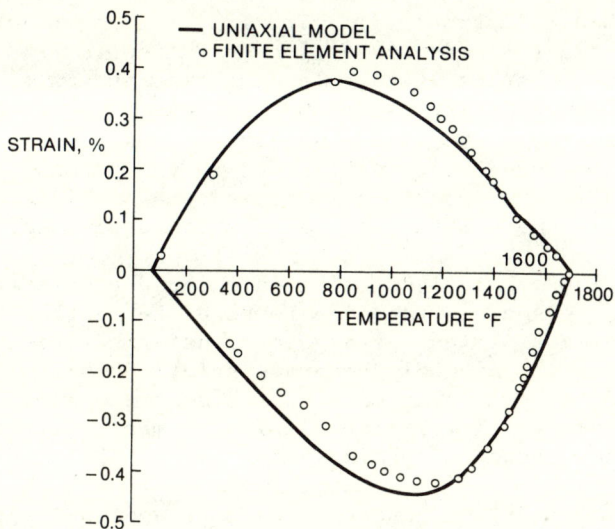

(a) Strain-temperature cycles.

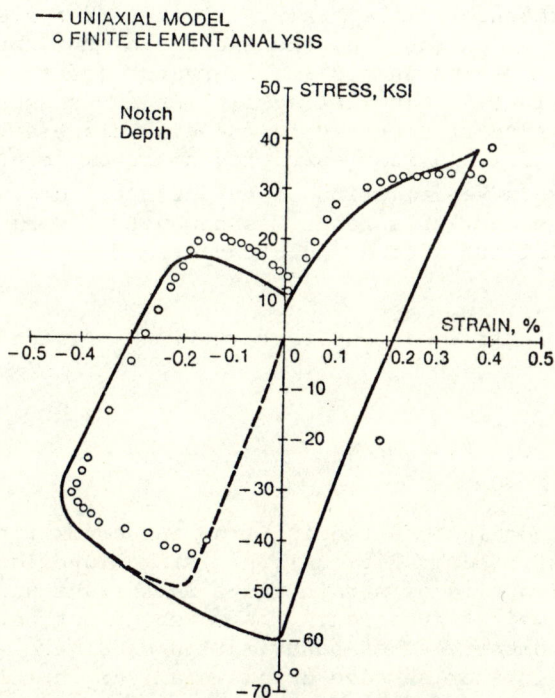

(b) Hysteretic loops.

Fig. 13.   Comparisons of calculated results using the uniaxial
           model and FE methods of analysis.

fracture mechanics approach.[4,5] The fracture mechanics approach
assumes that the rate of propagation of fatigue cracks is controlled by the elastically calculated stress intensity factor, K.
Normally the range of stress intensity factor, ΔK, is considered
the first order variable, and mean biases of stress intensity
factor level a second order variable. In the present case, the
crack propagation is described in terms of a strain intensity
factor based on the elastically calculated total strain fields.

Application of an elastically calculated parameter may not
at first seem reasonable for high temperature thermal fatigue
behavior. However, it can be argued[11] that the strain intensity
factor produces a reasonably accurate solution for the crack
driving force for nonlinear material in a strain limited loading
condition. Some further limited finite element analysis was
carried out to demonstrate this fact.

The weight function concept[12] is employed to calculate strain
intensity factor levels in the high gradient strain fields arising
in the disk thermal shock test. In using this concept the following assumptions are made: (1) the temperature field is not dis-
turbed by the presence of the crack, (2) the effective crack
driving forces are described by the average of the through-
thickness radial strains and (3) the curvature of the disk periphery can be neglected. The latter implies the assumption that
the crack remains short compared to the disk diameter. For this
reason the data analysis was restricted to crack lengths less
than ~ 0.3 inches. Verification of the accuracy of the weight
function solution over this range was demonstrated by comparison
with a detailed finite element solution of one of the test
geometries.

The range of strain intensity factor (ΔK_ε) is defined in the
weight function method by

$$\Delta K_\varepsilon = \sqrt{2/\pi} \int_0^a p(x)\, m(x)\, dx \tag{5}$$

where x is the coordinate extending from the crack tip to the
periphery (illustrated in Fig. 14), p(x) is a function representing
the radial distribution of strain range in the absence of a crack,
and m(x) is the weight function. The weight function is dependent
only upon the geometry. The approximation used here incorporates
the weight function for an edge crack in a semi-infinite body:

$$m(x) = 1/\sqrt{x}\,[1.0 + 0.615(x/a) + 0.250(x/a)^2]\ . \tag{6}$$

The strain distribution function was represented by a second
order polynomial fit of the calculated radial strain distributions:

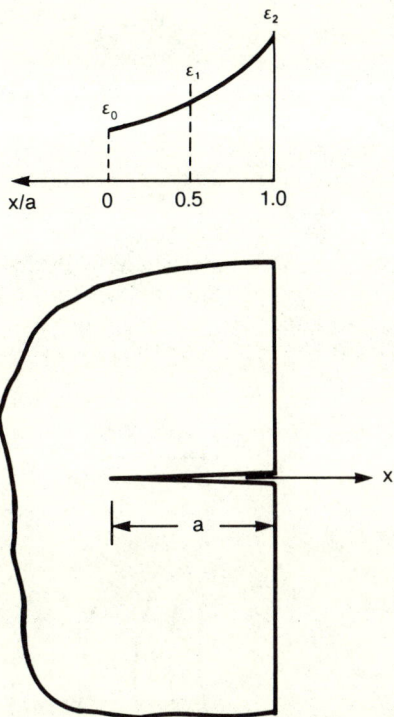

Fig. 14. Definition of crack tip coordinates and applied strain.

$$p(x) = C_0 + C_1(x/a) + C_2(x/a)^2, \tag{7}$$

where

$$C_0 = \Delta\varepsilon_0$$

$$C_1 = 4\Delta\varepsilon_1 - \Delta\varepsilon_2 - 3\Delta\varepsilon_0 \tag{8}$$

$$C_2 = 2\Delta\varepsilon_2 + 2\Delta\varepsilon_0 - 4\Delta\varepsilon_1 .$$

$\Delta\varepsilon_0$, $\Delta\varepsilon_1$ and $\Delta\varepsilon_2$ are the strain range levels obtained for the un-cracked disk corresponding to the tip, mid-point and end (peri-phery) portions of the crack, respectively (Fig. 14). Substitu-tion of Eqs. (6), (7) and (8) into (5) yields the following for ΔK_ε:

$$\Delta K_\varepsilon = \sqrt{a} \; (0.655 \; \Delta\varepsilon_0 + 1.126 \; \Delta\varepsilon_1 + 0.222 \; \Delta\varepsilon_2) \qquad (9)$$

Figure 15 shows examples of the calculated strain intensity factor as a function of crack length for several of the test configurations. ΔK_ε varies with crack length in the fashion expected. For example, starting at the shortest crack lengths, ΔK_ε increases with increasing crack length until achieving a relatively constant value, and then decreases at the longer crack lengths. This behavior duplicates the trends exhibited by the crack growth rates as illustrated in Fig. 16 for one geometry and test condition. The range of strain intensity factor, therefore, appears a reasonable choice as the controlling variable.

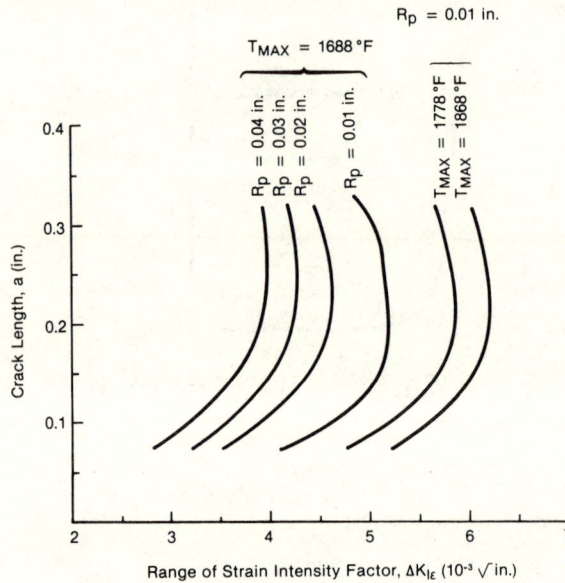

Fig. 15.   Strain intensity factor - crack length relationships for several test conditions.

To provide a check on the accuracy of the approximate weight function used in the calculations, finite element solutions for the strain intensity factor were obtained for the R_p = .02 in. disk and T_{max} = 1688°F test condition. The solutions were based on the strain energy release rate principle[13] and assumed plane stress behavior. The calculated ΔK_ε versus crack length results are compared with the weight function in Fig. 17. The two methods of analysis yield curves of identical shape, and for a given crack length the results differ by only a few percent. Hence, based on these results it was concluded that the approximate weight function analysis was sufficiently accurate, and because

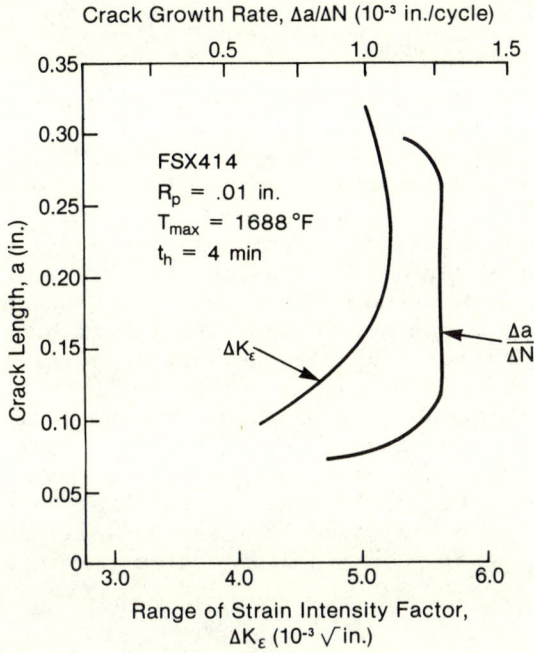

Fig. 16.   Variation of strain intensity factor and crack growth
rate with crack length.

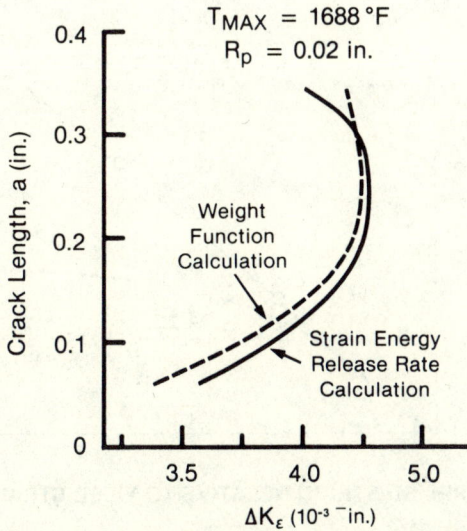

Fig. 17.   Comparison of strain intensity factor – crack length
relationships calculated by the weight function and
strain energy release rate methods.

of its simplicity, was used to construct ΔK_ε - solutions for all other loadings.

The ability of the elastically calculated strain intensity factor to describe the crack driving force in loadings where the crack is propagating through a field of plastically strained material was examined through nonlinear finite element analysis. A nonlinear solution was obtained for a uniformly strained edge crack specimen approximating a semi-infinite body. J-integral calculations[14] were performed to assess the crack driving force in the highly yielded state. The material stress-strain curve was represented by a Ramberg-Osgood type law with parameters chosen to be typical of a high temperature superalloy.

A version of the ADINA[15] computer program modified to permit the J-integral to be evaluated by contour integration was employed in the calculations. The model was loaded to gross strain levels approximately twice the yield strain. Comparative results are shown in Fig. 18 on coordinates of K and normalized applied strain level. The J-integral determined by contour integration was con-

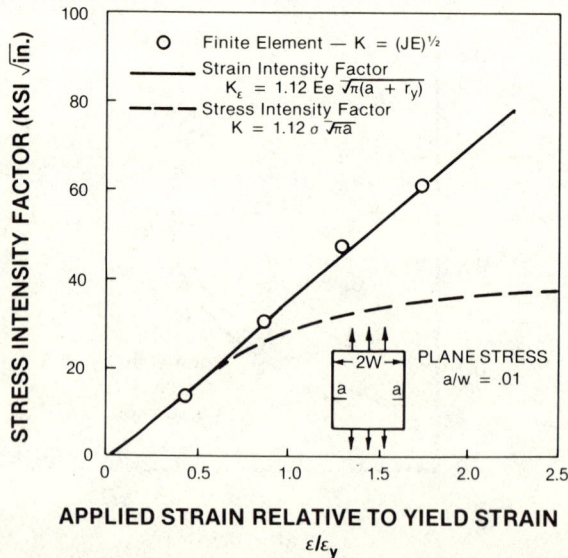

Fig. 18.   Comparison of crack driving force calculated by linear and nonlinear methods for a cracked-plate.

verted to K by the relationship $K = \sqrt{J} \, E$. Observation of the
figure shows that the nonlinear finite element solution deviates
from the linear elastic stress intensity factor solution at applied
strain levels of approximately 60% of the yield strain. This deviation increases roughly exponentially with increasing applied
strain. The solution defined by the strain intensity factor
($K_\varepsilon = 1.12 \, E \, \varepsilon \sqrt{\pi a}$) follows the J-contour integration solution very
closely, with no more than a few percent deviation at any applied
strain level examined. Hence, this result indicates that the
strain intensity factor should define the crack driving force very
well in plastic strain fields of a strain limited nature.

ANALYSIS RESULTS

A. Crack Initiation

     Examples of the calculated stress-strain behavior at the disk
periphery are illustrated in Fig. 19 for a variety of test conditions. Part A of the figure shows the effect of varying R_p at
constant T_{max} and t_h, part B shows the effect of varying T^P_{max} with
R_P and t_h held fixed, and part C the difference between $t_h = 1$ and
4 minutes at constant T_{max} and R_p. Loops calculated for exposure
times in the hot bath greater than 4 minutes have the same size
and shape as those shown for the 4 minute time.

     One can observe in Fig. 19 that stresses achieved in each half
of the cycle vary little between the test conditions investigated.
The major effect that each test variable has on the loop size is
the loop width or inelastic strain components.

     The calculated strain components and creep damage per cycle
(ϕ) are listed in Table 1 together with the crack initiation lives
(N_i). The creep damage per cycle is based on a linear summation,
defined by

$$\phi = \sum_0^t \frac{\Delta t}{t_r} \tag{10}$$

where Δt is the time increment in the numerical calculational procedure and t_r the creep-rupture time corresponding to the average
stress and temperature within the time increment. After Spera,[3]
the creep in compression was assumed additive to that in tension.

     Examination of each of the calculated inelastic variables in
Table 1 as the mechanical variable controlling fatigue damage does
not lead to completely satisfactory correlations of all the results,
but there are reasonable trends present. For example, a plot of
total inelastic strain range (ε_{ti}) versus cycles to crack is shown

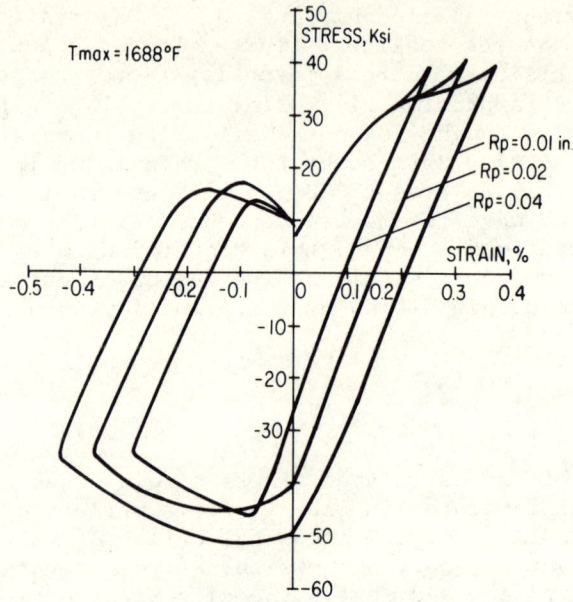

(A)   Effect of R_p.

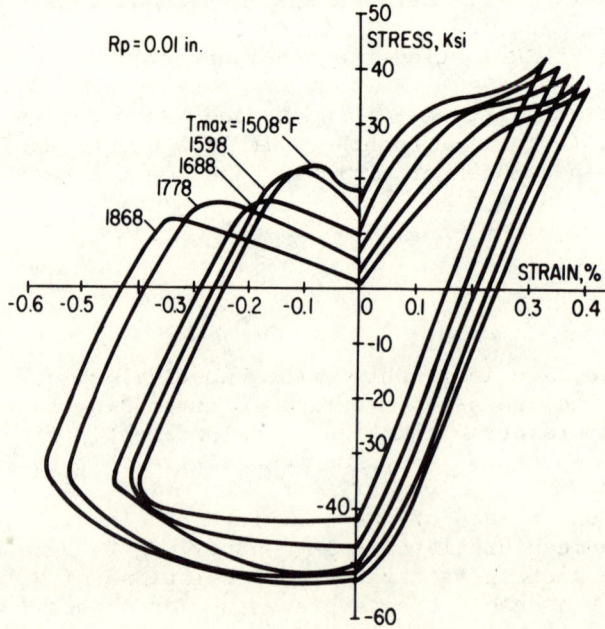

(B)   Effect of T_{max}.

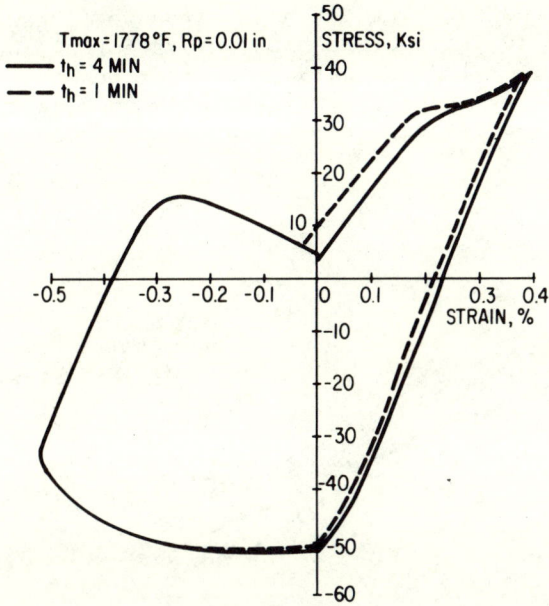

(c) Effect of t_h.

Fig. 19. Calculated hysteretic loops for various test conditions (uniaxial model method).

Table 1. Summary of calculated strain and life data for FSX 414.

T_{max}, °F	R_p, in.	t_h, min.	ε_t, %	ε_{ti}, %	ε_p, %	ε_c, %	ϕ	N_i
1508	.01	4	.722	.402	.309	.093	.081	230
1598	.01	4	.770	.451	.332	.119	.123	45
1688	.01	4	.819	.502	.340	.162	.112	49
1688	.02	4	.688	.372	.253	.119	.119	70
1688	.04	4	.552	.249	.174	.075	.089	116
1778	.01	4	.914	.608	.272	.336	.126	20
1778	.01	1	.900	.598	.329	.269	.097	60
1868	.01	4	.969	.690	.278	.412	.125	15
1688 BC	.02	4	.652	.283	.194	.089	.085	89
1688 SC	.01	4	.576	.354	.196	.158	.105	64
1688 SH	.01	4	.742	.329	.280	.049	.075	144

BC - bore cooled
SC - slow cooled
SH - slow heat

in Fig. 20 for tests having hot bath exposure times of 4 minutes.
A reasonable correlation of results exists for all but the lowest
temperature (T_{max} = 1508°F) data. In order to use such a correlation in a predictive sense, similar plots would have to be constructed for other exposure times, and the frequency sensitivity
established.

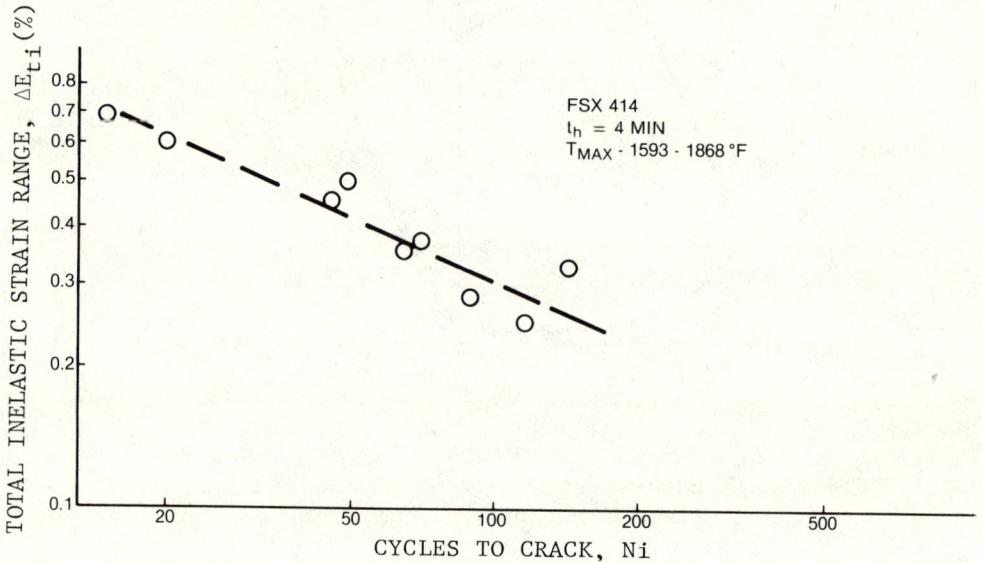

Fig. 20. Correlation of crack initiation data based on the total
inelastic strain.

The creep damage summation appears to discriminate well between the test variables, but the damage per cycle summed to a
damage state of 1.0 would predict a gross underestimate of life.
Also, the strain range partitioning approach may have applicability. The life data in the table appear to be correlatable by some
function of the two inelastic strain components as utilized in the
strain range partitioning approach. Use of this function

$$f(\Delta\varepsilon_p, \Delta\varepsilon_c) = \frac{1}{A} \Delta\varepsilon_p^\alpha + \frac{1}{B} \Delta\varepsilon_c^\beta \qquad (11)$$

would require first determining the material properties, α, β, A
and B from separate experiments.[10]

B. Crack Propagation

The analysis of crack growth rate data involved determining

growth rates from the a versus N curves such as shown in Fig. 2.
This was accomplished by fitting polynomial expressions to the
data and differentiating these to obtain da/dN as a function of
crack length.   Equation (9) and the appropriate radial strain
distribution were then used to estimate ΔK_ε values at selected
values of crack length for each of the test conditions.   The results are summarized in Fig. 21 on coordinates of da/dN versus
ΔK_ε.   The exposure time effect is illustrated on the left side
of the figure for T_{max} = 1688°F, and the temperature effect on
the right for t_h = 4 minutes.

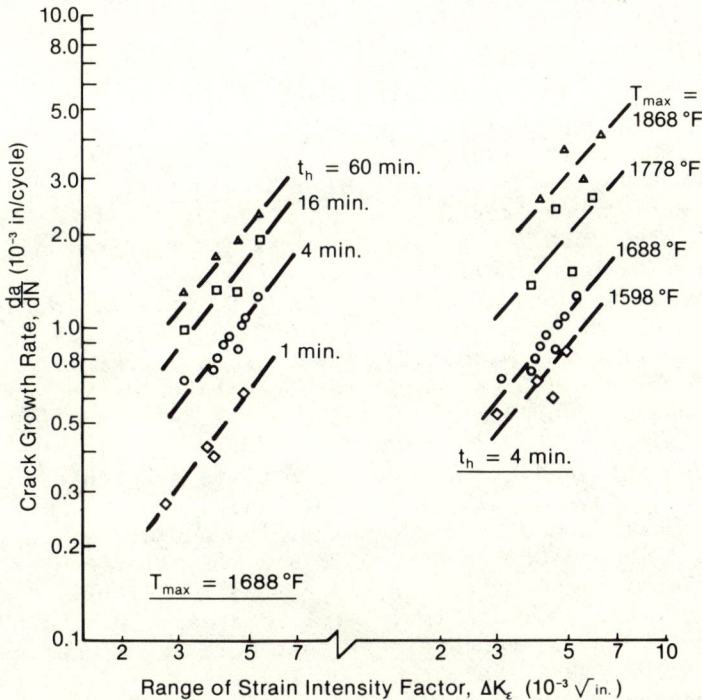

Fig. 21.   Fracture mechanics correlation of FSX 414 crack growth
           rate data.

      The correlations of data obtained (Fig. 21) based on using
the strain intensity factor as the crack driving force appear satisfactory.   The specimen peripheral radius or applied strain level
appears to be normalized correctly and a very clear ordering of
the exposure time and maximum temperature effects result.   Unfortunately, the data do not cover a very broad range of growth
rates, but this limitation appears to be a function of the test
conditions and not the method of analysis.

SUMMARY

Some simplified analytical procedures are developed for treating the complex thermal fatigue problem associated with the rapid temperature changes of gas turbine start-up and shutdown cycles. The analysis procedures are based on considering the fatigue damage process to consist of crack initiation and crack propagation phases. The local inelastic stress and strain behavior and the crack tip strain intensity factor are considered the damage controlling variables for each of these phases. Correlation of crack initiation and crack propagation data obtained on tests of tapered disk specimens of high temperature superalloy are demonstrated based on the calculated quantities.

REFERENCES

1. S. S. Manson, Thermal Stress and Low Cycle Fatigue, McGraw-Hill, New York, 1966.
2. P. W. H. Howe, in Thermal and High Strain Fatigue, The Metals and Metallurgy Trust, 1973, pp. 242-254.
3. D. A. Spera, NASA TMX 52558, National Aeronautics and Space Administration, 1971.
4. D. F. Mowbray, D. A. Woodford, and D. E. Brandt, in Fatigue at Elevated Temperatures, ASTM STP 520, American Society for Testing and Materials, 1973, pp. 416-426.
5. D. F. Mowbray and D. A. Woodford, Institute for Mechanical Engineers, Conference Publication 13, 1973, pp. 179.1-179.11.
6. D. A. Woodford and D. F. Mowbray, Materials Science and Engineering, Vol. 16, 1974, pp. 5-43.
7. D. F. Mowbray and J. E. McConnelee, in Thermal Fatigue of Materials and Components, ASTM STP 612, American Society for Testing and Materials, 1977, pp. 10-29.
8. A. D. Foster and C. T. Sims, Metal Progress, July, 1969.
9. L. F. Coffin, Jr., Proceedings, Institute for Mechanical Engineers, 1974, Vol. 188, pp. 109-127.
10. S. S. Manson, G. R. Halford, and M. H. Hirschberg, NASA TMX-67838, National Aeronautics and Space Administration, May, 1971.
11. A. S. Tetelman and A. J. McEvily, Fracture of Structural Metals, John Wiley & Sons, New York, 1967, p. 375.
12. H. Buechner, in Methods of Analysis and Solution of Crack Problems, G. C. Sih, Editor, Wolters-Noordhoff, 1972.
13. G. R. Irwin, in Structural Mechanics, Pergamon Press, New York, 1960.
14. H. G. deLorenzi and C. F. Shih, in ADINA Conference, Report No. 822448-6, Massachusetts Institute of Technology, August, 1977.
15. K. J. Bathe, ADINA, Report No. 82448-1, Department of Mechanical Engineering, Massachusetts Institute of Technology, 1977.

LIFE PREDICTION FOR TURBINE ENGINE COMPONENTS

T. Nicholas and J. M. Larsen

Air Force Wright Aeronautical
Laboratories, AFWAL/MLLN
Wright-Patterson AFB, OH 45433

ABSTRACT

An alternate approach to life management of turbine engines is
being considered by the U.S. Air Force. Whereas most major structural components are currently limited by low cycle fatigue and are
retired from service after their design life has been reached, a
"Retirement for Cause" approach would keep components in service
until a fatigue crack has been detected. The approach is based
on non-destructive inspection and prediction of fatigue crack
growth behavior under engine operating conditions. This paper
discusses the concept of retirement for cause and reviews the
problems associated with the prediction of crack growth. Several
aspects of crack growth under engine spectrum loading including
creep crack growth and crack retardation are discussed. Recommenda-
tions for future research efforts are presented.

INTRODUCTION

Ever increasing performance requirements for U.S. Air Force
turbine engines have placed stringent demands on material capability. A need to increase the ratio of engine thrust to weight
has forced design stresses higher and higher, approaching the
material yield strength, while engine operating temperatures have
continued to increase. In the hot section of the engine, turbine
airfoils operate under extreme thermal and mechanical conditions
which require strength at high temperature and dimensional stability. Components such as disks which do not lie directly in the
path of combustion gases are exposed to somewhat less severe

conditions. Many other components such as those found in the fan
and compressor regions of the engine are exposed to a cooler environment of inlet air. The variety of operating conditions within
the engine dictate diverse material capability requirements. In
the 1960's material strength and creep resistance were the primary
design limitations for turbine components. However, by 1975 low
cycle fatigue (LCF) performance had become the life limiting factor
for over 75% of the major structural components in advanced engines
such as the F-100 which powers the Air Force's F-15 and F-16 aircraft. Many of the LCF limited components are subjected to low
frequency stress cycles that are primarily due to centrifugal
loads related to engine speeds. A typical loading cycle is comprised of engine start-up, flight, and engine shutdown. In military engines these major cycles are altered by superimposed minor
cycles resulting from tactical maneuvers.

As disks and disk spacers represent the most critical individual turbine components, extreme design conservatism is exercised
in order to insure their structural reliability. The present design philosophy is based entirely on the concepts of fatigue crack
initiation.[1,2] Any additional component life due to subcritical
crack propagation following initiation is not considered. According to the initiation philosophy, one in 1000 of these components
will have developed a measurable flaw, while the remainder of the
population has not exhausted the initiation life. The difference
between the design life of a disk and the mean lifetime of test
data may be an order of magnitude or more. This is illustrated
in Fig. 1 which shows the average fatigue behavior of Inconel 718
and the 1/1000 lower bound used as the design curve. Throughout
the range of data presented, there exists a 10 fold difference
between the design curve and the average fatigue behavior. Thus,
the forced retirement policy for this disk material requires the
elimination of 999 statistically sound disks in order to remove
the one lower bound disk which is expected to have initiated a
measurable flaw. In addition, the disk which has initiated a
measurable flaw may have some useful crack propagation life remaining. Although this design approach is extremely conservative,
it is also very safe for major structural components whose failure
would be catastrophic.

The conservatism in this design approach is due in part to
the variability in material properties as well as uncertainty in
the actual stresses and thermal gradients imposed upon the components. The complex loading conditions active at critical disk
locations result from the superposition of stresses due to centrifugal forces, thermal gradients, and attachment constraints. In
turn, these individual loading components depend upon the specific

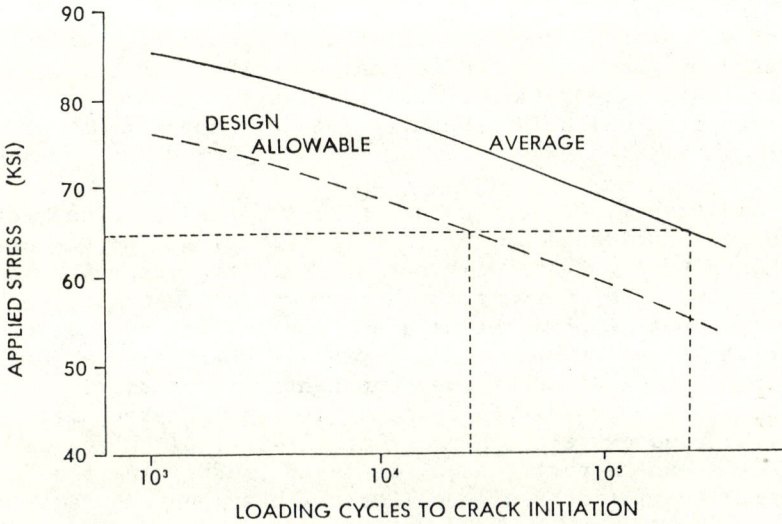

Figure 1.   S-N Curves for Alloy 718.

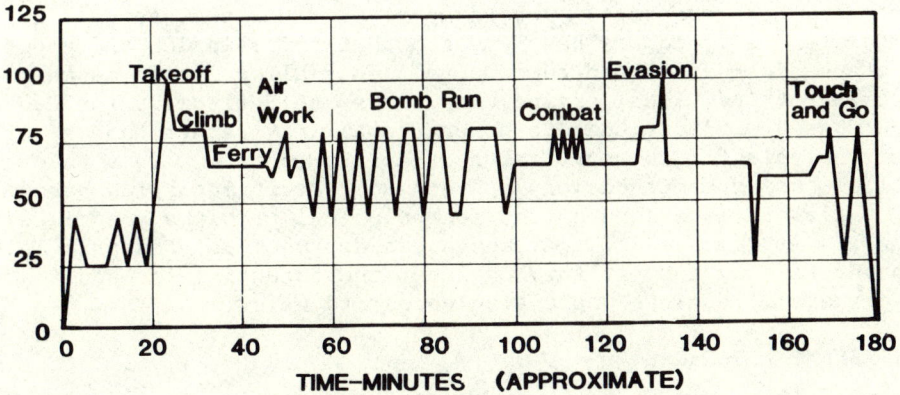

Figure 2.   Schematic of Typical Engine Load Spectrum.

mission or load spectrum to which the engine is operated. A typical mission imposed on turbine disks is presented in idealized form in Fig. 2. It can be seen that there is a variety of mission activities which includes a mix of fatigue and sustained load segments which may interact synergistically. In the hot region of the engine component response to such a mission loading spectrum is further complicated by the associated thermal spectrum. That is, throttle excursions which generate the mechanical spectrum also produce a thermal spectrum which is generally not in phase with the stress history.

In addition to mission considerations, the low cycle fatigue analysis is complicated by the fact that critical fatigue crack initiation sites are generally stress concentrators which produce local yielding. For many of the high temperature components, this time independent deformation is also coupled with time dependent deformation, i.e., creep. The addition of environmental effects extends the complication of the component life prediction process. In light of the uncertain fatigue conditions, inherent material variability, and extremely high reliability requirement associated with turbine engine performance, it is not surprising that a lower bound crack initiation design philosophy has been adopted.

Turbine engine components, however, are becoming increasingly costly to replace and, in addition, contain materials which have limited domestic supply. Therefore, from an economic and strategic viewpoint, it would be most desirable to fully utilize the available component life without sacrificing safety. To accomplish this, an alternative approach to life management of engine components has been proposed. According to this new philosophy, component life prediction is based upon both the initiation and growth of cracks in LCF limited components. This allows performance tracking of individual engine components and Retirement for Cause (RFC) based on individual component capability rather than a worst case analysis for the population of all components. This paper will discuss the methodology of such a fracture mechanics based RFC philosophy and some of the technological problems associated with its implementation. Although the approach is presently being considered only for military engines, the potential for application to commercial engines would also appear to exist.

RETIREMENT FOR CAUSE

The replacement of a safe life design policy, or low cycle fatigue life limited design, by one of "fly to a safe crack" has been termed Retirement for Cause[3,4] in this country or "on-condition lifing"[5] in the United Kingdom. This approach, which is being considered for use in military engines by the U.S. Air Force, is based on statistical considerations that most disks have no detectable crack and thus some useful life remaining when they reach the design lifetime. At present, each component is retired from service when it has reached its design life, whether or not any cracks are found. Under RFC, components would be kept in service and inspected at predetermined intervals for fatigue cracks. Components which did not have any cracks would be kept in service for another inspection interval. The fracture mechanics analysis would have to show that a crack smaller than the detectable size would not grow to a catastrophic size within one or more inspection intervals as illustrated in Fig. 3. If A_O is the size of the crack

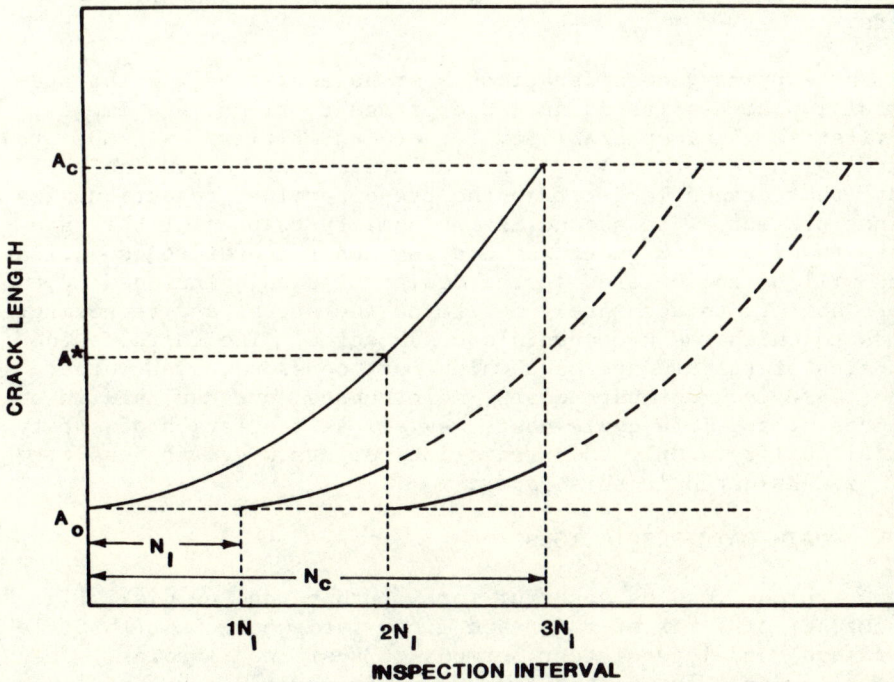

Figure 3.   Schematic of Crack Growth Considerations in Retirement for Cause

which is detectable by Nondestructive Evaluation (NDE) techniques, the crack growth analysis indicated by the solid line shows that such a crack will grow to a critical size, A_c, in three inspection intervals. The choice of three intervals is arbitrary, it could be less or more depending on economic, statistical, and reliability considerations. It is apparent from the figure that a crack less than A_o will not grow to critical size before the next inspection interval, N_1. In fact, for the hypothetical case shown, the crack will have grown to a size A* in two inspection intervals. Thus, there should be two opportunities to find a crack greater than A_o if the diagrammed procedure is followed. Components which do not have cracks greater than A_o are kept in service and inspected at each interval until a crack of A_o or greater is detected at which time the component is retired from service for cause. It is obvious from the procedure that the reliability of both the inspection technique and the crack growth predictions are vital aspects of RFC. The difficulty of both of these tasks is accentuated by the complexity of the components and crack geometries. Fig. 4 presents a schematic of typical types of LCF limited components, and common crack geometries are illustrated in Fig. 5. As shown flaws generally originate as surface or corner cracks at stress risers.

Three primary considerations must be addressed for the successful implementation of an RFC approach to fleet management. The first is the inspectability for cracks or flaws. An NDE procedure which is both reliable and accurate must be available to assure that cracks larger than the pre-determined rejection size are not present. The second is the ability to predict the rate of growth of a crack under the loading and temperature conditions which will be encountered in the design mission. Included in this is the ability to accurately determine the stress and temperature fields to which the components are subjected. The third is the economics of the RFC approach which must consider, among other factors, replacement part costs, maintenance, overhaul and inspection costs, and life cycle costs as well as statistics of safety and reliability. Only the predictions of crack growth behavior will be considered in this paper.

CRACK GROWTH RATE PREDICTIONS

A typical loading spectrum for a turbine engine disk, Fig. 2, is composed of a mix of major and minor fatigue cycles (throttle excursions) and intermittent periods of sustained loading. The fatigue loadings are in the frequency regime of 0.1 Hz or slower, and hold times are on the order of minutes. At elevated temperatures crack propagation produced by mission loading is quite

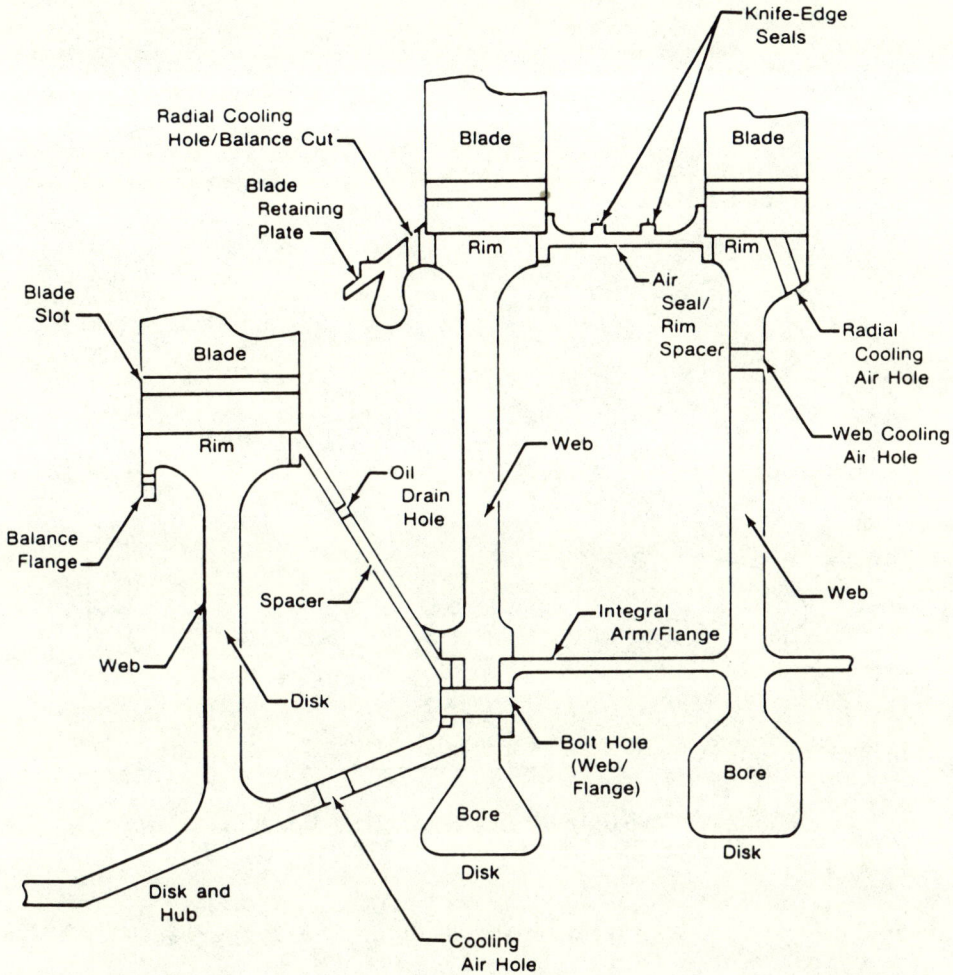

Figure 4.   Typical Life Limited Structural Components in an
            Engine 4

complex, requiring the capability to predict the effects of fatigue,
creep, environment, and the interaction of these variables. Fortunately, the degree of difficulty of the component life prediction problem is lessened by the foreseeable, approximately repeat-
able, nature of the mission spectra. Each mission contains major
cycles which occur regularly and are generally definable. Extreme

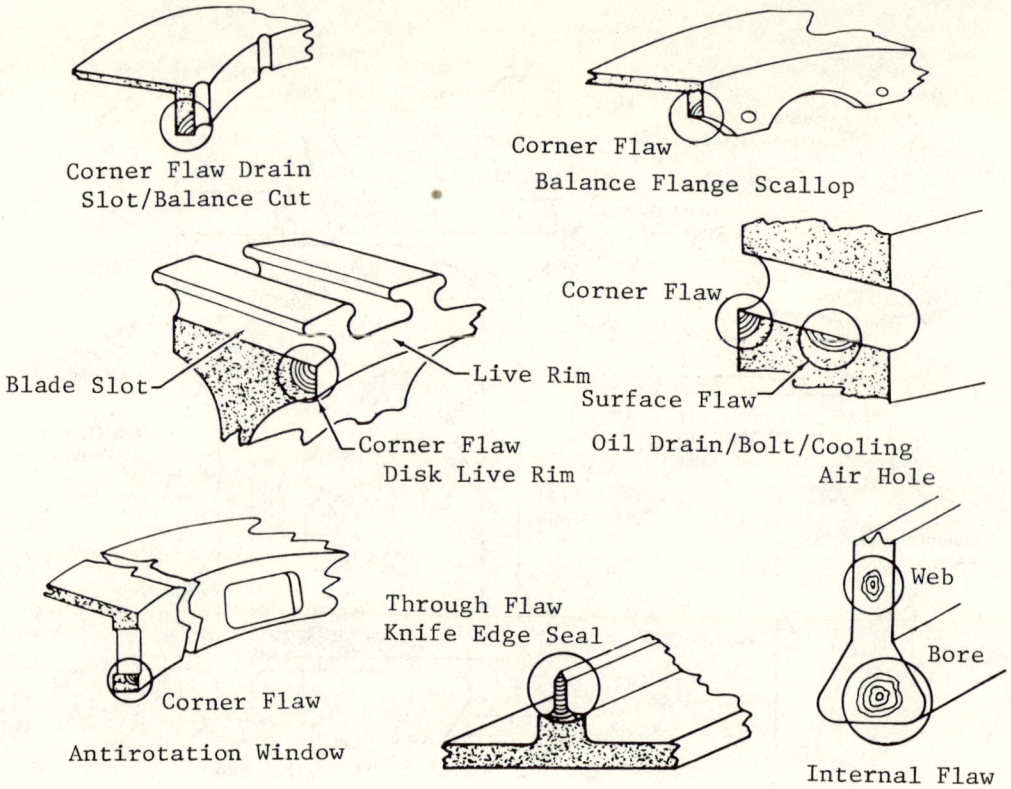

Figure 5.   Typical Flaws in Engine Components 4

variety in low cycle fatigue spectra is not characteristic of engine operation.

Superimposed on the primary LCF loading are fatigue cycles resulting from vibrations of turbine blades and other disk attachments. The secondary cycles are of relatively low amplitude and in a frequency range several orders of magnitude above that of the primary cycles. There has been essentially no research addressing the effect on crack propagation of superimposed high cycle and low cycle fatigue. Therefore, this aspect of turbine disk fatigue will not be discussed.

The prediction of crack growth in turbine engine components has been approached in two ways. The more common approach has been to study the basic phenomena of crack growth associated with cyclic and sustained loading under conditions of temperature and environment which are representative of engine operation. With the addition of interaction effects, the results of these fundamental studies may then be collected to address crack growth in the specific engine spectrum. The second approach has been to decompose the turbine spectrum into groupings of cycles and hold-times and to model the crack growth empirically. The discussions to follow will focus primarily on the former approach.

Linear elastic fracture mechanics (LEFM) has been successfully applied in predicting the crack growth rate behavior in turbine engine materials under constant amplitude cyclic loading. Using either ΔK or K_{max} as a correlating parameter, the crack growth rate curve has the traditional sigmoidal shape commonly found in other structural materials. The primary variables influencing the shape and location of the basic fatigue crack growth curve are temperature, frequency, stress ratio (R = minimum/maximum stress), and environment. Since most high temperature disks are surrounded by an atmosphere of bleed air used to cool the turbine blades, considerations of environmental variability as applied to disks are limited. Temperature, frequency, and stress ratio effects on crack growth may be described by a number of analytical methods. One example of such a method is the SINH model[6,7,8] which provides the capability to interpolate across crack growth data generated over a wide range of temperature, frequency, and stress ratio. The use of this model has been extended to describe load sequence effects representative of engine disk operation.[9]

As discussed earlier, engine operating histories are complex but limited in variety. Typically, engine operating profiles differ significantly from the loading spectra which are characteristic of airframe usage. Analysis of turbine disk missions has revealed that major load excursions (overloads), such as occur during takeoff, occur on a relatively frequent basis. The significance of this observation may be appreciated by considering the phenomenon of fatigue crack retardation associated with a single overload. A schematic illustrating the various features of this process is presented in Fig. 6. Typically, constant load amplitude fatigue, interrupted by a single overload, experiences a deceleration in crack growth (delayed retardation), minimum crack growth, and crack acceleration until the unretarded rate of crack growth is resumed. The increase in crack life due to the application of the overload, defined as N_D, has been the focus of many crack retardation studies. Depending primarily upon the overload ratio (OLR = $\sigma_{overload}/\sigma_{max}$), N_D may be many thousands of cycles or even infinite (crack arrest) under some conditions. However,

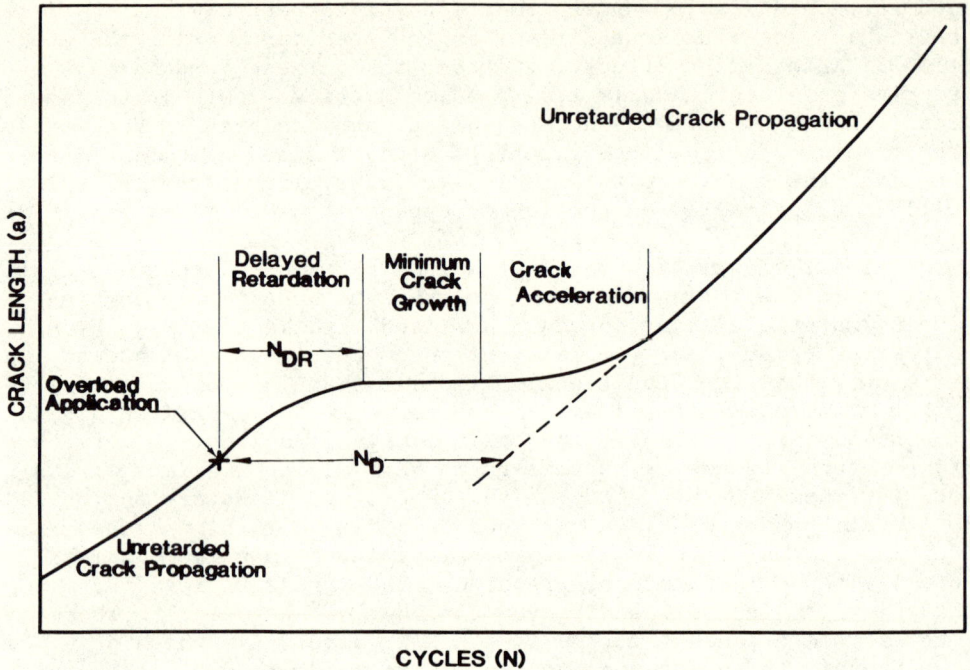

Figure 6.   Schematic of Crack Growth Behavior After a Single
             Overload Cycle

for single overloads typical of disk operation, its magnitude is
commonly on the order of hundreds or a few thousand. This number
is still much greater than the approximate 50 or less cycles which
typically separate successive overloads in the spectrum for military engine operation. As a result of these frequent major load
excursions, the immediate post-overload fatigue behavior is of increased significance, while the long-term effect of each overload
is truncated by the succeeding overload. From a physical point
of view, crack growth between overloads is not sufficient to remove the crack tip from the overload plastic zone. For the superalloys IN100* and Waspaloy subject to frequently occurring overloads characteristic of mission disk usage, it has been shown that
the fatigue crack retardation process is dominated by delayed retardation.[10]   That is, both materials exhibited continuous post-
overload deceleration in crack growth until the recurrence of an

---

*All references to IN100 in this paper refer to the P/M superalloy
GATORIZED[R] IN100, described in Ref. 7.

overload truncated and restarted the load interaction process.
This behavior deviates significantly from the general long term
effects of an overload on fatigue crack propagation.   Therefore,
mission crack growth in these alloys is difficult to predict by
any retardation model which is developed from experimental measurements of the total period of retardation, N_D.   The interaction effects between cyclic loading, hold-times, and overloads all have
to be considered in modeling crack growth rate behavior of engine
alloys.

A second significant feature of engine spectra is a limitation on the magnitude of the overload ratio. Peak stresses are
related to maximum engine speeds, which are limited by ultimate
engine performance. Since military engines are designed to operate
near peak performance (and near the material yield strength), the
major load cycles (overloads) are on the order of the intervening
fatigue cycles, and overload ratios rarely exceed 1.5. As a result of the absence of higher overload ratios, crack propagation
in disks under engine spectra does not exhibit the extreme crack
retardation or crack arrest which is observed under other types
of fatigue spectra containing overloads ratios of 2 or 3.

At some locations on engine disks, the combination of mechanically and thermally induced stresses may result in the occasional
occurrence of an overload-underload sequence. An investigation[11]
focusing on the period of retardation, N_D, has shown that, in IN100,
massive umderloads ($\sigma_{compressive}/\sigma_{max} < -1$) are necessary in order to
significantly alter the N_D produced by the single overload alone.
Since underloads of this magnitude do not occur during engine operation, the presence of the underload appears unimportant. However,
additional work is necessary to define its effect under frequent
overload-underload occurrences.

Another feature of engine spectrum loading which must be
modeled is the effect of sustained loads or the associated creep
crack growth. Although stress intensity factor, K, has been applied in a number of investigations involving creep crack growth,
numerous questions have been raised about the applicability of
LEFM to creep crack growth modeling.[12] Other parameters such as
Rice's J - integral[13] or the C* parameter of Landes and Begley[14]
have also been proposed as correlating parameters. The question
can be asked legitimately as to whether any single parameter can
be used to correlate creep crack growth because of the complex
phenomenology involved. Fig. 7 is a schematic of the effect of a
sustained load interspersed within a block of fatigue cycles.
Letting a_c denote the amount of creep crack growth observed on the

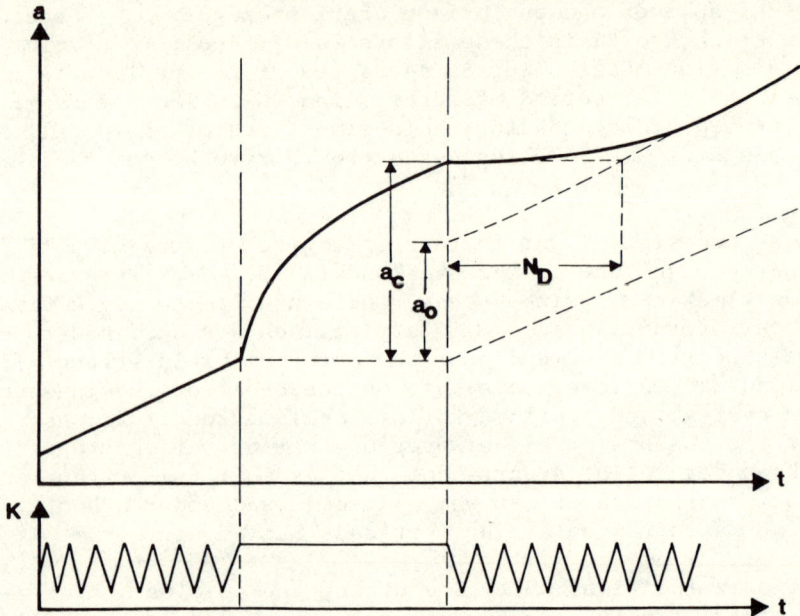

Figure 7.   Schematic of Crack Growth Behavior Due to Hold Time
            Interrupting Fatigue Cycling

fracture surface, and a_o the offset to the constant amplitude
fatigue crack growth curves before and some time after the sustained load interval, it has been observed that a_c and a_o are not
necessarily identical. The difference, $a_c - a_o$, is an increment
of crack extension retardation, which can be related to the apparent number of delay cycles through

$$N_D = \frac{a_c - a_o}{da/dN}$$

where da/dN is the constant amplitude fatigue crack growth rate.
Using K as the correlating parameter, it has been found that the
initial creep crack growth rate is higher than the rate after the
creep crack has been growing for some length of time at the same
value of K.

Donat et al.[15] showed that in compact (CT) specimens of IN100 at 732°C covering a range of thicknesses from 5.5 to 18 mm, the initial portion of the curves of a vs. K was consistently higher than the average curve. In constant K cracked ring geometry specimens, this effect was even more noticeable. Data on the same material for a thickness of 7.5 mm, Fig. 8, show the same general

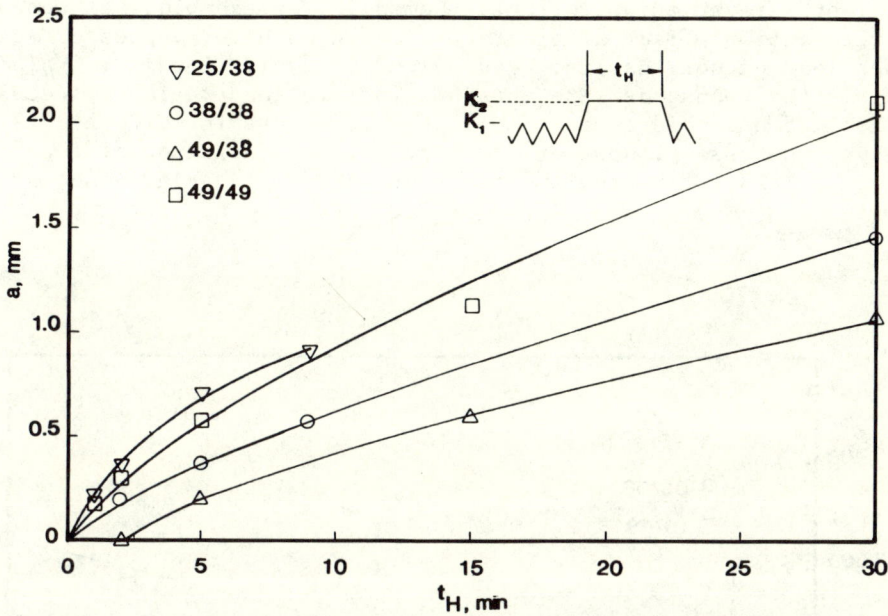

Figure 8. Crack Growth Due to Hold Times Following Fatigue Cycling IN100, 732°C, R = 0.1, ν = 2.5Hz (K_1/K_2 in MPa$\sqrt{m}$)

effect for two values of K, 38 and 49 M Pa$\sqrt{m}$ in the cases where the prior fatigue cycling was at the same K_{max} (R = 0.1) level as the sustained load portion, indicated by 38/38 and 49/49, respectively. In the data of Donat et al., however, the prior fatigue cycling was at a much lower level, which approaches the condition shown as 25/38 in Fig. 8. On the other hand, if the prior fatigue cycling is at a higher K_{max} level, an incubation period can be observed. In essence, the prior "overloading" in fatigue causes a retardation of subsequent creep crack growth. The curve indicated as 49/38 in Fig. 8 shows such an effect, where an incubation period of approximately 2 min. is observed. In other

tests, for cycling at K_{max} = 38 M Pa$\sqrt{m}$, followed by sustained load
at K = 25 M Pa$\sqrt{m}$, no creep crack growth was observed for hold-times
up to 10 min. Creep crack growth experiments on specimens previously subjected to creep crack growth have also shown crack arrest at subsequent lower K levels which had previously produced
creep crack growth in fatigue precracked specimens.

In addition to the non-uniform creep crack growth behavior,
the fatigue cycling after creep crack growth generally shows a
delay, or a retardation, effect. Gemma,[16] for example, has shown
the retardation effect after creep crack growth for values of sustained load higher than the peak values in cyclic loading. Values
of N_D in the thousands were obtained for overload ratios between
1.2 and 1.5 and hold-times of the order of hours in IN100 at 649°C.
It is not necessary, however, to increase the amplitude of the
sustained load to observe the retardation effect. Fig. 9 shows

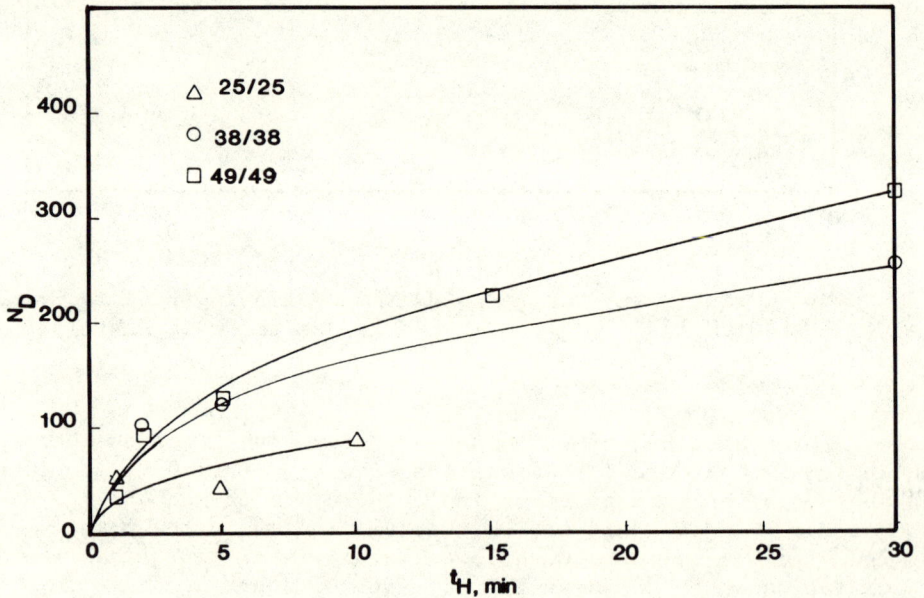

Figure 9.   Delay Cycles (N_D) Due to Sustained Load; IN100, 732°C,
R = 0.1, ν = 2.5Hz (K_i/K_i represents Value of K_{max} in
MPa$\sqrt{m}$ for the Sustained Load and Cyclic Loading)

data on IN100 at $732^{\circ}C$ on the number of delay cycles, N_D, against
hold-time for three different K values. In each case, the value
of K during the creep crack growth portion was the same as the
K_{max} value during fatigue both before and after the hold-time.
For hold-times less than 30 min. N_D values of the order of hundreds
were obtained. It is important to note that the values of N_D had
to be computed from the measured creep crack extension (a_c in Fig.
7) and the observed offset, a_o, to the constant amplitude fatigue
cycling. In performing experiments of this nature, therefore, it
is important to observe both the effect of the creep crack growth
and the subsequent retardation in modeling the overall behavior.

In the work of Gemma[16] primary attention was paid to the effect of hold-time on the subsequent retardation of crack growth
under fatigue loading. Although not specifically mentioned, it is
assumed that the amount of crack growth during sustained loading
was negligible. Macha et al.[11] on the other hand, performed
similar experiments looking for the amount of crack growth during
sustained loading. Using measured Δa offsets obtained from the
steady state fatigue cycling curves (a vs. N) taken before and
after the sustained load portion, they attributed this offset entirely to creep crack growth. We have subsequently re-examined
the data and fracture surfaces and have noted that the actual
amount of creep crack growth was substantially greater than that
determined by the offset method. The reason for this is the subsequent retardation of fatigue crack growth after creep crack
growth. Thus, these experiments which demonstrated an apparent
incubation period for creep crack growth, in actuality demonstrated
an initial higher creep crack growth rate when replotted using
actual crack growth values.

Another feature which these and other data reported in the
literature point out is the lack of a constant crack front geometry
when going from fatigue crack growth to creep crack growth. Creep
crack growth experiments consistently show a propensity for higher
growth rates at the center of a specimen than at the free surfaces,
or a variation from nearly plane strain stress states to ones of
plane stress. Crack tunneling is a direct result of this variation
in growth rates and makes both definition and measurement of crack
length a difficult task. Thus, the same phenomenological behavior
can be interpreted as an incubation period based on surface crack
measurements or as an initial crack acceleration based on measurements at the center from fracture surface measurements. In our
investigations, we have chosen to define crack length based on
compliance measurements and LEFM analysis, or as the average length
from nine equally spaced points across the fracture surface.

Other investigations have addressed simple spectrum loading repeated at regular intervals. Various combinations of cycles and hold-times have been used to study the general phenomenology and the interaction effects. The effect of a hold time on cyclic load-ing can either increase or decrease the effective crack growth rate when interspersed within fatigue cycles. Shahinian and Sadananda[17] have observed that hold times at very low load levels increase the rate of crack growth in Inconel 718. This appears to be an environ-mentally induced phenomenon in that the same effect could not be reproduced in vacuum. The implications of such an observation are that hold times at temperature under zero load conditions might play a role in the overall crack growth rate behavior. In an actual engine mission, the time after engine shut down correspond-ing to zero load, which precedes the temperature drop, might con-tribute to subsequent crack growth through some combination of oxidation and residual stress effects. Thus, in all of these cases we are dealing with a combination of mechanical and environmental effects with generally strong interactions.

Load interaction effects characteristic of engine operation may also be examined by performing crack growth testing under simplified repetitive missions of a single overload followed by a block of fatigue cycles. A series of such missions is simply con-stant load amplitude fatigue interrupted by periodic overloads. Crack growth data[9] generated in IN100 subject to such periodic overload fatigue are presented in Fig. 10. Under these representa-tive elevated temperature test conditions, average fatigue crack growth rate decreases significantly as the number of base line fatigue cycles between successive overload (N_{OL}) increases. Con-stant load amplitude fatigue crack growth under the base line cycle is approximately equivalent to crack propagation under the condition with 5 base line cycles separating periodic overloads. Over the range shown, increasing N_{OL} above 5 results in a generally retarded crack growth rate. As N_{OL} becomes larger still, a minimum crack propagation curve should result. Thereafter, average crack growth rates should increase, approaching the constant P_{max} fatigue curve, as N_{OL} becomes very large. However, as discussed earlier, only crack growth behavior associated with much more frequent over-loads is applicable to disk life prediction in military gas turbine engines.

Another aspect of crack growth behavior in engine materials at elevated temperatures which must be considered is the effect of three-dimensional stress fields or the effect of thickness in test samples as one goes from states of plane stress to plane strain with increasing thickness. A state of plane stress is approached as thinner and thinner samples are used because stresses in the

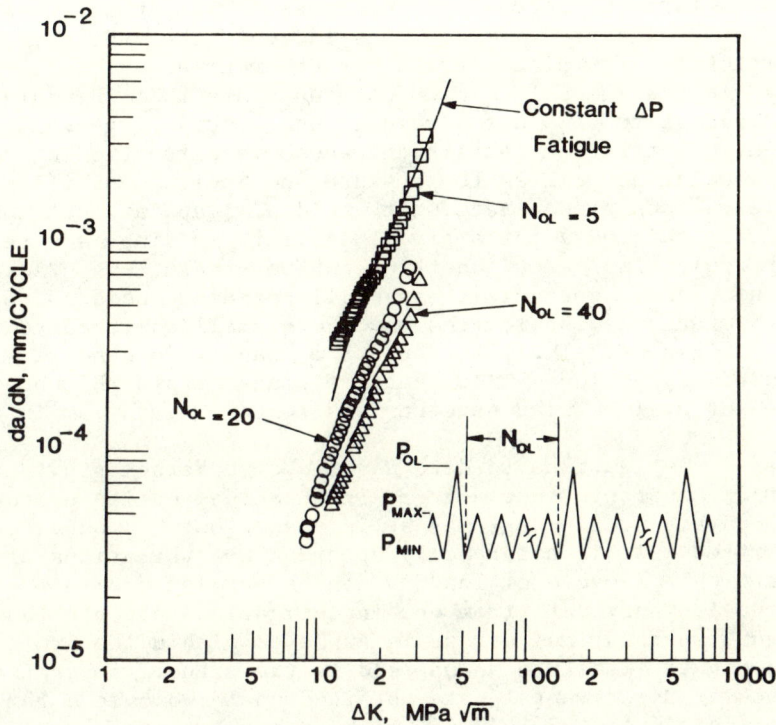

Figure 10. Crack Growth Behavior Due to Fatigue Cycling With Intermittent Overloads; IN100, R = 0.5, ν = 0.167Hz, OLR = 1.5

thickness direction are absent at the two free surfaces. We define the specimen thickness for plane strain conditions as that for which further increases in thickness provide no change in crack growth rate behavior. Investigations have shown that under fatigue cycling, especially at higher frequencies, the thickness required to reach a state of plane strain is considerably smaller than the thickness to reach plane strain under sustained loading conditions. In IN100, for example, plane strain conditions are

achieved under cyclic loading at thicknesses of the order of 2 mm
whereas for creep crack growth thicknesses close to 10 mm are necessary before no further change in crack growth behavior is observed with further increase of thickness.[7] Recent data based on
results using side grooved specimens indicate that the creep crack
growth thickness for plane strain may be even larger. Thus, there
is no single thickness for which one can state that the material
will be in a state of plane strain unless a very large value is
used. Rather, the plane strain thickness is a function of the
type of loading as well as temperature and environment if these
also affect crack growth rate. It would also appear that the
plane strain thickness for a given material, loading, and test
specimen would also be dependent on parameters such as crack
length and plastic zone size, especially if one considers the
growth of cracks having lengths which are small compared to the
specimen thickness. Thus, for spectrum loading in general, it is
not intuitively obvious as to what the plane strain thickness
might be for a given load spectrum and test specimen combination.

     Most investigations to date have involved crack growth rate
behavior at constant temperature. In an actual engine component
the temperature and loading may be in phase, out of phase, or
phase shifted. It is not readily apparent how variations of temperature within a cycle of loading can be handled from a predictive
basis. Preliminary data from one investigation indicate that the
temperature when maximum stress is achieved within the cycle may
be the dominating feature as opposed to the actual temperature
variations.[9] Experimental data obtained on IN100 over a temperature range from 427° - 704°C with temperature and load out
of phase, show that the out of phase data match most closely with
the isothermal crack growth data at the lower temperature extreme
in the cycle. On the other hand, preliminary unpublished data[18]
on Inconel 718 and P/M Rene 95 indicate little difference between
in and out of phase thermal-mechanical cycling crack growth rates,
although behavior in the threshold region appears to be different.
There are insufficient data at this time to draw any conclusions,
and no models have been developed to address this type of problem
in general.

SUPPORTING TECHNOLOGY

     Some of the major considerations for life prediction in engine
components using a Retirement for Cause approach have been addressed above. It is apparent that a thorough technology base has
not yet been developed, although great progress has been made in
the areas of understanding and predicting crack growth rate behavior in engine materials under typical operating environments.

Numerous questions remain to be answered before RFC can be applied on a large scale to both military and commercial engines. We will attempt to highlight some of the technical areas which remain to be addressed below.

One of the major questions which has arisen in the application of fracture mechanics to the prediction of the growth of flaws in engine disks concerns the applicability of the modeling to small flaws such as corner cracks or surface cracks in bolt holes. Successful implementation of RFC on modern engines will require NDE capability for ever decreasing flaw sizes which, in turn, will require crack growth predictive capability for initially smaller and smaller flaws. Most of the laboratory data generated on crack growth rate behavior of engine materials have been obtained on conventional test specimens such as the compact type where the flaw lengths are typically an order of magnitude larger than those which we will seek to predict in actual components. For these test specimens with long cracks, we have a high degree of confidence in the LEFM analysis. For short cracks, however, the stress intensity analysis becomes more questionable because we are approaching the region where plastic zone sizes at crack tips may no longer be small compared with the crack length, i.e., the fundamental assumptions of LEFM are violated. Further complicating the situation is the fact that many of the regions where cracks initiate are regions of stress concentrations such as at holes or notches. Since stresses in the component can approach the yield strength of the materials, it is not unusual to expect inelastic deformation in the regions where cracks initiate. Thus, we must face the problem of the prediction of the growth of small cracks in regions of stress concentrations. The applicability of LEFM to such an analysis must certainly be questioned, or verified, for this situation. A companion problem is the ability to obtain accurate and reliable experimental data in such a situation. Conventional crack measurement techniques such as those based on optical or compliance methods lose accuracy when applied to very small cracks because of lack of sensitivity. It would appear that both analytical and experimental work are needed in this area.

Another area where further development seems necessary is that of cumulative damage, i.e., the prediction of crack growth under engine spectrum loading. As discussed previously concepts such as total period of retardation (number of delay cycles) which are applicable in airframe analysis are of little value in engine spectra because the number of cycles between overloads or hold times is small compared to the number of delay cycles, and overload ratios are generally small. Thus, although the concepts of retardation or cumulative damage and interaction effects of creep

and fatigue are quite important, the approaches must be different
for engine materials. Cumulative damage and spectrum loading
modeling concepts for crack growth in engine materials have received little attention outside of some highly empirical approaches,
and could provide an opportunity for significant technical advances
of the state-of-the-art in engine life prediction.

The lack of an experimental data base and the apparently con-
flicting preliminary results in thermal-mechanical fatigue crack
growth indicate that this is another area where future work and
understanding is needed. The key problem here is that the temperature may vary within a given stress cycle, so interpolation
between isothermal curves is not strictly valid. This is another
example of the need for cumulative damage concepts applied to
crack growth behavior.

The investigations to date on crack growth in engine alloys
clearly demonstrate that environment plays a significant role in
the behavior. If mathematical or mechanistic models are to be
developed, this fact should not be overlooked. Thus, not only
stress, strain, or stress intensity are to be considered, but
time will play an important role. From a mechanisms viewpoint, it
is important to determine the roles which surface phenomena and
diffusion play in environmental effects.

Another aspect of crack growth behavior which is not clearly
understood is thickness effects or the role of three-dimensional
stress states. As pointed out above, the thickness required to
achieve plane strain conditions depends on the type of loading,
for example high frequency fatigue versus sustained load creep
crack growth. It is to be expected that the plane strain thickness will also depend on frequency and temperature. Furthermore,
crack length will probably play a role, especially as shorter
cracks are considered. The question as to what equivalent thickness should be used for edge cracks or corner cracks in actual
components is not readily answered. The use of plane strain data
for all cases would certainly be safe, but it might be unduly conservative. Until a rationale can be developed for explaining
thickness effects, this conservative approach may be the only
method for handling life prediction.

One of the aspects of crack growth behavior or material behavior in general which must not be overlooked in present or
future research is that of statistical variability. It is extremely tempting to draw general conclusions based on limited data.
In dealing with expensive materials, high machining costs, and
complicated tests, we are often forced into drawing conclusions
based on very limited data. Our experience has shown that a

large amount of material scatter can be expected. Creep crack growth tests, for example, exhibit significantly more scatter than high frequency fatigue crack growth tests in the same material. Factors such as frequency, temperature, stress ratio, and maximum stress can also affect the amount of scatter in material data, i.e., one cannot simply assign a scatter band to crack growth rate data for a given material. Furthermore, much of this scatter may be due to test techniques. Any such variability, whether associated with material properties or testing methods, must be accounted for in the component life prediction.

A final technical consideration which will influence the future applicability of a Retirement for Cause philosophy to fleet management pertains to the area of alloy development. In recent years, as low cycle fatigue performance has become life limiting for many turbine engine components, alloy development has been tailored toward improved LCF properties. Refinement of alloy chemistry and processing methods has resulted in increased crack initiation lives. In engine disks, achievement of smaller grain sizes and the use of powder metallurgy alloys have improved initiation performance allowing an increase in the LCF design lives without structural modification. However, since the LCF design philosophy did not include a crack propagation life, the crack growth behavior of the developing materials was of secondary importance. In some cases crack propagation life was reduced in order to improve the initiation life. With the implementation of RFC, total life including both initiation and crack propagation has to be considered in evaluating a material for potential application. Material development should therefore move in the direction of alloys which have greater resistance to fatigue crack propagation. This may be in conflict with the desire to improve initiation life, since the two are not necessarily compatible. For example, Ti-6Al-2Sn-4Zr-6Mo, which has excellent crack initiation resistance, has relatively poor crack growth characteristics. Under RFC, both characteristics have to be considered, along with the economics and NDE capabilities. The implementation of RFC, therefore, could have a significant impact on the criteria for new materials development and selection.

ACKNOWLEDGEMENTS

The support of the U. S. Air Force under In-house Project 2307 P1 02 at the Materials Laboratory is greatly appreciated. The assistance of Mr. Douglas Deaton and Mr. Stanley Flagel, Systems Research Laboratories (SRL), Dayton, OH, in the conduct of the experiments and in the data reduction, respectively, is gratefully acknowledged. Special thanks are also extended to Dr. Noel Ashbaugh, SRL, for his contributions.

REFERENCES

1. "Analysis of Life Prediction Methods for Time-Dependent Fa-
     tigue Crack Initiation in Nickel-Base Superalloys,"
     National Materials Advisory Board, Publication NMAB-347,
     National Academy of Sciences, Washington, D.C., 1980.
2. Cruse, T. A. and Meyer, T. G., "Structural Life Prediction
     and Analysis Technology," Air Force Aero Propulsion Labora-
     tory Report AFAPL-TR-78-106, Wright-Patterson AFB, OH, 1978.
3. Hill, R. J., Reimann, W. H., and Ogg, J. S., "A Retirement-
     for-Cause Study of an Engine Turbine Disk," Unpublished
     Air Force Wright Aeronautical Laboratories Report.
4. Harris, J. A., Jr., Sims, D. L., and Annis, C. G., Jr.,
     "Concept Definition: Retirement for Cause of F100 Rotor
     Components," Air Force Wright Aeronautical Laboratories
     Report, AFWAL-TR-80-4118, Wright-Patterson AFB, OH, 1980.
5. Duggan, T. B., "Philosophy of Safe Life Design," Department
     of Mechanical Engineers, Portsmouth Polytechnic, Portsmouth,
     England, Unpublished Report, 1980.
6. Annis, C. G., Wallace, R. M. and Sims, D. L., "An Interpola-
     tive Model for Elevated Temperature Fatigue Crack Propaga-
     tion," Air Force Materials Laboratory Report, AFML-TR-76-
     176, Part I, Wright-Patterson AFB, OH, 1976.
7. Wallace, R. M., Annis, C. G. and Sims, D. L., "Application of
     Fracture Mechanics at Elevated Temperature," Air Force
     Materials Laboratory Report, AFML-TR-76-176, Part II,
     Wright-Patterson AFB, OH, 1976.
8. Sims, D. L., Annis, C. G. and Wallace, R. M., "Cumulative
     Damage Fracture Mechanics at Elevated Temperature," AFML-
     TR-76-176, Part III, Wright-Patterson AFB, OH, 1976.
9. Larsen, J. M., Schwartz, B. J., and Annis, C. G., Jr.,
     "Cumulative Damage Fracture Mechanics Under Engine Spectra,"
     Air Force Materials Laboratory Report, AFML-TR-79-4159,
     Wright-Patterson AFB, OH, 1980.
10. Larsen, J. M. and Annis, C. G., "Observation of Crack Retard-
     ation Resulting from Load Sequencing Characteristics of
     Military Gas Turbine Operation," Presented at ASTM Sympo-
     sium, San Francisco, CA., May 1979, to be published in
     ASTM-STP 714.
11. Macha, D. E., Grandt, A. F., Jr., and Wicks, B. J., "Effects
     of Gas Turbine Engine Load Spectrum Variables on Crack
     Propagation," presented at ASTM Symposium, San Francisco,
     CA., May 1979; to be published in ASTM-STP 714.
12. Fu, L. S., "Creep Crack Growth in Technical Alloys at Elevated
     Temperature - A Review," Engineering Fracture Mechanics,
     Vol. 13, pp. 307-330, 1980.

13. Rice, J. R., "Mathematical Analysis in the Mechanics of
     Fracture," in Fracture, Vol. II, H. Liebowitz, ed.,
     Academic Press, New York, pp. 191-311, 1968.
14. Landes, J. D. and Begley, J. A., "A Fracture Mechanics Ap-
     proach to Creep Crack Growth," Mechanics of Crack Growth,
     ASTM STP 590, pp. 128-148, 1975.
15. Donat, R. C., Nicholas, T., and Fu, L. S., "An Experimental
     Investigation of Creep Crack Growth in IN100," presented
     at 13th Nat'l. Symposium on Fracture Mechanics,
     Philadelphia, PA., June 1980; to be published in ASTM-STP.
16. Gemma, A. E., "Hold Time Effect of a Single Overload on Crack
     Retardation at Elevated Temperature," Engr. Fr. Mech., Vol.
     11, pp. 763-774, 1979.
17. Shahinian, P. and Sadananda, K., "Effects of Stress Ratio
     and Hold-Time on Fatigue Crack Growth in Alloy 718,"
     Trans. ASME, J. Engr. Mat. Tech., Vol. 101, pp. 224-230,
     1979.
18. General Electric Co., AF Contract F33615-71-C-5193, un-
     published data.

A KINETIC MODEL OF HIGH TEMPERATURE FATIGUE CRACK GROWTH

J.J. McGowan and H.W. Liu[†]

Department of Aerospace Engineering, Mechanical
Engineering and Engineering Mechanics
University of Alabama
[†]Department of Chemical Engineering
and Materials Science, Syracuse University
Syracuse, New York

ABSTRACT

The increased fatigue crack growth rate at high temperature
is primarily caused by environmental effects. The amount of the
environmental effect depends on the penetration rate of the detrimental chemical specie relative to the crack growth rate. Depending upon the relative chemical penetration rate, there can either
be no, partial, or full environmental effects. A kinetic model is
formulated to predict the environmental effects. The kinetic
model is applied with good success to the high temperature fatigue
crack growth of IN-100, a nickel base superalloy.

HIGH TEMPERATURE FATIGUE CRACK GROWTH: CREEP VS. ENVIRONMENT

The fatigue crack growth rate in air at high temperature can
be considerably higher than the rate at room temperature, with the
increase in growth rate heavily dependent upon the frequency. The
increase in growth rates could be caused by either environmental
effects (corrosion) or by creep cracking. The results of some
early high temperature fatigue crack growth experiments by Coffin
and Solomon [1,2,3,4] on A286 in air and vacuum are shown in
Figure 1. The crack growth rate in air is substantially higher
than the rate in vacuum. From these results they conclude that
the primary cause of the accelerated crack growth was due to the

377

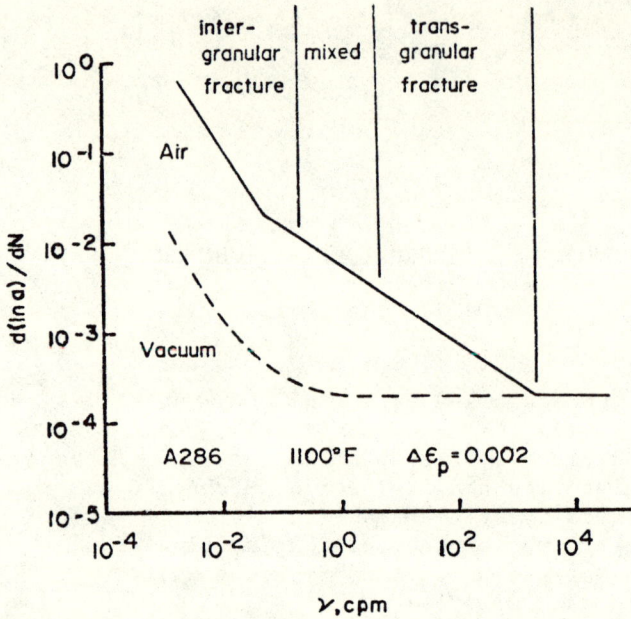

Fig. 1.   Schematic comparison of the air and vacuum fatigue crack
          growth behavior in A286 at 1100°F (ref. 4).

environment.   The figure also shows that (in the air environment)
as the frequency is increased from 0.01 cpm to 10 cpm, the fracture
morphology changes from intergranular to transgranular, with a corresponding decrease in growth rate.

James [5] tested 304 stainless steel at temperatures, 70°F,
800°F, and 1000°F.   The fatigue crack growth rates in air at elevated temperatures were much higher than those in room temperature
air, with an intergranular fracture mode.   Furthermore, the growth
rates in inert environments of liquid sodium and vacuum at both of
these two elevated temperatures were approximately equal to that in
room temperature air with a transgranular fracture mode.   It was
concluded that the accelerated crack growth rate in air at the elevated temperatures were caused by environmental effects.

Mahoney and Paton [6], Hill [7], White [8], and Nachtigal [9]
have also shown that the detrimental effects of high temperature
on fatigue behavior were primarily caused by the chemical attack
rather than by creep effects.

James [5] and Solomon and Coffin [4] have suggested that creep
effects might be important if the test temperature exceeds half
of the melting temperature or it the cyclic frequency is extremely
low.   However in the majority of the cases, the increased crack
growth rate is caused by environmental effects.

HIGH TEMPERATURE FATIGUE: FREQUENCY AND TEMPERATURE EFFECTS

James [10] investigated the effect of frequency on the fatigue crack growth of type 304 stainless steel at 1000°F. As shown in Figure 2, the growth rate approaches a limit at low ΔK and is in-dependent of frequency. A family of parallel lines (one for each frequency) intersect the "frequency-independent" limit line on the log-log plot. James [10] found that an equation of the form

$$da/dN = A(\nu)(\Delta K)^n$$

correlated the growth rate well in this frequency-dependent region to the right of the limit line.

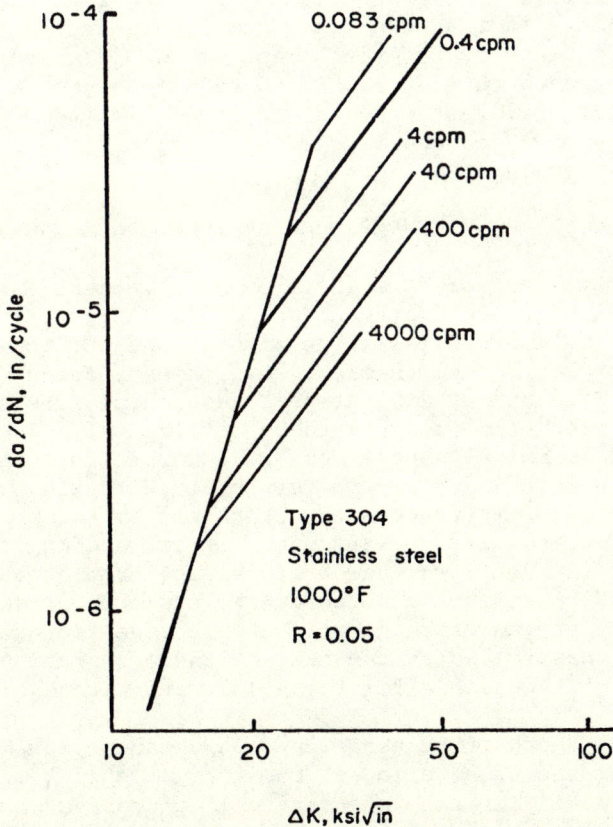

Fig. 2.   Frequency effect on fatigue crack growth in type 304 stainless steel at 1000°F (Ref. 10).

In a recent study Bamford [11] reviewed the fatigue crack growth in pressure vessel steels in a light water reactor environment (2000 psi, 550°F, H_2O). A summary of frequency and R-ratio effects are shown in Figure 3. As with the stainless steel results [10], the crack growth rate has a frequency independent limit line at low ΔK with branching parallel "frequency-dependent" lines at a given R-ratio. However, the pressure vessel steels merge to another "frequency-independent" line at high ΔK. This high ΔK limit line is the air fatigue curve (70°F - 550°F). As shown in this figure, increasing the R-ratio does not seem to change the general behavior, only the low ΔK, frequency independent limit line is moved.

Sadananda and Shahinian [12] have studied fatigue crack growth rate data in IN-718 at 800°F, 1000°F and 1200°F, as shown in Figure 4. The fracture mode at 800°F is entirely transgranular, and the fracture mode at 1200°F is entirely intergranular with considerably higher growth rate than the 800°F data. However, at 1000°F the fracture mode at low ΔK (20 - 50 ksi$\sqrt{\text{in}}$) is intergranular and at high ΔK (> 60 ksi$\sqrt{\text{in}}$) the fracture mode is transgranular. Accompanying the change in fracture mode from intergranular to transgranular, the fatigue crack growth rate at 1000°F approaches the transgranular growth rate of 800°F. Note that also the general fatigue behavior at 1000°F at moderate ΔK levels (20 < ΔK < 50 ksi$\sqrt{\text{in}}$) is comparable to the 1200°F behavior, and the log (da/dN) vs. log (ΔK) curves are parallel.

A KINETIC MODEL FOR HIGH TEMPERATURE FATIGUE CRACK GROWTH

For a given material at a given ΔK level, there exists a corresponding fatigue crack increment per cycle, Δa_f, which is caused purely by the fatigue process alone without the environmental effects. In a detrimental chemical environment, during the time period per fatigue cycle, Δt, the pertinent chemical specie must move with the crack tip in order to be detrimental. In gaseous environments, a chemical specie must be adsorbed onto the metal surface. The specie may reach the crack tip directly by the adsorption process, or it reaches the crack tip by the adsorption process followed by surface diffusion. After reaching the crack tip, additional kinetic process or processes, such as bulk or grain boundary diffusion, might be necessary to cause the accelerated crack growth. Fatigue crack growth rate in a detrimental chemical environment is controlled by the rate of the penetration of the chemical specie responsible for the accelerated crack growth.

A certain "depth of penetration", Δa_p, can be reached by the chemical during the time period of one fatigue cycle, Δt. For example, if adsorption is the rate controlling process, Δa_p is given by the number of monolayers of oxygen formed during the time period Δt. It is the amount of the crack increment during Δt, which will be fully covered by oxygen by the absorption process. If diffusion

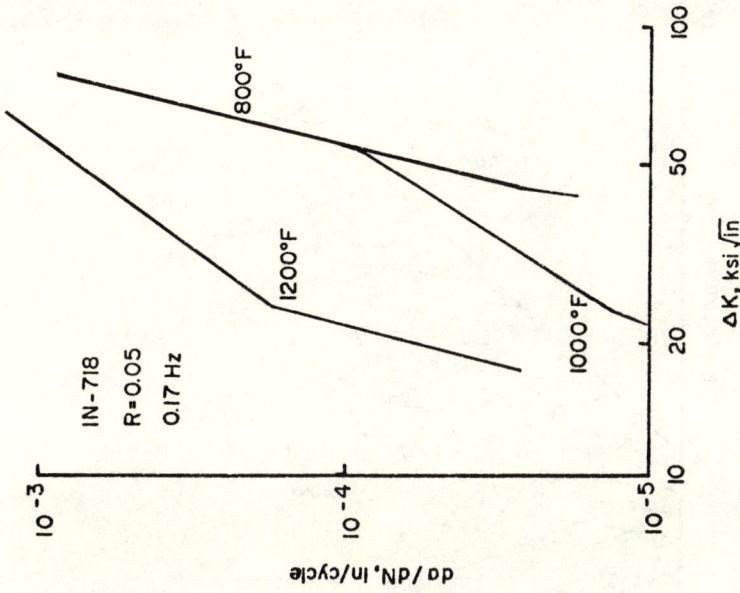

Fig. 4. Temperature effect on fatigue crack growth in alloy 718 (Ref. 12).

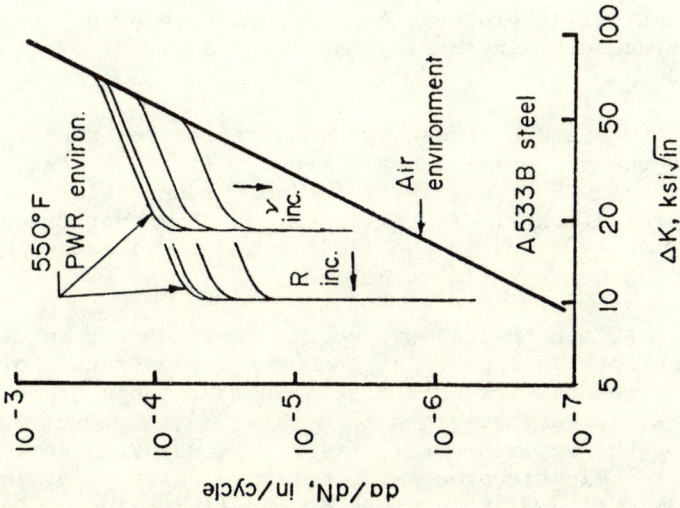

Fig. 3. Schematic of frequency and R-ratio effects – pressure vessel steel in PWR environment (Ref. 11).

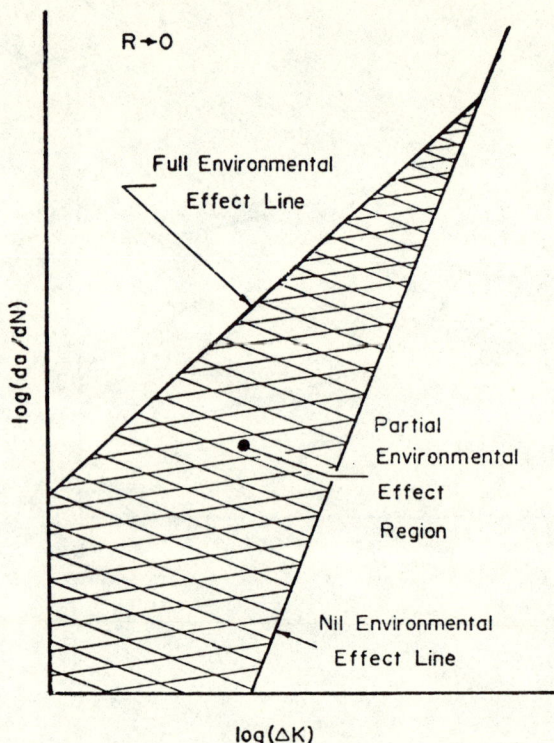

Figure 5.   Kinetic model schematic.

is the rate controlling process, Δa_p is the depth of diffusion. Δa_p can be increased by increasing the temperature or by reducing the frequency.

The relative sizes of these two quantities, Δa_f and Δa_p, determine the degree of environmental effect. If $\Delta a_f \gg \Delta a_p$, fatigue crack growth is caused primarily by fatigue process alone without any environmental effects. In this case, da/dN is not frequency dependent, and it is given by the limit line of nil environmental effect as shown schematically in Figure 5.

If $\Delta a_p \overset{>}{-} \Delta a_f$, the environment has its full effect on da/dN. Any further increase in Δa_p by increasing the temperature or by reducing the frequency will not increase da/dN. When $\Delta a_p \overset{>}{-} \Delta a_f$, a fatigue crack, in essence grows in a material, which is fully "embrittled" by the environmental effect, and da/dN is no longer controlled by the kinetic process. Therefore da/dN is again not frequency dependent, and it is given by the limit line of full environmental effect as shown in Figure 5. The data of James [10], Bamford [11], and to a certain degree that of Sadananda and

IN-100
R→0

Full Environmental Effect Line

$$\frac{da}{dN} = 5.53 \, E\text{-}9 \; (\Delta K)^{3.1}$$

T inc.

γ inc.

$$\frac{da}{dN} = f(\gamma) \, g(T) \, (\Delta K)^{1.7}$$

Partial Environmental Effect Region

Nil Environmental Effect Line

$$\frac{da}{dN} = 1.27 \, E\text{-}13 \; (\Delta K)^{5.5}$$

log(da/dN)

log(ΔK)

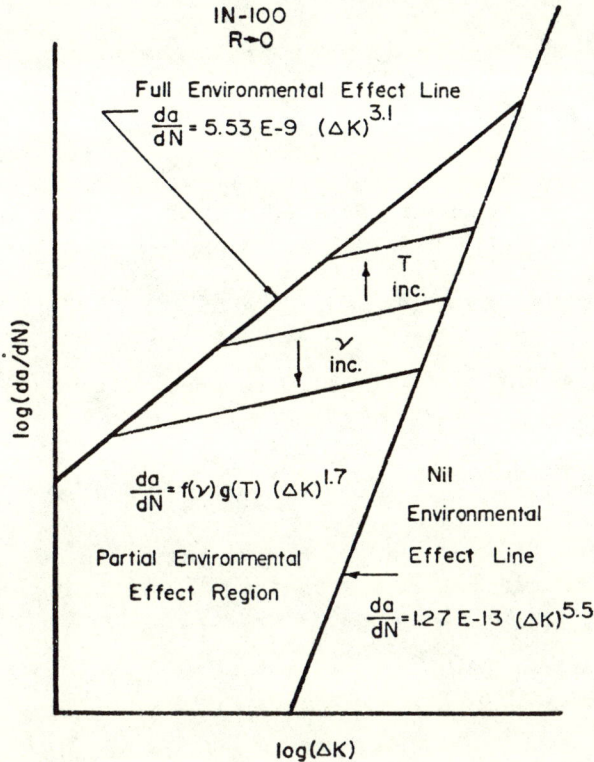

Figure 6.   Kinetic model as applied to IN-100.

Shahinian [12] support the dual limit line model. One may consider these two limit lines as fatigue crack growth lines in two differ- ent materials: one in the material without any environmental effect, and one in the material which is fully embrittled by the environment. The crack growth mode usually changes from transgran- ular without environmental effects to intergranular with environ- mental effects as observed by Solomon and Coffin in A286 [4], James in 304 stainless steel [5] and Sadananda and Shahinian in alloy 718 [12].

In the region between these two limit lines, Δa_p is comparable to Δa_f, the rate controlling kinetic process becomes important, and da/dN is temperature and frequency dependent. The da/dN lines in this region vary with temperature and frequency. At a given tem- perature, the lines for different frequencies are parallel. Simi- larly, the lines for different temperatures at a given frequency are also parallel as shown by James [10], Bamford [11], and Sadananda and Shahinian [12]. The parallel lines are shown sche- matically in Figure 6.

The parallel straight lines in a log da/dN vs. log ΔK plot

indicate that

$$\frac{\Delta a}{\Delta N} = F(\nu, T) \Delta K^n \tag{1}$$

where ν and T are cyclic frequency and temperature. The function F is related to the rate controlling kinetic process. At a given value of ΔK, as the frequency and temperature change, the crack growth rate can be written as

$$\frac{\Delta a}{\Delta N} = \frac{\Delta a}{\Delta t} \frac{1}{\nu} \tag{2}$$

If $\Delta a / \Delta t$ is controlled by a kinetic process, we have

$$\frac{\Delta a}{\Delta t} = A \exp\{-\frac{Q}{kT}\} \tag{3}$$

where Q is the activation energy; k, the Boltzmann constant; T, absolute temperature; and A, a proportional constant. With the data at T_o, and ν_o, as a reference,

$$\frac{\Delta a}{\Delta N} = (\frac{\Delta a}{\Delta N})_{T_o, \nu_o} (\frac{\nu_o}{\nu}) \exp\{-\frac{Q}{k} (\frac{1}{T} - \frac{1}{T_o})\} \tag{4}$$

In Equation (2), $\Delta a / \Delta t$ is the average rate over the time interval Δt. During Δt, the crack growth rate is not constant, therefore, the crack increment per cycle is

$$\Delta a = \int_o^{\Delta t} \frac{da}{dt} dt \tag{5}$$

da/dt is an unknown function of local stresses and strains at the crack tip and the rate controlling kinetic process. Very likely an interaction exists between the crack tip stresses and strains and the kinetic process. Therefore da/dN may not be linearly proportional to the frequency ν. With the assumption of a simple power function for ν, Equation (4) is modified to the form

$$\frac{da}{dN} = (\frac{da}{dN})_{T_o, \nu_o} (\frac{\nu_o}{\nu})^m \exp\{-\frac{Q}{k} (\frac{1}{T} - \frac{1}{T_o})\} \tag{6}$$

$$= A(\frac{1}{\nu})^m \exp\{-\frac{Q}{kT}\}$$

where

$$A = (\frac{da}{dN})_{T_o, \nu_o} \nu_o^m \exp\{\frac{Q}{kT_o}\}$$

The ΔK dependence is implicitly contained in $(da/dN)_{T_o, \nu_o}$. Including the ΔK dependence, we have

$$\frac{da}{dN} = B \; (\frac{\nu_o}{\nu})^m \; \exp\{- \frac{Q}{k} \; (\frac{1}{T} - \frac{1}{T_o})\} \; \Delta K^n \qquad (7)$$

With an equation of this type, where the effects of temperature and frequency are separable, James [13], successfully correlated fatigue crack growth data, for type 304 stainless steel. An
equation of this type is also employed to describe fatigue crack
growth data in the Nuclear Systems Materials Handbook [14], for a
number of alloys.

APPLICATION OF THE KINETIC MODEL TO HIGH TEMPERATURE BEHAVIOR OF
IN-100

To illustrate the use of the kinetic model in high temperature fatigue, the behavior of a nickel base super alloy (IN-100)
will be studied. The fatigue crack growth rates for this material have been studied at various frequencies, temperatures, dwell
times and R-ratios by Wallace et al. [15], Macha [16] and Larsen
et al. [17].

The fatigue crack growth rate at 1200°F and R = 0.1 for
various frequencies is shown in Figure 7. The fatigue crack
growth rate at 70°F is transgranular and independent of frequency
[15]. The fatigue crack growth rate at 1200°F shows frequency
independence at both the high and low ΔK limits, with frequency
dependence at moderate ΔK levels. In the frequency dependent
region the log (da/dN) vs. log (ΔK) data is essentially parallel
for the three frequencies tested. The fracture mode at 1200°F
is intergranular except at high ΔK levels, where a transition to
transgranular fracture is suspected as the data merges with the
70°F line.

The fatigue crack growth rate at 2.5 Hz and R = 0.1 for various temperatures is shown in Figure 8. The fatigue crack growth
rate at moderate ΔK levels is temperature dependent except at
high ΔK levels, where a transition to the 70°F line occurs. Study
of the 1200°F fracture surface indicated that the behavior at

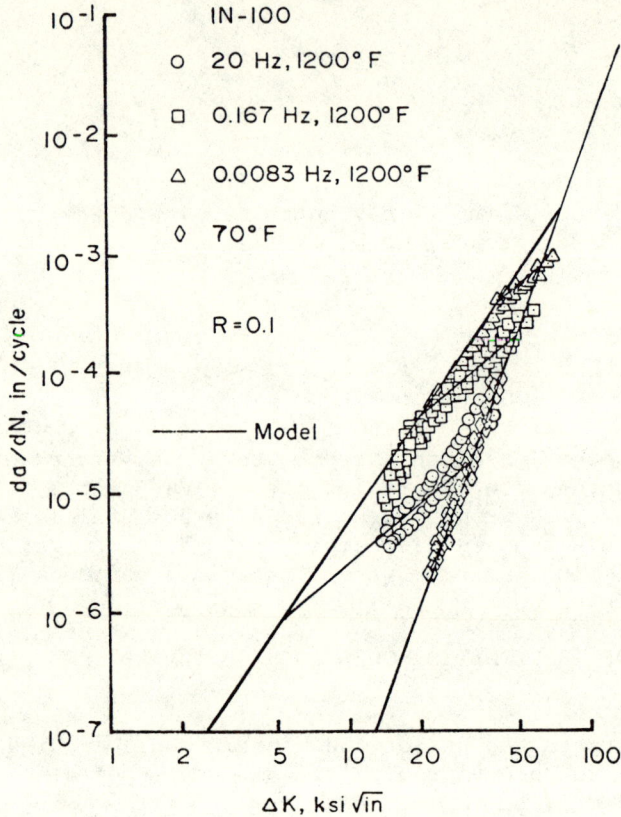

Fig. 7. Frequency effect on fatigue crack growth in IN-100,
         1200°F, R = 0.1 (Ref. 15, 17).

moderate ΔK levels was intergranular with a transition to transgranular at high ΔK levels. No study of the 1350°F fracture surface was attempted due to the heavy oxide coating. No data was
generated in the low ΔK region to establish the left limit line.
As before, the log (da/dN) vs. log (ΔK) data appears to be parallel
for the different temperatures in the moderate ΔK region. The
fatigue crack growth rate at 0.167 Hz and R=0.1 for various temperatures is presented in Figure 9. The fatigue crack growth rate
shows temperature independence at both high and low ΔK limits, with
temperature dependence at moderate ΔK levels. The fracture mode
is intergranular for the high temperature data except in the high
ΔK levels, where a transition to transgranular fracture is suspected
as the data merges with the 70°F line. As mentioned previously,

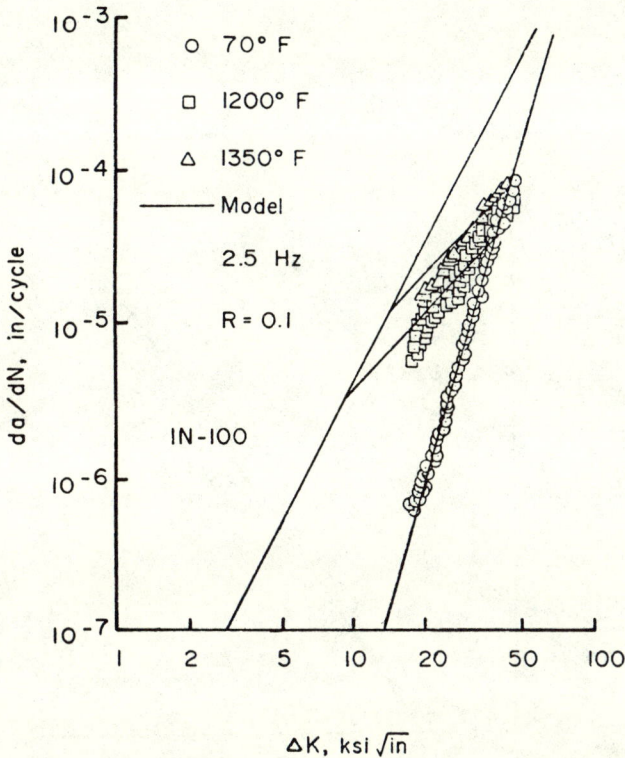

Fig. 8.   Temperature effect on fatigue crack growth in IN-100,
          R = 0.1, 2.5 Hz (Ref. 16).

the log (da/dN) vs. log (ΔK) data appears to be parallel for the
different temperatures in the moderate ΔK region.

The kinetic model is composed of two limit lines (independent
of temperature and frequency) bounding a region with temperature
and frequency dependence as shown in Figure 6. The low ΔK limit
line has the complete effect of the environment (full environmental effect line) and the high ΔK limit line has no effect of
the environment (nil environmental effect line). The equation for
the two limits were determined from the data in Figures 7, 8 and 9:

$$\frac{da}{dN} = 5.53 \times 10^{-9} \ (\Delta K)^{3.1} \ \text{(Full environmental effect)} \qquad (8)$$

$$\frac{da}{dN} = 1.27 \times 10^{-13} (\Delta \dot{K})^{5.5} \ \text{(Nil environmental effect)} \qquad (9)$$

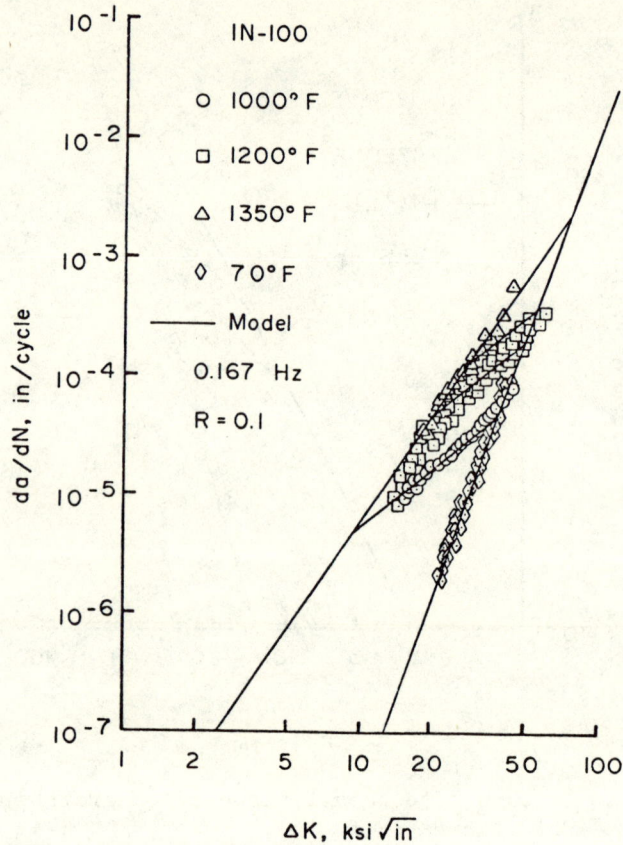

Fig. 9.   Temperature effect on fatigue crack growth in IN-100,
          R = 0.1, 0.167 Hz (Ref. 15,17).

Study of the data in Figures 7, 8, and 9 for the frequency and
temperature dependent area (partial environmental effect region)
yields the fatigue crack growth equation of the form

$$\frac{da}{dN} = 2.8 \times 10^{-7} \left(\frac{\nu_o}{\nu}\right)^{0.33} e^{-11500\left(\frac{1}{T} - \frac{1}{T_o}\right)} (\Delta K)^{1.7} \tag{10}$$

where T is in °R, T_o is 1660°R, ν is in Hz and ν_o is 0.167 Hz.
A summary of the equations is shown in Figure 6. These equations
are also plotted in Figures 7, 8, and 9 as solid lines. The comparison of the kinetic model prediction with the data in these
figures is very good.

SUMMARY

The increased fatigue crack growth rate at high temperature in metals is caused primarily by environmental effects. The fatigue crack growth rate is controlled by the rate which a detrimental chemical specie can penetrate the crack tip region. If the crack growth rate is significantly larger than the chemical penetration rate, then the fatigue crack growth rate has no environmental effect (virgin material behavior). If the crack growth rate is less than or equal to the chemical penetration rate, then the crack growth rate is fully affected by the environment (fully embrittled behavior). At both of these limiting states the fatigue crack growth rate is independent of frequency and temperature. In the intermediate region when the fatigue crack growth rate and chemical penetration rate are comparable, the fatigue crack growth rate is controlled by a kinetic process. Thus the fatigue crack growth rate is temperature and frequency dependent. The formulated kinetic model is applied to analyze the fatigue crack growth data of a nickel-base super alloy (IN-100). The kinetic model fits the data very well.

ACKNOWLEDGEMENT

The authors wish to acknowledge the support of Dr. T. Nicholas and the staff of the Metals Behavior Branch, Metals and Ceramic Division, Air Force Materials Laboratory, Wright-Patterson Air Force Base. The technical discussions with Drs. T. Nicholas and A. F. Grandt, Jr., and Messrs. D. E. Macha and J. M. Larsen were especially helpful. They gratefully acknowledge the financial support of the Air Force Materials Laboratory.

REFERENCES

1. L. F. Coffin, Jr., "The Effect of High Vacuum on the Low Cycle Fatigue Law," Met. Trans., No. 7, 3:1777-1788 (1972).
2. H. D. Solomon, "Low Cycle Fatigue Crack Propagation in 1018 Steel," Jour. of Mats., 7:299-306, Sept. (1972).
3. H. D. Solomon, "Frequency Dependent Low Cycle Fatigue Crack Propagation," Met. Trans., 4:341-347, Jan. (1973).
4. H. D. Solomon and L. F. Coffin, Jr., "Effects of Frequency and Environment on Fatigue Crack Growth in A286 at 1100°F," Fatigue at Elevated Temperature, STP 520, ASTM, Phil. (1973).
5. L. A. James, "Some Questions Regarding the Interaction of Creep and Fatigue," Jour. of Eng. Mats. and Tech., Transactions of ASME, 98:235-243, July (1976).
6. M. W. Mahoney and N. E. Paton, "The Influence of Gas Environments on Fatigue Crack Growth Rates in Types 316 and 321 Stainless Steel," Nuclear Tech., No. 3, 23:290-297,Sept. (1974).

7. G. J. Hill, "The Failure of Wrought 1% Cr-Mo-V Steels in Reverse-Bending High Strain Fatigue at 550°C," Thermal and High-Strain Fatigue, The Metals and Metallurgy Trust, London, 312-327 (1967).

8. D. J. White, "Effect of Environment and Hold Time on the High Strain Fatigue Endurance of 1/2 Percent Molybdenum Steel," Proceedings Inst. of Mech. Engineers, Part 1, No. 12, 184:223-240 (1969-79).

9. A. J. Nachtigal, S. J. Klima, J. C. Freche, and C. A. Hoffman, "The Effect of Vacuum on the Fatigue and Stress-Rupture Properties of S-186 and Inconel 550 at 1500°F," Tech. Note D-2898, NASA, June (1965).

10. L. A. James, "The Effect of Frequency Upon the Fatigue Crack Growth of Type 304 Stainless Steel at 1000°F," Stress Analysis and Crack Growth, ASTM, STP 513, 218-229 (1972).

11. W. H. Bamford, "Application of Corrosion Fatigue Crack Growth Rate Data to Integrity Analysis of Nuclear Reactor Vessels," Jour. of Eng. Mats. and Tech., Transactions of ASME, 101:182-190, July (1979).

12. K. Sadananda and P. Shahinian, "Crack Growth Behavior in Alloy 718 at 425°C," Jour. of Eng. Mats. and Tech., 100: 381-387, Oct. (1978).

13. L. A. James, "Frequency Effects in the Elevated Temperature Crack Growth Behavior of Austenitic Stainless Steels - A Design Approach," Trans. ASME, Jour. of Pressure Vessel Tech., 101:171-176, May (1979).

14. Nuclear Systems Materials Handbook, Part 1, Group 1, Section 2, Property Code 2431, Vol. 1, Report TID-26666, Westinghouse Hanford Co. (1976).

15. R. M. Wallace, C. G. Annis, Jr., and D. L. Sims, "Application of Fracture Mechanics at Elevated Temperatures," AFML-TR-76-176, Air Force Materials Laboratory, Wright Patterson Air Force Base, Ohio 45433.

16. D. E. Macha, "Fatigue Crack Growth Retardation Behavior of IN100 at Elevated Temperature," Engineering Fracture Mechanics, 12:1-11.

17. J. M. Larsen, B. J. Schwartz, and C. G. Annis, Jr., "Cumulative Damage Fracture Mechanics Under Engine Spectra," AFML-TR-79-4159, Air Force Materials Laboratory, Wright Patterson Air Force Base, Ohio 45433.

DESIGN-,OPERATION-,AND INSPECTION-RELEVANT FACTORS OF FATIGUE

CRACK GROWTH RATES FOR PRESSURE VESSEL AND PIPING STEELS

W. H. Cullen and F. J. Loss

Naval Research Laboratory
Washington, D. C.   20375

INTRODUCTION

To assure continued safe operation of nuclear reactors used
for power generation by the public utilities, a comprehensive and
industrious program of inspection and monitoring is carried out.
If a flaw indication is found, the reactor owner may proceed in one
of two ways: the flaw may be repaired, usually through a weld repair technique, or a fracture mechanics analyses may be carried out
resulting in computation of future inspection intervals during which
the flaw indication will be carefully monitored to determine whether
extension of the flaw has occurred. In addition to knowledge of the
stress levels in the neighborhood of the flaw, the fatigue crack
growth rates for the material in question must also be known or determined. These fatigue crack growth rates should take into account
not only the product form, but also the environment, waveform, temperature and irradiation levels which pertain to the vicinity of the
flaw. If satisfactory growth rates are not available, the Appendix
A of Section XI of the Boiler and Pressure Vessel Code[1] details a
set of default rules, providing crack growth rates which may be used
in lieu of actual crack growth rates for the material and conditions
which actually pertain.

Typically the transients which occur during operation of a
power reactor are the result of brief, temporary shutdowns prompted
by an indication of an equipment problem, or more simply, pressure
adjustments which are made in response to power demands. Start-ups
and shut-downs also account for a few hundred loading cycles in a
reactor lifetime. These cycles have periods ranging from a few

seconds to a few minutes and waveshapes that are basically triangular
or sinusoidal. The loads which are realized in the vicinity of a
flaw are generally a combination of deadweight loads of the reactor,
its internals and the water, the water pressure, torques and vibrations due to pumping equipment, residual stresses near welds and
some stresses induced by thermal gradients. The reactor coolant
environment is relatively well-characterized in a pressurized water
reactor (PWR) but the lack of buffering solutes and varying dissolved
oxygen content of boiling water reactor (BWR) coolants make the
latter environment more difficult to typify.

A brief account of the history of these code default lines is
given in Ref. 2. The default line for air environment growth was
adopted in 1971; for reactor-grade water environment growth - 1972.
Since that time, additional research involving an ever-widening scope
of external variables has shown that there are some reactor-typical
conditions for which growth rates may substantially exceed the 1971/
1972 guidelines. Both the reactor vendors and the Nuclear Regulatory
Commission (NRC) have sponsored, and continue to sponsor, research
aimed at determining fatigue crack growth rates and the impact and
understanding of the various factors which influence these growth
rates.

From 1976 through 1978, the Naval Research Laboratory, together
with Westinghouse Corporation, carried out a cooperative test plan,
sponsored by the Nuclear Regulatory Commission, and termed the "preliminary test matrix". A follow-on program, the "main test matrix"
was conducted in 1979 and 1980. The object of the first program was
to determine the influence of waveform, temperature, and load ratio
on a single pressure vessel steel - A508, Class 2. Then, utilizing
the two most illustrative, reactor-typical, waveform conditions, as
determined by the preliminary matrix, the main test matrix was directed at determination of fatigue crack growth rates for a wide
variety of reactor pressure vessel steels, welds and heat-affected
zones (HAZ). Testing of a limited variety of irradiated steels is
also within the scope of the main test matrix.

APPROACH

The approach to this research, as embodied in the concept of
the preliminary test matrix, was to determine fatigue crack growth
rates for a variety of test waveforms consisting of ramp and hold
components of varying lengths. As examples, one second, and one,
five and thirty minute ramp times were combined with one, three,
thirty and sixty minute hold times. To avoid confusing results, tests
were conducted from start to finish with a single waveform selection,
composed of one ramp time component and one hold time component from
the above list. As adjuncts to the set of preliminary matrix tests,
other waveforms were selected for additional tests, with the rationale
for these choices explained in the following section.

In order to provide the most germane data, the testing is conducted using simulated pressurized water reactor primary loop coolant. Crack growth rates are determined in autoclaves, or small pressure vessels using the water chemistry given in Table 1. The environment is monitored very carefully, with continuous online sampling of dissolved oxygen contents, specific conductance, and pH. Batch samples for more complete elemental analysis by classical wet chemistry methods are taken once each week, or more often, depending on test schedules and progress. Temperature is held constant to within two centigrade degrees, and the water circulates to prevent stagnation.

Tests were conducted at both 93^0C (200^0F) at essentially atmospheric pressure, and 288^0C (550^0F), with constant amplitude loads at a load ratio of 0.1 to 0.2. Each specimen was instrumented with a displacement gage, which measured crack mouth opening (δ). Since the ratio of δ to load (P) is proportional to the crack length, this provided a method to determine the latter without visual observation. The empirical relationship derived for this purpose accounted for gage position, temperature, and specimen geometry. Over the three-year duration of this test series, steadily improved methods of data acquisition were employed, culminating in a fully-automated computer-controlled data acquisition and processing system. The details of autoclave hardware, specimen instrumentation and data acquisition methods may be found in Refs. 3-5.

A508-Class 2 forging steel was selected as the one material to be investigated in this test matrix. Specimens were cut from the nozzle drop-outs by first removing about 40 mm of the outer surfaces, including the inside cladding, and blanking out specimens from the remaining material. All specimens were oriented so that the tensile stress at the crack tip was parallel to the hoop stress direction of the drop-out.

Table 1.   Typical pressurized water reactor coolant chemistry

Boron (as boric acid)	1000 ppm
Lithium (as lithium hydroxide)	1 ppm
Dissolved oxygen	1 ppb
Dissolved hydrogen	30-50 cc/kg (H_2O)
Chloride	0.10 ppm
Florine	0.10 ppm

All other elements should be held to trace levels; a small amount of iron, both solid and soluble, will be detected, and is the inevitable result of corroding specimens.

The samples were either 25.4 mm thick (1T) compact specimens, or 50.8 mm thick (2T) WOL specimens. In the case of the compacts, the machined notch extended to an a/W of 0.25 and these were pre-

cracked to an a/W of 0.30. The WOL specimens had a machined notch
25.4 mm (a/W = 0.19) and were precracked an additional 6 to 50 mm
(0.3 to 2 in.) depending on the initial ΔK level desired. The
maximum stress intensity applied during precracking was less than or
equal to the initial stress-intensity levels for the test. All precracking was performed at room temperature in an air environment.
During the latter phases of testing, additional precracking in the
test environment was performed. This was done in an attempt to
promote environmental equilibrium at the crack tip before initiating
the actual test.

RESULTS – UNIRRADIATED STEELS

     In the following presentation of the data, the first four
figures are paired, for certain similar, or identical, characteristics
of the waveform, with the lower temperature results on the left, and
the higher temperature results on the right. There appear to be
three factors which strongly influence the crack growth rates of
A508-2 in the reactor-grade water environment:

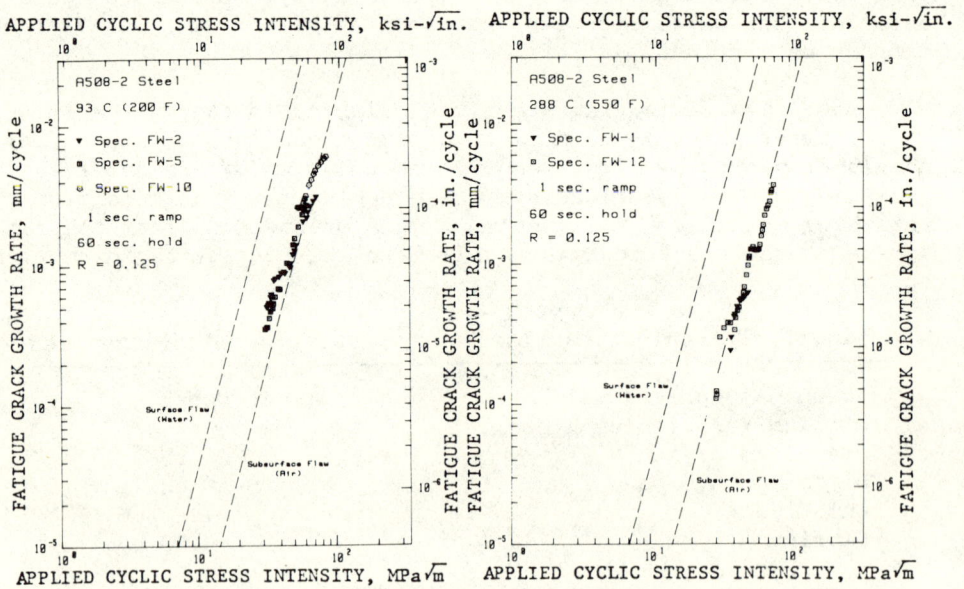

Fig. 1.   Fatigue Crack Growth Rate Data vs Applied Cyclic Stress-
          Intensity Factor for all the Tests With a Very Short (1
          sec) Rise Time. In This and the Following Figures, the
          Lower Temperature Test Results are on the Left. Both
          These Temperatures Yield Growth Rates Close to the ASME
          Section XI Air Default Line.

ramp time, hold time, and temperature. An interrelationship among
the three determines the particular crack growth rate; it is im-
possible to isolate the effects of one variable without fixing the
other two.

Figure 1 shows test results for waveforms with a one second
ramp time component, coupled with substantially longer hold times.
At either temperature, this produced essentially identical data,
located on, or very near the ASME air default line, which is included
(together with the 1972 water default line) on all the plots in this
article. For these short ramp time components, during which the
environment has an insufficient opportunity to attack the crack tip
enclave, there appears to be no environmental assistance.

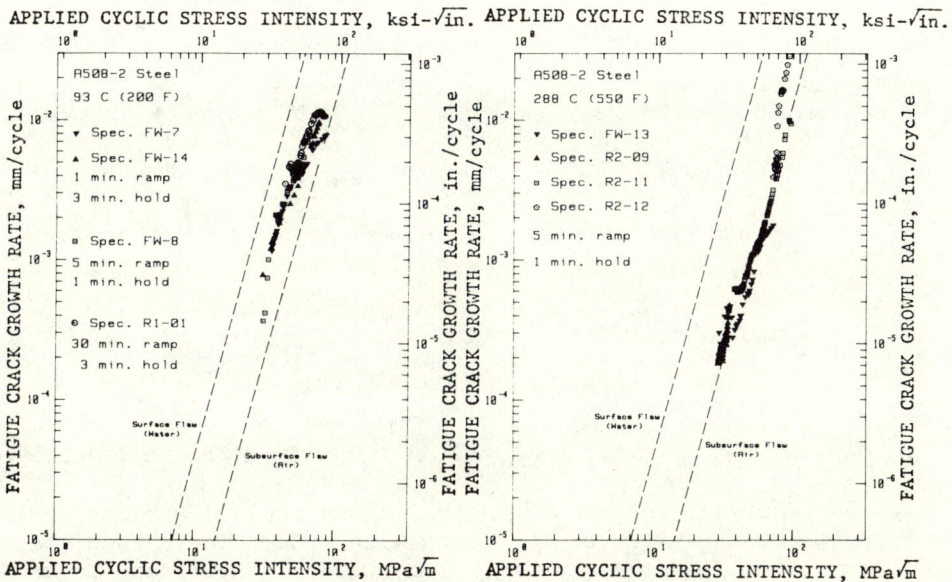

Fig. 2. Fatigue Crack Growth Rate Data vs Applied Cyclic Stress-
         Intensity Factor for all Tests With Both Significant Ramp
         Times and Hold Times (1 Min or Greater for Each). In the
         Left-Hand Data Sets, the Effect of the Increase in Crack
         Growth Rates Due to the Lower Temperature is Clearly Seen.

Figure 2 shows test results for waveforms with longer ramp
times (from one to thirty minutes), coupled with one to three minute
hold times. In this case, the lower temperature results reside
midway between the ASME air and water default lines, which represents
a substantial increase over the results shown in Fig. 1a. However,

for the higher temperature, the test results, as in Fig. 1b, reside
on or near the ASME air default line. Thus, a substantial ramp time
component, together with a low temperature, produces environmentally
assisted crack growth rates.

To determine whether the hold time component was influencing
the results, tests involving only ramp components were conducted,
and the results shown in Figs. 3 and 4. The ramp time only tests

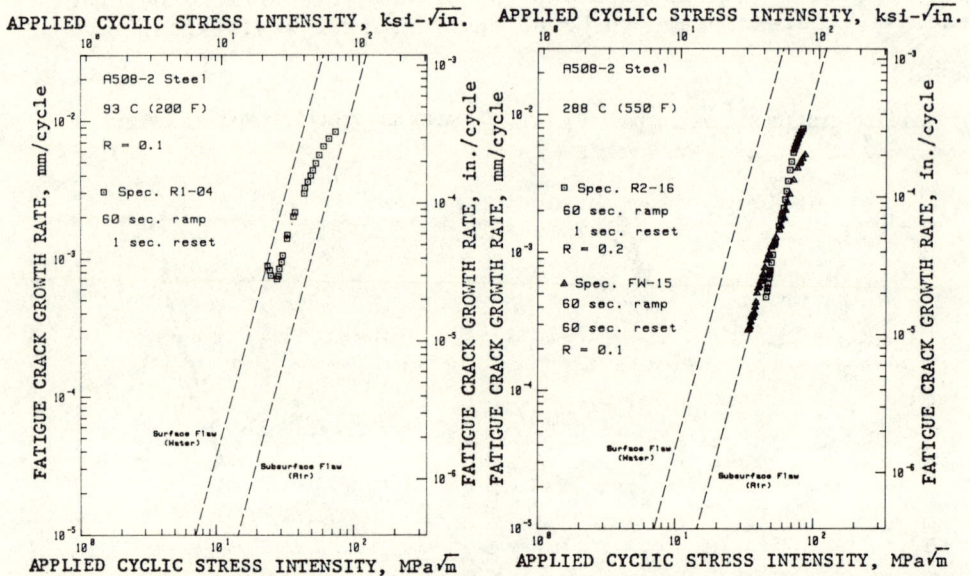

Fig. 3.   Fatigue Crack Growth Rate Data vs Applied Cyclic Stress-
          Intensity Factor for Tests With a One Minute Ramp, but No
          Hold Time. As in Fig. 2, the Lower Temperature Tests Re-
          sulted in the Higher Fatigue Crack Growth Rates.

at low temperature produced environmentally accelerated data, shown
in Fig. 3 for 60 sec ramp times, while the high temperature results
remain located near the air default line, showing no environmental
assistance. Figure 4 shows results of a similar test, but with a
22 sec ramp time component. Even with the shorter ramp time, the
results are similar – higher growth rates for the lower temperature,
and no clear cut environmental assistance for the higher temperature
test. For both of these cases (Figs. 3 and 4), it should be noted
that the results are nearly identical to those of Fig. 2, for the
respective temperatures, indicating that the presence or absence of

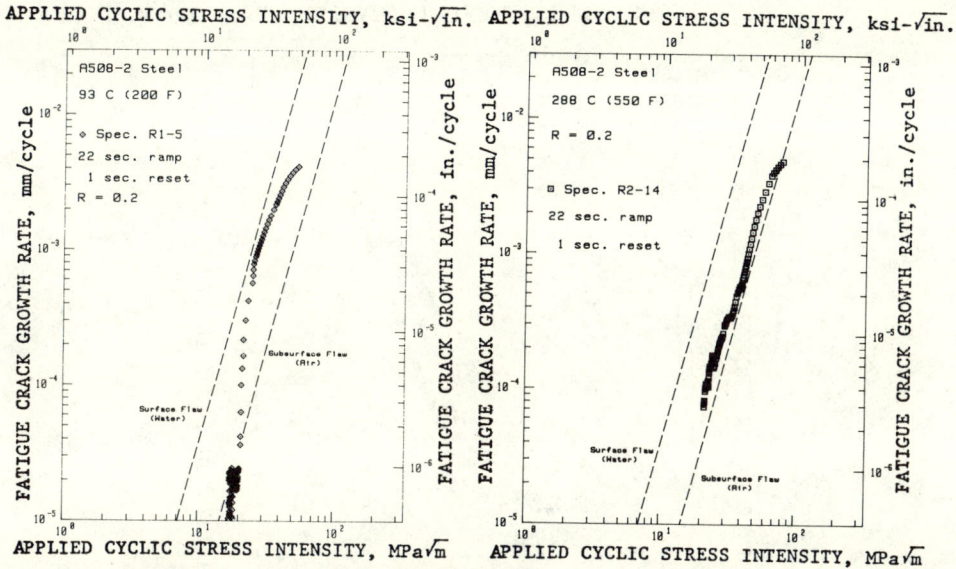

Fig. 4.   Fatigue Crack Growth Rate Data vs Applied Cyclic Stress-
Intensity for Tests With a 22 Second Ramp and No Hold Time.
The Results are Very Similar to Those Shown in Fig. 3.

a hold time component does not affect the degree of environmental
assistance.

About midway through the series of tests represented by the data
in the previous four figures, it was realized that a 17 mHz sinusoidal waveform, together with a high temperature (288°C), produced
crack growth rates which were significantly higher than any of the
ramp/hold or ramp/reset combinations which had been tested at the
288°C temperature. Figure 5 shows results of one test which bears
out this fact. Attempts were made to reproduce these data using a
ramp/hold combination (22 sec ramp, 4 sec hold) which is the trapezoidal approximation to the positive slope of a sinusoidal waveform.
The results of two tests (both shown in Fig. 5) fail to reside as
high as the 17 mHz sinewave test. Consideration of similar tests
conducted at other facilities[6] indicates that sinusoidal waveforms
with periods ranging from 0.2 to 2 minutes may all produce crack
growth rates which reside midway between the ASME air and water default lines. In short, for 288°C temperatures, there are no ramp
only, or ramp/hold waveforms, which yield environmentally-assisted
fatigue crack growth rates, but a substantial range of sinusoidal
waveforms do result in enhanced growth rates. The environmental
mechanism which produces this result is not understood at the present

APPLIED CYCLIC STRESS INTENSITY, ksi-√in.

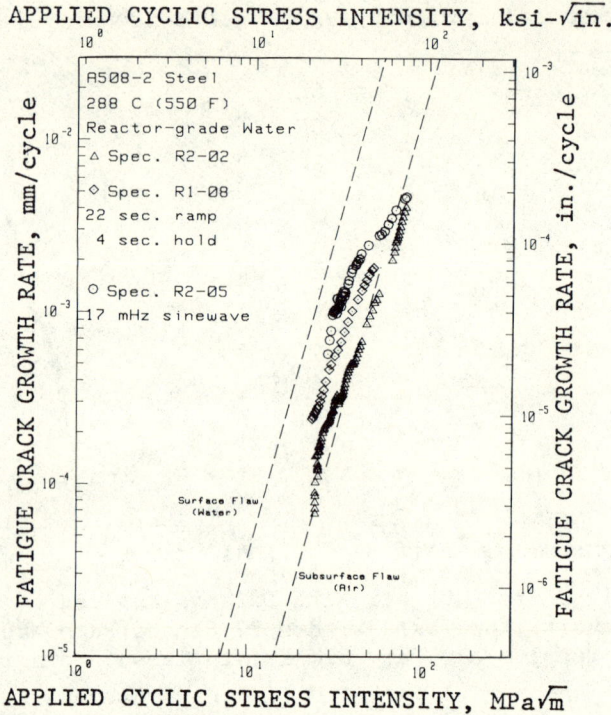

Fig. 5.   Fatigue Crack Growth Rate Data vs Applied Cyclic Stress-
Intensity Factor for a Test With a 17 mHz Sinusoidal
Waveform. Also Shown are the Results of Two Tests With a
22 Second Ramp, 4 Second Hold Time, Which is the Trapezoidal
Equivalent of the Positive Slope of a Sinusoidal Waveform.

time. It is possible that the constantly varying ramp rate resulting
from the sinusoidal waveform may alter the potentio-kinetic processes
at the crack tip in a way that is substantially different from that
for fixed ramp rate waveform. Another speculation is that the decreasing load portion of the sinusoidal waveform, or the dwell time
around the minimum load may enhance the environmental assistance.

During the initial phases of this research, when test systems
were being implemented and improved, and outages and other failures
were frequent, common test practice was to choose an initial value
of applied cyclic stress intensity such that relatively easily measurable crack extensions would occur over rather short times. Typically, for R = 0.2 and waveforms of about 60 sec periods, a ΔK
value of 25 MPa √m was selected. While the research carried out
under this proviso produced the desired amount of scoping data on
various materials[6] and enabled the understanding of the temperature
and waveform effects described above, the measurement of fatigue
crack growth rates for lower stress intensity factor ranges (10 to
25 MPa √ m for R = 0.2) has not been carried out. Now that the

reliability of test and data acquisition systems has substantially
improved, recent research has extended our understanding of fatigue
crack growth rates into the lower ΔK ranges.

On the basis of these recent tests, it is becoming evident that
fatigue crack growth rates for these reactor pressure vessel materials, at least for sinusoidal waveforms and low load ratios, follow
the conventional three-region behavior which has been amply discussed in the classical presentation of stress-corrosion cracking[7]
and corrosion-fatigue crack growth[8]. This is shown schematically
in Fig. 6a, which has been adapted in a simplified way, to these
results. In terms of ΔK, the exact onset of the transition from
Region I to Region II behavior is generally a function of the
material and the environment. In terms of crack growth rate, the
level of Region II is primarily a function of test frequency, all
other things being equal. The reader should note that Region I
growth behavior should not be construed as an indication of
"threshold", or lower limit behavior of these fatigue crack growth

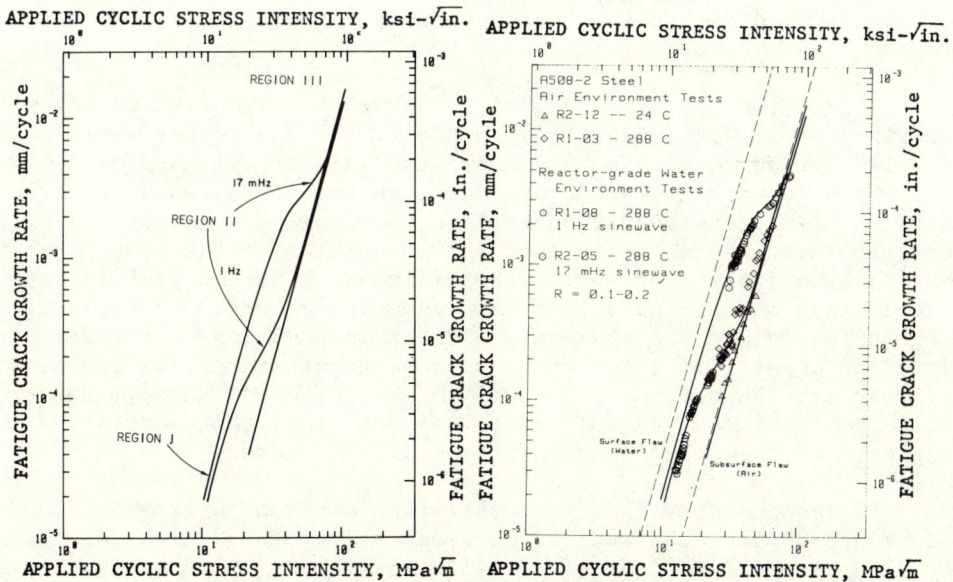

Fig. 6.   Illustration of the Three-Region Behavior of Fatigue Crack
Growth Rates in the Reactor-Typical Environments (a) the
Trend Lines for 1 Hz and 17 mHz Data in 288°C Reactor-
Grade Water, and for Baseline, Air Environment Data (b)
Actual Fatigue Crack Growth Data Superposed on the Trend
Lines of (a).

rates. The location of Region I for these data sets is simply a function of the initial value of ΔK chosen for these tests; true threshold studies on these materials have only been conducted for air environments, with the results detailed in Ref. 8. The onset of Region II, the value of the growth rates throughout Region II, and the transition of Region II into Region III are established by the test frequency.

In illustration of this conclusion, fatigue crack growth rates for A508-2 steel are shown in Fig. 6b, together with the trend lines shown in the accompanying figure. At the present time, these data describe the limits of the research completed on the extension of data into the lower ΔK regime. While this set of trend lines adequately describes the behavior of this material, there remain some aspects of this description for which additional research would be helpful. The exact position of the trend lines for lower values of ΔK, especially for longer period waveforms, is not well established. At the Naval Research Laboratory, efforts are now underway to measure fatigue crack growth rates of 17 mHz (or comparable frequency) waveforms on tests for which the initial stress intensity factor was about 15 MPa $\sqrt{m}$. Similar efforts, in constant ΔK tests, with ΔK = 16.5 MPa $\sqrt{m}$, are underway at Babcock and Wilcox Research Laboratories, Alliance, Ohio. The data resulting from the above two efforts should assist in a more complete definition of the actual trends.

At the time this contribution was written, testing of the main matrix program was essentially complete, and the results were essentially as expected. Figure 7 shows two shaded regions, for $R \approx 0.7$ and $R \approx 0.2$, which for the most part, encompass the data generated for the respective conditions. Also shown are the new surface flaw default lines as contained in the 1980 edition of the ASME Code. Much of the justification for the bilinear shape and positioning of these lines may be found in a reference by Bamford.[9] Briefly, a regression method was employed to determine a 95 percent confidence level on upper limits for fatigue crack growth rates for two sets of load ratio conditions, $R \approx 0.2$ and $R \approx 0.7$. At the present time, this choice of code default lines seems to represent a conservative estimate of the upper bound of the available data.

It seems likely, however, that the rather high growth rates described by these default lines especially for lower ΔK values, may lead to calculations of unrealistically short inspection intervals, or reactor lifetimes. In these cases, many utilities may chose, as the code indicates they may, to determine crack growth rates for the particular product form equivalent to that in which the suspected flaw resides. In some cases, the data may be available; in other cases, it may have to be generated – either case may prove to be an alternative less expensive than a costly repair job, or a nearly continuous inservice inspection procedure.

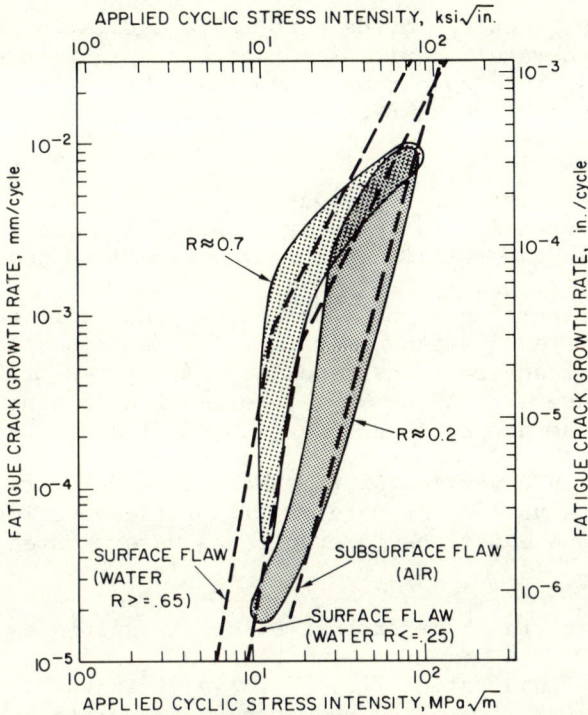

Fig. 7.   A Summary Plot of the Available Data for Fatigue Crack
          Growth Rate Tests in High-Temperature, Pressurized,
          Reactor-Grade Water, for Both High (≈0.7) and Low
          (≈0.1-0.2) Load Ratios. The 1980 Versions of the ASME
          Air and Water Default Lines are Also Shown.

An incorporation of the Section XI computational methodology
in Section III of the ASME Code[10] is also under consideration by
the code writing committees of the ASME. This makes the selection
of and rules for application of default crack growth rate laws even
more critical, since flaws which are discovered during the construction phase of a reactor pressure vessel will face an entire
reactor lifetime of applied stress cycles, irradiation and environmental attack.

With this preliminary matrix data as a background, a main test
matrix was subsequently configured. The object of the main test
matrix was to determine crack growth rates under simulated pressurized
water reactor conditions for a wide variety of reactor pressure
vessel steels, submerged arc welds, and the associated heat-affected
zones, but for a limited selection of two waveforms and two load
ratios. A 60 sec ramp/reset and 17 mHz sinusoidal waveform were
selected as being both reactor-typical, and from the preliminary

matrix results, should provide a valuable indication of the range
of crack growth rate behavior for reactor-typical materials and test
conditions. At this same time, a smaller, but otherwise similar,
matrix was constructed to define a series of tests of irradiated
materials.

There are several conclusions drawn from this main matrix data,
which is presented in its entirety in Ref. 11. These conclusions
are detailed in the reference, but are summarized below:

(a) There are no significant categorical differences in fatigue
crack growth rates between any of the RPV materials tested, which
included submerged-arc welds deposited with three different fluxes
(Linde 80, Linde 124, Linde 0091), heat-affected zones associated
with each of the above, A508-2 forging steel and A533-B plate steel.

(b) The 1980 version of the Section XI default lines is a
generally conservative estimate to all of these various materials,
even through the equations for the lines were derived only for A508
and A533 data.

(c) There are three strong, external influences on the data:

(1) Load ratio effect - for a given value of ΔK, decreasing
the load ratio, R, increases the crack growth rate. This effect has
been well-documented for many alloy-environment systems and this
evidence confirms that pressure vessel steels behave similarly.

(2) Waveform effect - while changes in crack growth rate
are known to stem from changes in the waveform, RPV steels in the
reactor-typical environment have a rather unique dependence on
waveform, as outlined in a preceeding section of this paper.

(3) Effect of sulfur content - while a significant collection of data only exists for several heats of A533-B, it appears
that increasing sulfur content leads to higher corrosion fatigue
crack growth rates. Crack growth rates for steels with .014 to
.021 percent sulfur were grouped distinctly above growth rates for
steels with lower (.008 to .012 percent) sulfur contents.

RESULTS - IRRADIATED STEELS

Testing of irradiated specimens has been performed in the same
autoclaves as were some of the unirradiated material tests described
above, although these autoclaves are now located in hot cells, and
have been refixtured for the remote handling of specimens. In
essentially all other respects, the tests are conducted in the same
way as before. The results of this research, which is in its infant
stages and which does not yet cover a wide scope of external variables such as waveform, temperature, etc., are given below. In all

cases, the total fluence was 2 to 4 x 10^{19} neutrons/cm^2, (> 1 MeV) which is representative of the end-of-life condition at the quarter-wall thickness in the beltline region of a pressurized water reactor.

Data for A533-B steel in both the irradiated and unirradiated condition are shown in Figs. 8a and b. Although the materials (codes L83 and HT) are from different heats, and the initial ΔK for the HT-material tests is somewhat higher, the trend of the results is very similar. The 1 Hz test produce results residing on or near the ASME air default line over the higher portion of the ΔK range

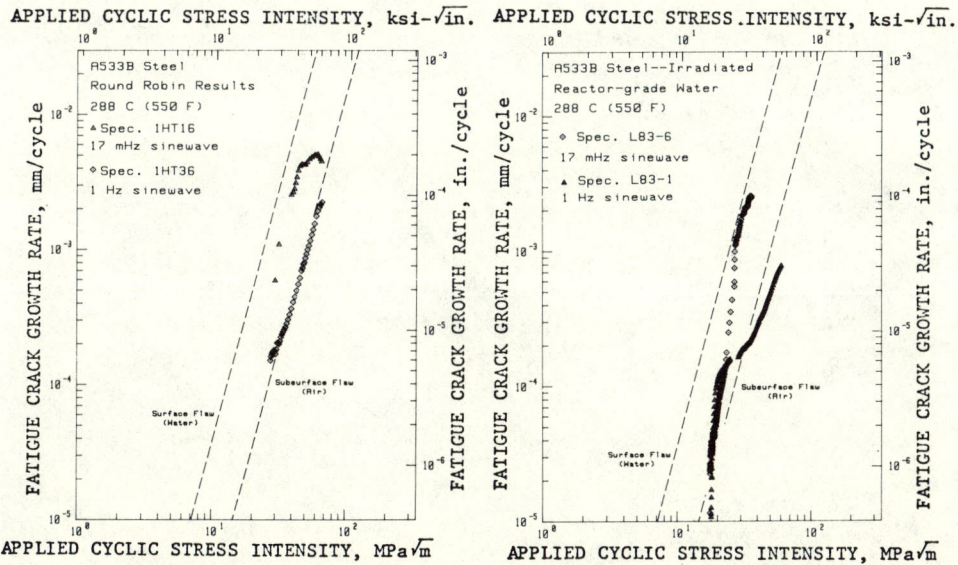

Fig. 8. Fatigue Crack Growth Rates vs Applied Cyclic Stress-Intensity Factor for (a) Irradiated and (b) Unirradiated A533-B Steel in High-Temperature, Pressurized Reactor-Grade Water. Comparison of These Two Graphs Shows the Parallel Behavior of the Irradiated and Unirradiated Specimens. Since the Unirradiated Test Began at a Rather High Initial Value of ΔK, the Three-Region, Corrosion-Fatigue Behavior, Seen Clearly in Irradiated Specimen L83-1, Does not Manifest Itself. Specimens L83-1 and L83-6 were Irradiated to a Total Fluence of 3.4 x 10^{19} Neutrons/cm^2 (> 1 MeV).

which was tested. The 17 mHz results show a substantial increase
in growth rate over the 1 Hz results, but since the increase is
about the same for both irradiated and unirradiated material, it
appears to be a function of the environment and cyclic period, rather
than irradiation.

Figure 9 presents the results of an effort to separate, if
present, the effect of irradiation time at temperature from the

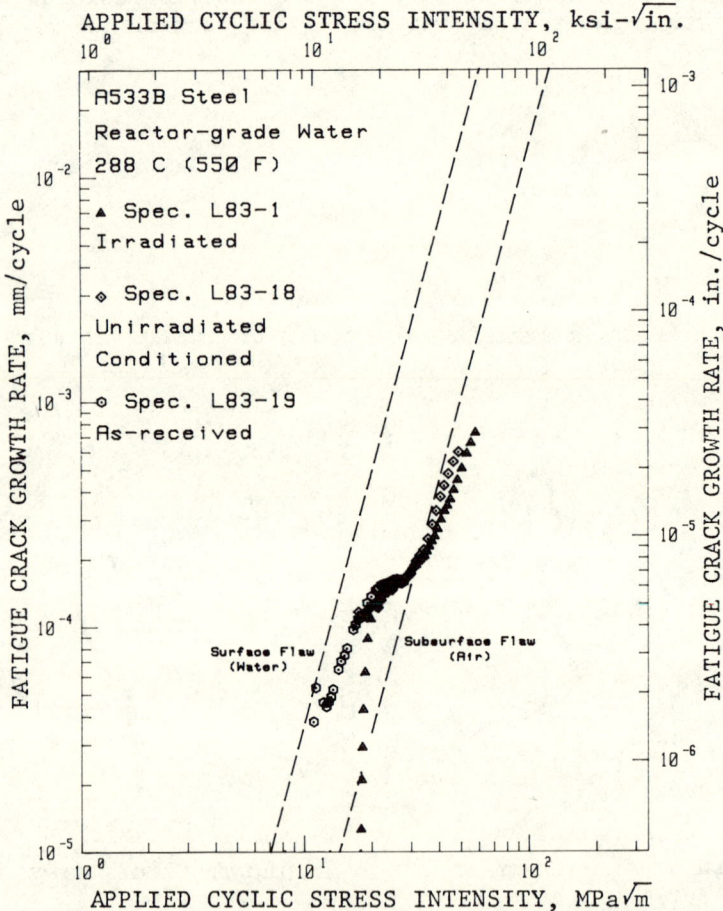

Fig. 9. Fatigue Crack Growth Rates vs Applied Cyclic Stress-
Intensity Factor for As-Received, Unirradiated but Con-
ditioned at Reactor Time and Temperature, and Irradiated
A533-B Steel in High-Temperature, Pressurized Reactor-
Grade Water. These Three Overlaping Data Sets Indicate
that Irradiation Under Conditions of This Initial Study
Has an Essentially Insignificant Effect on Fatigue Crack
Growth Rate. Specimens L83-1 was Irradiated to a Total
Fluence of 3.4×10^{19} Neutrons/cm^2 (> 1 MeV).

effect of irradiation damage itself. Specimen L83-19 was conditioned at 288°C for 1225 hours which, combined with residence time in the autoclave prior to and during the test, closely approximates the time for irradiations of the other L83 specimens described above. Specimen L83-18 was tested in the as-received condition. Results of 1 Hz tests in reactor-grade water environment of all three specimens (L83-1, -18, -19) are shown in Fig. 9, allowing the conclusion that the response of the material to fatigue is essentially unaltered by irradiation or by an equivalent time at temperature for these materials as evaluated in this study.

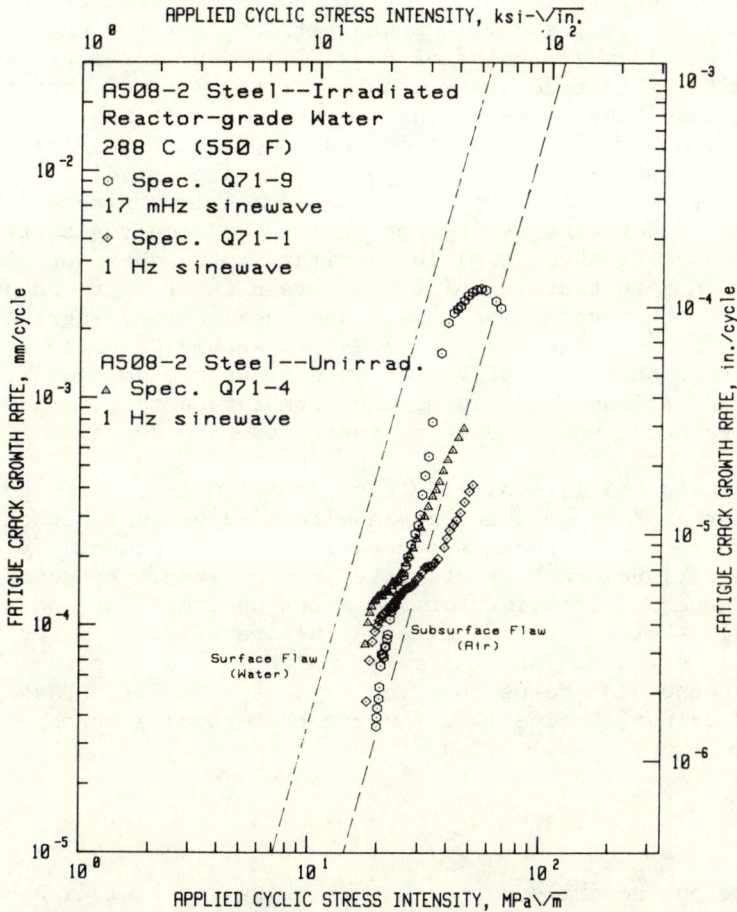

Fig. 10. Fatigue Crack Growth Rates vs. Applied Cyclic Stress-Intensity Factor for Irradiated and Unirradiated, but Reactor Time-and-Temperature Conditioned, A508-2 Steel. These Results are Very Similar to Those for A533-B Shown in Figs. 1 and 2. In This Case, the Irradiated Specimen Tested at 1 Hz Shows a Slight, but Measurable Decrease in Crack Growth Rates. Specimen Q71-1 was Irradiated to a Total Fluence of 3.4 x 10^{19} neutrons/cm^2 (> 1 MeV).

A similar set of tests was conducted on A508-2 material, code Q71. These results are shown in Fig. 10. As for A533-B, there is a significant increase in growth rates for the 17 mHz waveform as opposed to the 1 Hz waveform, but this increase is nearly identical to the increase shown in Fig. 8a, b and as before, is due to the influence of the environment during the longer period waveform. Note that the increase in growth rates between the 1 Hz and 17 mHz results is about a factor of twenty, while the increase in cyclic period is about sixty; the relationship is not, therefore, one-to-one. If irradiation has little or no effect, we can expect, on the basis of unirradiated results to date[3,6] that the 17 mHz waveform affords a near saturation of the environmental effect, for PWR environments, and thus, the 17 mHz results of Figs. 8 and 10 represent the probable upper limits of fatigue crack growth for the R = 0.2, constant-load-amplitude condition. The comparison of 1 Hz test results for irra-diated, and unirradiated but time-and-temperature-conditioned speci-mens also shown in Fig. 10, indicated that the irradiated results are somewhat lower than the unirradiated results, although the difference is not clearly greater than the commonly accepted scatter-band (a factor of about two) for similar, intralaboratory fatigue crack growth rate tests. While these results are very encouraging in the sense that radiation damage does not seem to aggravate corrosion-fatigue crack growth rates, it should be borne in mind that only a narrow selection of variables have been examined. There are higher load rations, additional waveforms, temperatures and water chemistries which may produce a synergistic and detrimental effect.

Specimens L83-1, -18, -19 (Fig. 8) and Q71-1, -4 (Fig. 9) were tested over a ΔK range which began with a value sufficiently low that the classical three-stage behavior of environmentally assisted corrosion fatigue crack growth rates can be seen. As described earlier, Region I behavior, or the start-up transient, occurs at the lowest values of the ΔK range, with the crack growth rates rising to a plateau, termed Region II growth, which is characterized by somewhat more ΔK-independent rates than for Regions I or III. Lastly, for the higher ΔK values, Region III, the growth resumes a strong ΔK-de-pendence.

CONCLUSIONS

Based on the above data, the following conclusions have been drawn.

1. For lower temperatures (93°C, 200°F) and low load ratios, (R = 0.1, 0.2), waveforms containing a ramp time component of greater than one second duration yield crack growth rates residing midway between the 1972 versions of the air and water environment default lines, or just under the 1980 version of the low load ratio (R $\leq$ 0.25) surface flaw default line.

2. For high temperature and low load ratios, all ramp and hold waveforms tested produced crack growth rates residing on or near the subsurface flaw (air environment) default line. Additionally, all low temperature and low load ratio tests with a one second ramp time fell on or near the air default line. This places all such data significantly below the ASME default line, and where such waveforms pertain, use of the default line growth rate law, rather than the actual data, will produce overly conservative computations of reactor lifetime or required inspection intervals. However, for high temperatures and low load ratios, 17 mHz sinusoidal waveforms produce significantly higher crack growth rates, residing about midway between the 1972 ASME air and water default lines.

3. For higher load ratios, there is more variance in the available data[6] and a general conclusion is not possible. There are apparently some product forms and compositions for which crack growth rates are substantially lower than the default line for high load ratio controlled growth rate.

4. For low load ratios, and for the limited number of materials and sinusoidal frequencies, irradiation damage does not appear to influence the fatigue crack growth rates of typical reactor pressure vessel steels.

5. When tested over a ΔK range which begins with a sufficiently low value ($\sim$ 15 MPa $\sqrt{m}$ for R = 0.2) and which extends over sufficiently large range (15 to MPa $\sqrt{m}$ or more) the classical three-region behavior of corrosion fatigue crack growth rates can be described.

CONSIDERATIONS FOR FUTURE RESEARCH

The above results were developed for a pressurized water reactor environment and irradiation conditions. Boiling water reactor (BWR) environments are most notably different in that during normal operation, the coolant contains 200 to 300 ppb dissolved oxygen, as opposed to the negligibly small (< 1 ppb) level in a PWR. Consideration of the available data relevant to BWR environments[12,13] indicates that the data points fall in the same two zones as the PWR-type data (Fig. 6), however, the trends of the data, as functions of waveform and temperature, are not as clearly established as in the above description for A508-2 in a PWR environment.

Fatigue crack growth rate research for reactor safety use is far from a completed task, for there are many areas which have not been explored, or in which only slight progress has been made. A look at the data in this report, or in review papers[6,14] shows that there is no data in the near-threshold, or very low growth rate regime ($\sim 10^{-7}, 10^{-8}$ m/cycle). It is in this area that flaw extension will require millions of cycles in order to produce measurable extensions. Thermal cycling and pump vibrations do, however, provide alternating stresses at frequencies high enough to result in millions of cycles. Thus, it is important to gain a better understanding of this topic. The current ASME default rules do not provide a threshold, or lower stress-intensity factor limit, below which fatigue crack growth will not occur. Thus, in calculations using the code default lines, any flaw will grow, regardless of its size or the stress level, if the requisite number of cycles are fed into the calculations.

Another important consideration in the application of these data is that reactor-typical waveforms involve cycles of a variety of amplitudes and periods, while the available crack growth data pertains only to constant amplitude, constant frequency testing. Furthermore, there is no indication of any predictive method which adequately enables computation of variable amplitude crack growth rates from constant amplitude test results. Variable amplitude testing of these materials in simulated nuclear coolant environments is in its infant states.[7,15]

Another closely related topic, which must be more thoroughly investigated, is stress-corrosion cracking (SCC), both as a mechanism by itself and in tandem with fatigue cycling involving hold times. Recent evidence indicates that these materials, formerly believed to be resistant to SCC, do show signs of cracking, albeit at rather high levels of applied stress-intensity factor.[16]

ACKNOWLEDGMENT

The investigations reported were sponsored by the Reactor Safety Research Division of the U. S. Nuclear Regulatory Commission. The continuing support of this agency is appreciated. The innovative experimental design and overall technical assistance of R. E. Taylor contributed to the success of this research. The assistance of J. R. Hawthorne and H. E. Watson, who handled the irradiation logistics, is also appreciated.

REFERENCES

1. "Rules for Inservice Inspection of Nuclear Power Plant Components", Section XI of the ASME Boiler and Pressure Vessel Code, ANSI-BPV-XI-1, issued annually, American Society of Mechanical Engineers, New York City.

2. W. H. Bamford, "A Review of Fatigue Crack Growth Studies of Light Water Reactor Pressure Boundary Steels", IAEA Meeting

on Time and Load Dependent Degradation of Pressure Boundary Materials, Innsbruck, Austria, November 1978.

3. W. H. Cullen, et al., "Fatigue Crack Growth of A508 Steel in High-Temperature, Pressurized Reactor-Grade Water NUREG/CR 0969, NRL Memorandum Report 4063, Naval Research Laboratory, Washington, DC., 28 Sep 1979.

4. W. H. Cullen, et al., "Operation of a High-Temperature, Pressurized Water Fatigue Crack Growth Facility", Closed Loop, MTS Corporation, Minneapolis, MN, October 1980.

5. W. H. Cullen, et al., "A Computerized Data Acquisition System for a High-Temperature Pressurized Water Fatigue Test Facility", in Computer Automation of Materials Testing, ASTM STP 10, pp. 127–140, American Society for Testing and Materials, Philadelphia, PA, 1980.

6. W. H. Cullen and K. Torronen, "A Review of Fatigue Crack Growth of Pressure Vessel and Piping Steels in High-Temperature, Pressurized, Reactor-Grade Water", NUREG/CR-1576, NRL Mem - orandum Report 4298, Naval Research Laboratory, Washington, DC, September 1980.

7. R. P. Wei, S. R. Novak, D. P. Williams, "Some Important Consideration in the Development of Stress-Corrosion Cracking Test Methods," Proceedings 33rd AGARD Conference on Structures and Materials, Brussels, Belgium, 1971.

8. P. C. Paris, R. J. Bucci, E. T. Wessel, W. G. Clark and T. R. Mager, "An Extensive Study on Low Fatigue Crack Growth Rates in A533 and A508 Steels", in Stress Analysis and Growth of Cracks, ASTM STP 513, American Society for Testing and Materials, Philadelphia, PA (1972), pp. 141–176.

9. W. H. Bamford, "Application of Corrosion Fatigue Crack Growth Rate Data to Integrity Analyses of Nuclear Reactor Vessels", Third ASME Congress on Pressure Vessels and Piping, San Francisco, CA, June 1979.

10. "Nuclear Power Plant Components: General Requirements", Section III of the ASME Boiler and Pressure Vessel Code, ANSI-BPV-III-1, issued annually, American Society of Mechanical Engineers, New York City, NY.

11. W. H. Bamford, W. H. Cullen, L. J. Ceschini, R. E. Taylor, and H. E. Watson, "Environmentally-Assisted Subcritical Crack Growth in LWR Materials", in Structural Integrity of Water Pressure Boundary Components, Annual Report, Fiscal Year 1980, NUREG-CR, NRL Memorandum Report, Naval Research Laboratory, Washington, DC (pending publication).

12. D. A. Hale, et al., "Task K-Effect of BWR Environment" in Reactor Primary Coolant System Pipe Rupture Study Progress Reports GEAP-10207-32 through GEAP-10207-40, General Electric Co., Nuclear Systems Division, San Jose, CA (1974-1978).

13. D. A. Hale, J. Yuen and T. Gerber, "Fatigue Crack Growth in Piping and RPV Steels in Simulated BWR Water Environment", GEAP 24098, General Electric Co., Nuclear Systems Division, San Jose, CA.

14. L. A. James, Fatigue Crack Propagation in Neutron-Irradiated
     Ferritic Pressure Vessel Steels, <u>Nuclear Safety</u>, 18:791-801
     (1977).

15. W. H. Cullen, H. E. Watson, V. Provenzano, "Results of Cyclic
     Crack Growth Rate Studies in Pressure Vessel and Piping Steels",
     in <u>Structural Integrity of Water Reactor Pressure Boundary
     Components</u>, Annual Report, Fiscal Year, 1979, F. J. Loss, (ed),
     NUREG/CR-1128, NRL Memorandum Report 4122, Naval Research
     Laboratory, Washington, DC, December 31, 1979.

16. W. H. Bamford, D. M. Moon and L. J. Ceschini, "Cyclic Crack
     Growth Rate Studies"Conducted by Westinghouse Nuclear Energy
     Systems Under NRL Contract, in "Structural Integrity of Water
     Reactor Pressure Boundary Components, Quarterly Progress
     Report, October-December 1979", NUREG/CR-1268, NRL Memorandum
     Report 4174, Naval Research Laboratory, Washington, DC,
     March 20, 1980.

Al-2024-T4 alloys
   accrued damage/failure predic-
      tion in, 76-80
   corrosion fatigue tests on,
      72-74, 80
   x-ray rocking curve studies
      on, 74-76
Alloy 718
   chemical composition/mechani-
      cal properties of, 165
   low cycle fatigue deformation
      behavior in, 166-179
   low cycle fatigue testing of,
      164-170
   plastic strain-life coeffi-
      cients for, 169
   time-dependent fatigue of, 257
Austenitic stainless steels
   deformation-microstructural
      relationships of, 184-
      192
   fatigue crack growth of welds
      on, 200-210
   fracture resistance of welds
      on, 200, 210-216
   time-dependent fatigue of,
      243-256
   types A533-B and A508-2, 403,
      405

Basquin equation, 322
Burger's vector
   relationship to dislocation
      density in x-ray analy-
      sis, 106
   relationship with plastic
      strain, 185

Ceramics
   combined-exposure/thermal cycl-
      ing effects on, 226-228,
      229
   high performance, 222, 229
   high temperature static fatigue
      in, 222-223
Coffin-Manson equation, 140, 168-
      169, *see also* Manson-
      Coffin equation
   limitations of, 174
Corrosion fatigue
   depth distribution of disloca-
      tion density induced by,
      74-76
   rate determining steps in, 41-
      43, 44-49
   tests on Al-2024-T4 alloys, 72-
      74, 80
Crack growth modeling, 293-294
Crack growth power law, 225
Crack initiation, *see also* Fatigue
      crack initiation
   relationshup to plastic strain,
      164
Crack retardation, 364-367
Crack tip opening displacement
      (CTOD)
   influence on fatigue crack
      growth, 99
Creep crack growth, 363-367
   applied mechanics considerations
      in, 151-156
   atomistics considerations in,
      148-152
   basic considerations in, 145-
      148

Creep crack growth (continued)
    composition/microstructural ef-
        fects in, 158–160
    correlation of rupture data,
        307
    environmental effects on, 156–
        159
    of Ni-base alloys, 153–154,
        159–160
Creep damage, 302, 326, 347, 350
    model for unbalanced loop test-
        ing, 16, 18–19
Creep-environment crack growth
        interaction, 34–37
Creep-fatigue, 9
    correlation of data, 307–310,
        320–323, 326
    strain-controlled testing in,
        246–249, 255
Creep-fatigue-environment inter-
        action
    high temperature effects, 129–
        140
Creep strain, 338

Default line
    ASME criteria in fatigue crack
        growth, 392–407
Deformation mechanism map, 184
    for pure titanium, 278
Diffusion-controlled crack
        growth, 149, 151
Dislocation density, 104–105,
        106, 115
    depth distribution by corro-
        sion fatigue, 74–76
    depth profiles in fatigued met-
        als, 109–110, 111
    depth profiles in unidirection-
        al stressed metals, 106–
        109
    effect on structural instabil-
        ity of metals, 110–112
Dislocation microstructure
    effect of hold-time on, 187,
        189
    effect of stress/temperature
        on, 185–189, 191
    induced deformation in stain-
        less steels, 184–192

Engineering plastics
    evaluation of mechanical pro-
        perties of, 231–232
    factors affecting fatigue life
        of, 232–239
Eutectic composites
    composition of, 302
    fatigue crack growth of, 324–
        326
    fatigue behavior at elevated
        temperatures, 302–320,
        326
    fractography in, 313–320
    metallography in, 313–320

Fast breeder reactor systems,
    austenitic stainless steels
        in, 243–248
    environmental interaction/ time-
        dependent fatigue in, 242–
        256
    Fe-$2\frac{1}{4}$ Cr-1 Mo steels in, 248–
        256
Fatigue, 1–2
    crack initiation and growth, 1–
        7
    damage determination by x-ray
        techniques, 113–114
    damage processes in time-de-
        pendent, 7–9
    high temperature, 379–380
    role of dislocation distribu-
        tion in, 112–113
    role of stacking fault energy
        in, 174
    time-dependent behavior of
        structural steels, 243–
        258
Fatigue crack growth, 41–42
    analytical approaches to, 331–
        337, 340–347
    ASME default line considera-
        tions in, 392–407
    comparison of gas transport vs.
        surface reaction con-
        trolled, 64
    effect of aging on, 286
    effect of delta ferrite con-
        tent on, 203

Fatigue crack growth, (continued)
  effect of environment on, 7–11,
      14–16, 43–44, 51–55, 85–
      87, 95, 99, 134–138, 238–
      239, 242–259, 282, 291–
      292, 295, 305–307, 394–
      407, 377–378
  effect of fatigue loading pat-
      terns on, 204–209, 369
  effect of frequency on, 11–14,
      233–237, 303–307, 379–
      381, 386
  effect of hold-times on, 24–
      32, 203–209, 243–244,
      250–251, 257–258, 310–
      313, 364–365
  effect of load ratio on, 84,
      96, 291–292, 400–402,
      406–407
  effect of microstructure on,
      11–14, 17–23, 52, 85–86,
      88–91, 120–129, 189, 281–
      282, 285–290, 295
  effect of sulphur content in
      steels on, 402
  effect of temperature on, 9–11,
      24–34, 92, 204–209, 234–
      236, 248–255, 257–258,
      277, 283–284, 290, 293–
      295, 377–378, 381, 388,
      395–396, 400–405
  effect of time/temperature on,
      332–337
  effect of unirradiated/radiated
      conditions on, 203, 204–
      209, 394–403, 403–406
  effect of waveform on, 395–407
  effect of yield strength on,
      93
  equation for, 383, 388
  influence of inhibitor gases
      in, 67–68
  kinetic model for high tempera-
      ture, 380, 382–385
  modeling in binary gas mix-
      tures, 65–68
  modeling in one component gas
      environment, 59–63
  non-destructive inspection/
      prediction of, 358–370

Fatigue crack growth, (continued)
  of ceramics, 222–229
  of engineering plastics, 232–
      239
  of eutectic composites, 324–326
  of Ni-base alloys, 120–141, 153–
      154, 159–160, 385–388
  of pressure vessels/piping
      steels, 393–394, 395–408
  of steels, 154, 156, 403, 405
  of superalloys, 24–32
  of titanium alloys, 83–96, 283–
      294
  of welds, 200–210
  rate controlling processes in,
      380, 382–385
  research need in reactor safe-
      ty, 407–408
  role of Knudsen flow parameter
      in, 61
  slip-dissolution model for, 44–
      51
  tests on stainless steel welds,
      200–203
  three-region behavior of, 399,
      407
  vs. applied cyclic stress inten-
      sity factor, 394–398, 403–
      405
Fatigue crack initiation, 3–7, 9–
      11
Fatigue life, see Fatigue crack
      growth
Fractography
  considerations for eutectic com-
      posites, 313–320
Fracture
  role of surface layer in, 106–
      113
  resistance of welds in steels,
      210–216
Fracture mechanics
  effect of metallurgical/engi-
      neering variables in, 155
Fracture mechanism map, 184
  for Ni-base alloys, 146
  for pure titanium, 278–279
Fracture resistance
  effect of delta ferrite con-
      tent in steels on, 213–
      215

Fracture resistance
  effect of elevated tempera-
      tures on, 213–216
  effect of unirradiated/irradi-
      ated conditions on, 200,
      211–216
  tests on stainless steel welds,
      200, 210–211
Fusion reactor first-wall sys-
    tems
  environmental interaction/
      time-dependent fatigue
      in, 258–259

Grain boundary damage, 9
Gross–Srawley equation
  for crack growth rate/stress
      intensity factor, 199,
      204–209

High-temperature gas-cooled re-
    actor systems
  environmental interaction/time-
      dependent fatigue in,
      257–258

Knudsen flow parameter
  in surface reaction controlled
      fatigue crack growth,
      61

Larsen–Miller parameter
  for titanium alloys, 277
  use in correlating creep-rup-
      ture data, 307
Low cycle fatigue, 120, 164, 176,
    259, 354–356
  deformation behavior of alloy
      718 by, 166–179
  effect of microstructure on,
      129–136
  effect of temperature on, 166–
      170, 177–180
  environmental/deformation mod-
      el for high temperature,
      134, 137–140
  high temperature behavior of
      Ni-base alloys, 129–138,
      142

in AISI 304 stainless steels,
    188
Low cycle fatigue (continued)
  of titanium alloys, 272, 279–
      282

Main test matrix
  crack growth study by, 392, 401–
      402
Manson–Coffin equation, *see also*
      Coffin–Manson equation
  modified form of, 320
Metallography
  considerations for eutectic com-
      posites, 313–320
Microtwin, 180, *see also* Twinning

Ni-base alloys
  creep crack growth of, 153–154,
      159–160
  fatigue crack growth of, 120–
      141, 153–154, 159–160,
      385–388
  fracture mechanics maps for,
      146
  low cycle fatigue of, 129–
      138, 142
Nuclear reactor systems
  environmental interaction/time-
      dependent fatigue in,
      see under types
  types of, 242, 257, 258

Paris's law
  fatigue crack propagation con-
      stants, 122
Plastic deformation
  effect of dislocation density
      on, 106, 108–109
  of titanium alloys, 271
  role of surface layer in, 103–
      105, 107–110
Plastic strain, 185
Preliminary test matrix
  crack growth rate study by, 392
Pressurized water reactor (PWR)
  coolant chemistry, 393

Reflection (Berg–Barrett) topo-
      graphs, 73–74

Retirement for cause, 120, 356–
358
life prediction management concept of, 353–356
supporting technology for, 370–
373

Shielded metal arc, 198
Short crack
definition of, 1–2
Single-edge-notch (SEN) fatigue
test specimen
design description of, 199
Slip, 172, 180
deformation mode of titanium,
271, 288
Slip-dissolution model
crack propagation rate study
by, 46–51
Stacking fault energy (SFE)
in fatigue deformation processes, 174
Static fatigue
of ceramic materials under
high temperatures, 222
Stepped-temperature stress-rupture (STSR) test, 223–
225, 226, 229
Strain distribution function,
343–344
Stress controlled crack growth,
150–151
Stress-corrosion fatigue
rate determining steps, 41–43,
51–55
Stress intensity factor, *see also* Gross-Srawley equa-
tion
vs. fatigue crack growth, 199,
204–209, 394–398, 403–
405
Stress-rupture tests, 225–226,
229

Thermal fatigue, 329–330
crack initiation analysis, 337–
340, 347–350
crack propagation analysis,
331–337, 340–347, 350–351
determination of, 330–331

Threshold ΔK_{th}, 92, 93, 94, 98,
99, 101
effect of environment on, 85,
86, 92–93
effect of load ratio on, 98
effect of microstructure on,
88–91
effect of short fatigue crack
on, 93
effect of temperature on, 88–
91, 92
effect of yield strength on,
99–91, 101
relationship with fatigue crack
growth, 85–87, 92–101
Time-dependent fatigue
of alloy 800H, 257
of stainless steels, 243–256
Titanium alloys, 270
creep deformation and rupture
of, 274–279
effect of zirconium on creep
strength of, 276
fatigue crack growth of, 83–
96, 283–294
Larsen-Miller parameters for,
277
low cycle fatigue of, 279–282
oxidation/oxidation kinetics
of, 272–274
plastic deformation of, 271
processing/microstructural
considerations of, 265–
271
slip deformation mode of, 271–
272, 288
twin deformation of, 271
yield stress and ΔK_{th} values
of, 100
Twin
deformation mode of titanium,
271
Twinning, 171–174, 178

Weight function method, 342
Wei-Landes superposition theory,
42
WEN diagram
for subcritical crack growth,
41–42

X-ray diffraction
  analyses of dislocation density
        vs. plastic deformation,
        106
  prefracture damage/failure pre-
        diction by, 76-80
X-ray double crystal diffracto-
        metry, 72
X-ray rocking curve, 72-74
  determination of, 76-79
  profiles of tensile deformed
        metals, 107, 109

Zirconium
  effect on creep strength of
        titanium alloys, 276